Oklahoma Notes

Basic Sciences Review for Medical Licensure
Developed at
The University of Oklahoma College of Medicine

Suitable Reviews for:
United States Medical Licensing Examination
(USMLE), Step 1

Oklahoma Notes

Physiology

Fourth Edition

Edited by
Roger Thies

Technical Editing by
Rita R. Claudet

With Contributions by
Kirk W. Barron Robert C. Beesley
Siribhinya Benyajati Robert W. Blair
Kenneth J. Dormer Jay P. Farber
Robert D. Foreman Kennon M. Garrett
Stephen S. Hull, Jr.
Philip A. McHale Y.S. Reddy
Rex D. Stith Roger Thies

Springer-Verlag
New York Berlin Heidelberg London Paris
Tokyo Hong Kong Barcelona Budapest

Roger Thies, Ph.D.
Department of Physiology
College of Medicine
Health Sciences Center
University of Oklahoma
Oklahoma City, OK 73190
USA

All contributors share the address above except:
Rex D. Stith
Assistant Dean and Director
Indiana University School of Medicine
Evansville Center, Box 3287
Evansville, IN 47732
USA

Library of Congress Cataloging-in-Publication Data
Physiology / edited by Roger Thies ; contributions by Kirk W. Barron
. . . [et al.]. – 4th ed.
p. cm. – (Oklahoma notes)
Includes bibliographical references.
ISBN 0-387-94397-8
1. Human physiology–Outlines, syllabi, etc. 2. Human physiology–Examinations, questions, etc. I. Thies, Roger. II. Barron, Kirk W. III. Series.
[DNLM: 1. Physiology–outlines. 2. Physiology–examination questions. QT 18 P5785 1995]
QP41.P52 1995
612′.0076–dc20
DNLM/DLC
for Library of Congress 94-45261

Printed on acid-free paper.

Production managed by Jim Harbison; manufacturing supervised by Jacqui Ashri.
Camera-ready copy prepared by the editor.
Printed and bound by Edwards Brothers, Inc., Ann Arbor, MI.
Printed in the United States of America.

9 8 7 6 5 4 3 2 1

ISBN 0-387-94397-8 Springer-Verlag New York Berlin Heidelberg

Preface to the *Oklahoma Notes*

In 1973, the University of Oklahoma College of Medicine instituted a requirement for passage of the Part 1 National Boards for promotion to the third year. To assist students in preparation for this examination, a two-week review of the basic sciences was added to the curriculum in 1975. Ten review texts were written by the faculty: four in anatomical sciences and one each in the other six basic sciences. Self-instructional quizzes were also developed by each discipline and administered during the review period.

The first year the course was instituted the Total Score performance on National Boards Part I increased 60 points, with the relative standing of the school changing from 56th to 9th in the nation. The performance of the class since then has remained near the national candidate mean. This improvement in our own students' performance has been documented (Hyde et al: Performance on NBME Part I examination in relation to policies regarding use of test. J. Med. Educ. 60: 439–443, 1985).

A questionnaire was administered to one of the classes after they had completed the Boards; 82% rated the review books as the most beneficial part of the course. These texts were subsequently rewritten and made available for use by all students of medicine who were preparing for comprehensive examinations in the Basic Medical Sciences. Since their introduction in 1987, over 300,000 copies have been sold. Obviously these texts have proven to be of value. The main reason is that they present a *concise overview* of each discipline, emphasizing the content and concepts most appropriate to the task at hand, i.e., passage of a comprehensive examination over the Basic Medical Sciences.

The recent changes in the licensure examination that have been made to create a Step 1/Step 2/Step 3 process have necessitiated a complete revision of the Oklahoma Notes. This task was begun in the summer of 1991 and has been on-going over the past 3 years. The book you are now holding is a product of that revision. Besides bringing each book up to date, the authors have made every effort to make the tests and review questions conform to the new format of the National Board of Medical Examiners. Thus we have added numerous clinical vignettes and extended match questions. A major revision in the review of the Anatomical Sciences has also been introduced. We have distilled the previous editions' content to the details the authors believe to be of greatest importance and have combined the four texts into a single volume. In addition a book over neurosciences has been added to reflect the emphasis this interdisciplinary field is now receiving.

I hope you will find these review books valuable in your preparation for the licensure exams. Good Luck!

Richard M. Hyde, Ph.D.
Executive Editor

Preface

This review covers major concepts of human physiology without being exhaustive. It assumes that students have completed a course in human physiology and wish to refresh their memory in preparing for Step 1 of the United States Medical Licensing Examination (USMLE). Students are encouraged to refer to a comprehensive textbook of physiology while using this review.

All chapters of this fourth edition have been revised. These changes include newer understandings of physiology and applications of physiology and pathophysiology. The endocrinology chapter and section on reproduction have been expanded with text and figures, since recent national examinations have greater numbers of questions on these topics. The endocrinology questions and the section of the final chapter on exercise have been completely rewritten. Aspects of aging have been included with each system rather than put in a separate section of the final chapter.

Review questions follow every few pages of text to monitor your understanding of the preceding material. Multiple choice questions conform to the styles in the USMLE examination, namely "single one best answer" type and matching from five or more choices. In addition, some chapters ask questions about clinical cases in a non-multiple choice format. Review questions are numbered consecutively within each of the eight chapters. Correct answers with brief explanations are given at the end of each chapter. There are more than 340 review questions in this book. Clinical laboratory values for answering some questions are on page 280.

The names of all contributors are listed on the title page. The writers of chapters are not identified, since this edition was a cooperative effort between all the faculty of our department. The editor appreciates Lula Rhoton for her capable assistance in preparing the typescript, Tamara Williams and Sharon Farber for drawing the figures, Rita Claudet for her efficient technical editing, and Springer-Verlag for their continued support of *Oklahoma Notes*.

Roger Thies

Contents

ELECTROPHYSIOLOGY

MEMBRANE TRANSPORT

The function of the human body depends upon the transport of substances across cell membranes. Transport across cell layers is the basis of respiratory, renal, gastrointestinal, and capillary functions. Many hormones control membrane transport of various substances, and neural activity is dependent upon control of ion movements across membranes.

Diffusion and Osmosis

Diffusion is the migration of molecules from a region of high concentration to one of lower concentration as a result of random motion. Lipid soluble substances, including fatty acids, steroids, oxygen, carbon dioxide and anesthetic gases, move across cell membranes down their concentration gradients by simple diffusion. The rate of solute diffusion is dependent on the electrochemical gradient, permeability of the membrane for the solute, the size of the solute, the area and thickness of the membrane (see flux equation on p. 100). **Osmosis** is the net diffusion of water across semipermeable (permeable to water, but not solutes) and selectively permeable (permeable to water and some solutes) membranes through **pores.** The **osmolarity** of a solution is determined by the total number of particles in the solution. The net movement of water across a semipermeable membrane is due to the concentration differences of the non-penetrating solutes. The **osmotic pressure** of a solution is the hydrostatic pressure that will prevent the flow of water across a semipermeable membrane. For solutions containing non-penetrating solutes, water diffuses from a low osmolarity solution (high water concentration) to a high osmolarity solution (low water concentration) in order to achieve equal water concentrations on both sides of the membrane. The normal osmolarity of body fluids in various compartments is 285 (±5) milliosmoles/kg of water. The **tonicity** of a solution is determined by the concentration of non-penetrating solutes. If a cell is placed in a hypertonic solution, water will pass out of the cell, and the cell will shrink. Conversely, if a cell is placed in a hypotonic solution, water will flow into the cell, and the cell will swell.

Mediated-Transport Systems

Many substances are hydrophilic or too large to cross the membrane by simple diffusion. Carrier proteins in cell membranes permit faster transport of lipid-insoluble substances than simple diffusion. The carrier binding sites are **specific** for a class of substances. Structurally related compounds can **compete** for the site and inhibit the binding of each solute to the carrier protein. The rate of carrier-mediated transport may show **saturation** at high solute concentrations, since the number of carrier proteins is finite, and the cycling of carrier proteins is limited. **Facilitated diffusion** is a carrier-mediated transport mechanism that moves solutes down their electrochemical gradients, so it does not use metabolic energy. An example of facilitated diffusion is the transport of glucose into red blood cells.

Active transport moves molecules against their electrochemical gradients and thus requires energy. Active transport is linked to energy metabolism either directly through use of ATP or indirectly through electrochemical potential gradients of another solute. **Primary active transport** uses ATP as the direct energy source, and these transport proteins are called ATPases. Primary active transport systems are especially sensitive to metabolic inhibitors. The Na^+, K^+-ATPase is a representative of this type of transporter. It maintains the low $[Na^+]_i$ and high $[K^+]_i$ by pumping Na^+ out of the cell and K^+ into the cell. **Secondary active transport** uses the potential energy stored in transmembrane electrochemical gradients of ions to transport substrates across the membrane (see Fig. 6-2 on p. 187). This category of transport is characterized by a physical coupling between the flow of the actively transported substrate and the "co-substrate" (the ion whose gradient is being tapped as the immediate energy source for active transport). Because co-substrate gradients are themselves the product of primary active, ATP-linked transport processes, there is an indirect link between cell metabolism and secondary active transport. In many instances the Na^+ electrochemical gradient is used as the energy source for secondary active transport. The physical coupling between the flow of the substrate and co-substrate can be in the same direction or in opposite directions across the membrane. In **co-transport**, the substrate and the co-substrate are transported in the same direction by a **synporter**. An example of this is the Na^+-dependent transport of amino acids into many epithelial cells, such as the gut. **Counter-transport** by **antiporter** proteins moves the substrate and co-substrate across the membrane in opposite directions. An example of this process is the Na^+/Ca^{2+} antiporter in cardiac muscle cells.

Endocytosis and Exocytosis

Endocytosis is the process of invagination of the plasma membrane to pinch off and become an intracellular vesicle enclosing a small volume of extracellular fluid. **Pinocytosis** is a type of endocytosis where water and solutes are taken up into vesicles and transported into the cell. **Phagocytosis** is a type of endocytosis where particulate matter, such as cell debris, is taken up by the cell. When molecules are too large or hydrophilic to diffuse across the plasma membrane, they are packaged in vesicles and secreted into the extracellular fluid by **exocytosis**. This process is also used for rapid, coordinated release of large numbers of small molecules, as in release of neurotransmitters or hormones. Under the appropriate stimulus, the vesicles migrate to the plasma membrane, fuse with it and release the substance into extracellular fluid. Both endocytosis and exocytosis require metabolic energy.

BIOELECTRIC PHENOMENA

Electrical events in the body are weak and fast, so they are measured in millivolts (mV) and milliseconds (msec). Typical amplitudes for recorded electrical signals are 1 mV for the QRS complex of the EKG, 1 mV for electromyograms (EMGs) and 0.1 mV for electroencephalograms (EEGs). These signals are only about 1% of the values generated across cell membranes, because potential differences are attenuated when recorded at a distance from the generator. The durations of nerve action potentials and of synaptic or neuromuscular delay are about 1 msec.

Ions are charged particles, so movement of ions is an electrical current. Ionic current from cations such as Na^+, K^+, Ca^{2+} and H^+ flows from an **anode** (relatively positive charge) to a **cathode** (relatively negative charge). The current (flow of charges) is directly proportional to the potential difference (usually concentration of ions) and inversely proportional to the resistance (Ohm's law). Cell membranes are selectively permeable to ions. Membrane permeability for ions is expressed as **conductance (g)**, the reciprocal of resistance. Consequently, flow of ions across membranes is the product of potential difference and conductance.

The lipid content of cell membranes makes them electrical insulators. Because there are charge differences across membranes, they have **capacitance**. This passive electrical property means that it takes time to move charges across membranes (discharging and recharging), and also that charge differences spread along membranes for some millimeters with exponential decay.

RESTING POTENTIALS

Equilibrium Potentials

Ion movement or flux is controlled by both concentration gradients and electrical gradients. If these gradients are equal but opposite in direction for a particular ion, then its total electrochemical potential is zero, and there is no net current flow. This is the definition of electrochemical **equilibrium.** Cells contain high concentrations of K^+ and low concentrations of Na^+ compared to extracellular fluid. The tendency for K^+ to flow out of cells and Na^+ to flow into cells due to concentration differences is affected by charges across the cell membranes. The **Nernst** equation calculates the electrical gradient for an ion which would exactly balance the force of a concentration gradient in the opposite direction. A simplified form of the Nernst equation at 37°C is

$$\text{Equilibrium potential or } E_{ion} = \frac{61}{z} \log_{10} \frac{[\text{ion}]_{out}}{[\text{ion}]_{in}}$$

where z is charge of the ion. A ten-fold difference of ion concentration produces a potential difference of 61 mV, and a 100-fold difference gives 122 mV. Typical values for ions are shown below, in the next to the last column of Table 1-1. The ratio of concentrations for potassium ions across the muscle membrane is 1:38, so its equilibrium potential (E_K) is -96 mV. The ratio of concentrations for sodium ions is 12:1, so its equilibrium potential (E_{Na}) is +66 mV. The ratio of concentrations for chloride ions is 29:1, but its equilibrium potential (E_{Cl}) is -90 mV, since it has the opposite charge. In fact, chloride is passively distributed across the membrane in skeletal muscle, so its concentration ratio has been determined by the membrane potential. "Free" calcium ions within the cell are at a very low concentration, so the ratio is very large. The values for E_K and E_{Na} are the limits for membrane potentials in excitable cells. These values are approached but never achieved, since membranes are never permeable to only one type of ion.

Table 1-1. Distribution of important ions in mammalian skeletal muscle extracellular fluid (ECF) and intracellular fluid (ICF) in mM.

Ion	ECF $[\text{ion}]_o$	ICF $[\text{ion}]_i$	Ratio $[\text{ion}]_o/[\text{ion}]_i$	Relative g	E_{ion} (Nernst)	Driving Force with RP = -90 mV
Na^+	144	12	12/1	4	+66	+144
K^+	4	152	1/38	100	-96	-6
Cl^-	116	4	29/1	1	-90	+0
Ca^{2+}	1.3	~ 100 nM	<10,000/1	0	>+122	>+212

The last column of Table 1-1 shows **driving forces** on ions. These are the differences between their equilibrium potentials and a resting potential of -90 mV. For a resting membrane the driving force on K^+ is small, because its large permeability brings the resting potential close to E_K. The driving force on Na^+ is large (both charge and concentration tend to move Na^+ into the cell), and its permeability is small. Ionic current is the product of driving force and conductance (fifth column of Table 1-1). The ionic currents for K^+ and Na^+ are equal and opposite at the resting potential.

Generation

Membrane potentials are due mainly to diffusion of ions. The resting potential is due to a high resting conductance to K^+. As K^+ ions diffuse out, they leave negative charges behind, so the cell interior is electronegative within 1 nm of the membrane. The resting potential approaches the equilibrium potential for K^+. The slight permeability to Na^+ ("a leak") is responsible for the membrane potential being less than the K^+ equilibrium potential. Exact values for membrane potentials can be calculated from the chord conductance equation, which is an average of equilibrium potentials weighted by ionic conductances.

The Na^+, K^+ pump maintains the steady state concentrations of ions within the cell in spite of Na^+ and K^+ leaks. The movement of charges across membranes by the Na^+, K^+-ATPase pump contributes up to a third of the membrane potential in some smooth muscle cells **(electrogenic potential)** but is trivial in nerve and striated muscle cells. The magnitude of the membrane potential under diverse conditions is determined by the **concentration differences** of various ions across the membrane (the physiochemical factor) and the membrane's **relative permeabilities** to these ions (the biological factor).

Review Questions

1. If a patient is poisoned with cyanide, they cannot utilize ATP. Which of the following transport processes would be most directly affected by this loss?

   A. Transport of Na^+ out of cells
   B. Transport of glucose into cells
   C. Transport of amino acids into cell
   D. Osmosis of water into cells
   E. Filtration of water out of capillaries

2. A membrane permeable only to sodium separates 10 M and 0.1 M solutions of NaCl. Which of the following is the equilibrium potential for this system?

   A. between 110 and 130 mV, 10 M side positive
   B between 110 and 130 mV, 0.1 M side positive
   C. between 70 and 110 mV, 0.1 M side positive
   D. between 50 and 70 mV, 10 M side positive
   E. between 50 and 70 mV, 0.1 M side positive

3. Compartments X and Y below are separated by a semipermeable membrane. After adding the indicated solutions to each side, compartment X is electrically negative with respect to compartment Y. The membrane is more permeable to

X	Y
10 mM KCl 100 mM NaCl	100 mM KCl 10 mM NaCl

A. potassium than to chloride
B. sodium than to chloride
C. potassium than to sodium
D. sodium than to potassium
E. chloride than to both potassium and sodium

4. Which of the following procedures will reduce the resting membrane potential to near zero instantaneously?

A. Poisoning the Na^+, K^+-ATPase pump with a metabolic inhibitor like cyanide
B. Replacing Na^+ with choline in extracellular fluid
C. Bathing the cells with isotonic KCl solution
D. Replacing Cl^- with SO_4^{2+}
E. Blocking conductance channels for Cl^-

5. A neurotoxin is applied to resting skeletal muscle cells, irreversibly increasing sodium conductance, gNa, but having no effect on potassium conductance gK. The transmembrane potential will

A. hyperpolarize
B. depolarize
C. not change
D. transiently hyperpolarize and then return to what it was before applying the neurotoxin
E. transiently depolarize and then return to what it was before applying the neurotoxin

ACTION POTENTIALS

Action potentials are the signals used for communication throughout the nervous system and excitable tissues. They are characterized by a rapid reversal of membrane potential (to + inside), an "all-or-nothing" response, and the ability to propagate. Recovery to the resting potential is rapid in nerve and muscle cells (about 1 msec) but slower in cardiac and smooth muscle cells. Action potentials are initiated by **graded potentials**, either **generator potentials** at sensory nerve terminals or **synaptic potentials** of neurons. Such graded potentials are local (do not propagate), graded (can be summated), have no threshold and no refractory period. They rise in about 1 msec and decay in a few msec, so their time course is slower than nerve action potentials. Synaptic potentials include endplate potentials, excitatory postsynaptic potentials and inhibitory postsynaptic potentials. Sensory nerve terminals may have receptor potentials in separate receptor cells or generator potentials (see p. 29).

Generation

The ability to generate action potentials is called electrical excitability. Excitable membranes contain voltage-sensitive Na^+ conductance channels. The changes of membrane potential (E_m) and associated ion conductances (gNa and gK) during an action potential are shown in Figure 1-1.

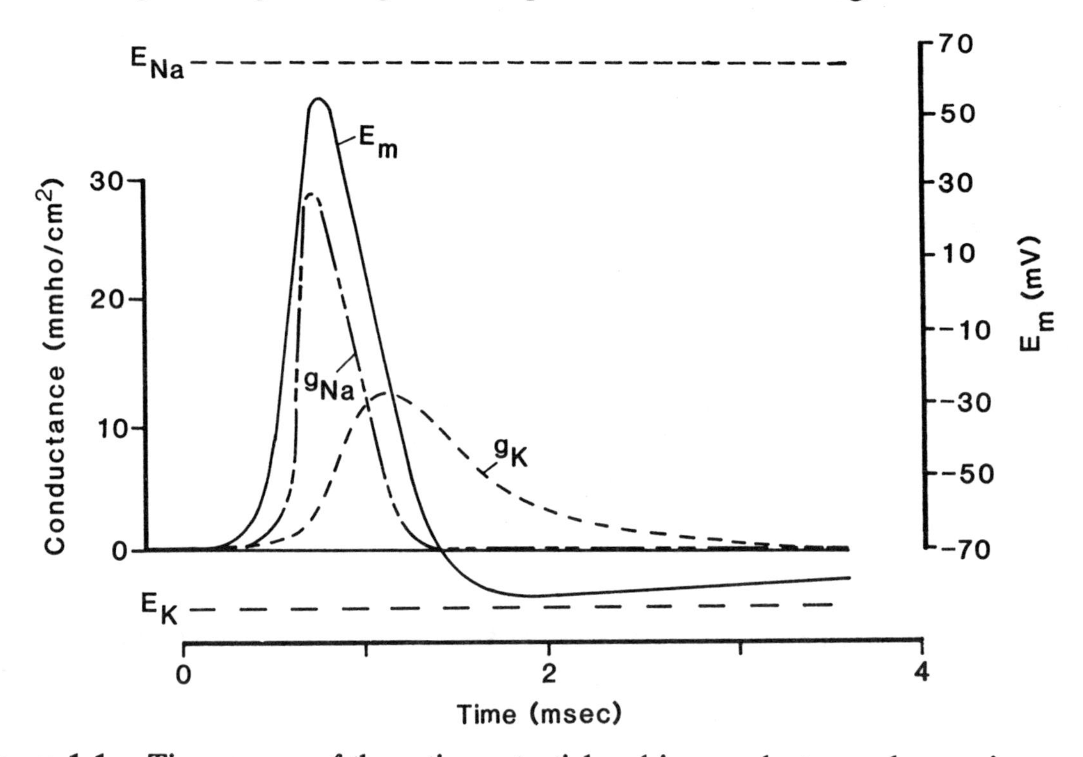

Figure 1-1. Time course of the action potential and ion conductance changes in a nerve.

Activation. Depolarization from a neighboring graded potential reduces the resting membrane potential. Suprathreshold depolarizations (about 10 mV) trigger a positive feedback sequence, since increased inflow of Na^+ causes more depolarization, which increases Na^+ conductance and then more inflow of Na^+. **Threshold** is the membrane potential level where action potentials are initiated. Subthreshold depolarizations may cause a **local response**, a slightly increased depolarization without sufficient Na^+ conductance channels being activated to sustain the positive feedback cycle. As activation continues, the relative conductance of the membrane to Na^+ increases to about 50 times that to K^+, so the membrane moves toward the Na^+ equilibrium potential (E_{Na}). This is responsible for the **rising phase** of the action potential. The action potential shows an **overshoot**, when the membrane potential passes zero and is reversed (+ inside the membrane). Artificially decreasing Na^+ outside the cell decreases the amplitude of the action potential and its rate of rise.

Inactivation. The Na^+ conductance channels close spontaneously within fractions of a msec after opening. Single ionic channels can be studied by "patch clamping". Such studies show that individual Na^+ conductance channels turn on for periods of 0.1 to 1.0 msec before switching off spontaneously.

The action potential approaches but never reaches E_{Na}. As the conductance to Na^+ decreases, the relative conductance of K^+/Na^+ increases, so the membrane **repolarizes** back toward the membrane potential, where gK is 35 times gNa. In addition, voltage sensitive K^+ channels open with depolarization (Fig. 1-1) but slower than gNa, so repolarization is faster than if gK were unchanged. Also, gK is still greater than at rest after repolarization to the original membrane potential, so a slight hyperpolarization, called the **undershoot**, is seen in some excitable cells (Fig. 1-1 at 2 msec).

Refractory Periods and Accommodation. The membrane demonstrates an **absolute refractory period** similar to the duration of the action potential spike. During this time the membrane cannot be stimulated to produce a second action potential, because gNa is entirely inactivated. This is followed by a **relative refractory period** that lasts up to 5-10 msec after the action potential peak. Significantly larger depolarizations are required for spike initiation during this period, because the prolonged increase in gK opposes depolarization and raises the threshold. Also, action potential amplitude is less than normal, since some Na^+ channels are still recovering from being activated. **Accommodation** is an increased threshold for spike initiation. Slight depolarization (less than threshold) causes a continuous activation and inactivation of some gNa channels and a sustained partial activation of gK. So less than normal numbers of Na^+ channels are available for activation.

Review Questions

6. At the peak of the action potential

   A. gNa is much less than gK
   B. the transmembrane potential is reduced in amplitude but unchanged in sign
   C. the driving force for Na^+ is greater than at the resting potential
   D. the driving force for K^+ is less than at the resting potential
   E. Na^+ conductance channels cannot be activated

7. If a nerve were placed in a solution containing one-half the normal $[Na^+]_o$, which of the following would be increased?

   A. Na^+ equilibrium potential (E_{Na})
   B. Overshoot of the action potential
   C. Rate of rise of the action potential
   D. Rate of fall of the action potential
   E. Resting potential

8. The relative refractory period of the neural membrane is characterized by

   A. partial recovery of gNa
   B. complete inactivation of gK
   C. a lower than normal threshold for action potential triggering
   D. a decreased potassium equilibrium potential (E_K)
   E. an increased sodium equilibrium potential (E_{Na})

Propagation

A propagating action potential depolarizes a length of nerve axon. This region with a reversed membrane potential (ie, + inside) is called the **active locus.** The movement of Na^+ in through the cell membrane at this region sets up a **local circuit current** that depolarizes the resting membrane ahead of the active locus. This depolarization is produced by outward flowing **capacitive** current across the inactive membrane in a local circuit, which removes positive charge from the outer membrane surface and adds positive charge to the inner membrane surface. Sufficient depolarization brings adjacent inactive membrane to threshold for a regenerative increase in gNa, and the active locus migrates smoothly along the unmyelinated axon. Behind the active locus repolarization occurs as high gNa is inactivated and gK increases. In myelinated axons the length of local circuits is greatly extended by the insulating layer of the myelin sheath. Local circuits can cross the membrane only at **nodes of Ranvier** to create active loci. Consequently, propagation of the action potential is discontinuous and jumps from node to node. This process, called **saltatory conduction**, significantly increases action potential conduction velocity. It also requires less energy for the Na^+, K^+ pump, since only nodes are depolarized for propagation.

Axon excitability and conduction velocity are directly proportional to the square of the axon radius. Consequently, larger axons have lower thresholds and faster conduction velocities than smaller axons. The **compound action potential** is the extracellular summation of thousands of action potentials, recorded from nerve trunks. It increases in amplitude with stronger stimulation of the nerve trunk, showing threshold differences among axons. **Temporal dispersion** is a function of differing conduction velocities among fiber groups. Compound action potential components increase their width and latencies at longer conduction distances.

Conduction velocity in patients is measured by stimulating a large subcutaneous nerve and recording the latency of the compound action potential at one or more locations on the same nerve either proximal or distal to the stimulation site. Such measurements are difficult, since nerve action potentials produce weak external signals. The latency of a muscle action potential after stimulating a motor nerve is inaccurate, since it includes neuromuscular delay. Consequently, the best method for measuring conduction velocity in motor nerves is to stimulate a nerve near the muscle and then farther away and record over the innervated muscle. Conduction velocity (m/sec) is calculated as the different in distance (mm) divided by the difference in latency (msec).

Agents that change the responses of excitable membranes are classified as membrane excitants or depressants.

Membrane excitants include

- Hypocalcemia - destabilizes resting membrane, lowers threshold, may provoke spontaneous spike initiation (eg, low calcium tetany in hyperventilation and respiratory alkalosis)
- Veratrum - antihypertensive alkaloid; prolongs active gNa time, promoting repetitive action potentials
- Hyperkalemia - depolarizes, moves resting potential towards threshold

Membrane depressants include

- Temperature - hypothermia slows membrane processes, acts as a local anesthetic
- Hypercalcemia - raises threshold by increasing membrane stabilization (loosely-bound Ca^{2+} on excitable membrane screens Na^+ from Na^+ conductance channels)
- Hypokalemia - hyperpolarizes resting potential, raises threshold
- Procaine - local anesthetics block activation of gNa (the effect is inversely proportional to axon diameter), affects small (eg, slow pain) fibers first
- Crush injury - traumatic crush effects are proportional to diameter; larger fibers are injured first, sparing smaller fibers
- Hypoxia - effect is proportional to axon diameter; larger fibers are depressed first, since more Na^+-K^+ transport sites are required per unit length to maintain the resting potential

Review Questions

9. The myelin sheath

   A. decreases conduction velocity
   B. decreases the relative refractory period
   C. increases the energy expenditure for membrane recovery
   D. is interrupted by nodes of Ranvier
   E. is found on C-fibers

10. A local anesthetic like procaine acts by

   A. facilitating sodium influx into excitable cells
   B. promoting potassium efflux from excitable cells
   C. blocking Na^+ conductance channels
   D. blocking the Na^+-K^+ pump
   E. depolarizing the membrane potential

11. The ulnar nerve of a patient is stimulated at the elbow and again 24 cm farther down the arm at the wrist. Both stimuli produce compound action potentials in the hypothenar muscle of the hand. What would be the latency difference for the two muscle action potentials if the nerve conduction velocity were 60 meters/sec?

   A. 0.2 msec
   B. 0.4 msec
   C. 2 msec
   D. 4 msec
   E. 40 msec

NEUROTRANSMISSION

Propagated action potentials carry information via axons over long distances, but they do not transfer electrical impulses directly to other neurons, glands or muscle. Neuronal communication is generally accomplished through **chemical transmission** at **synapses,** recognized by structural modifications in both pre- and postsynaptic cells. Neurotransmitter is released by presynaptic nerve terminals, binds to receptors on postsynaptic cells, and leads to changes in ion conductances in their cell membranes. Such changes alter the electric potential of postsynaptic membranes, resulting in either excitation (depolarization) or inhibition (hyperpolarization) of postsynaptic cells. The sequence of events in neurotransmission is as follows:

A. Presynaptic events **(excitation-secretion coupling)**

- Arrival of the action potential at the presynaptic nerve terminal
- Depolarization of the nerve terminal by the action potential
- Entry of Ca^{2+} through voltage-gated Ca^{2+} channels
- Ca^{2+}-induced fusion of synaptic vesicles with the presynaptic membrane
- Release of neurotransmitter into the synaptic cleft by exocytosis
- Diffusion of neurotransmitter across the synaptic cleft

B. Postsynaptic events

- Binding of neurotransmitter with receptors on the postsynaptic membrane
- Activation of receptors which causes changes in ion conductances
- Depolarization or hyperpolarization of the postsynaptic membrane
- Dissociation of the neurotransmitter from the receptor
- Removal of the neurotransmitter from the synaptic cleft via enzymatic degradation, reuptake or diffusion

There is a synaptic delay of approximately 0.5 msec between the arrival of the action potential at the nerve terminal and the generation of the action potential at the postsynaptic cell. Most of this delay is the time for entry of Ca^{2+} into the nerve terminal and the Ca^{2+}-induced release of neurotransmitter.

Neuromuscular Transmission

Acetylcholine (ACh) is the neurotransmitter at skeletal neuromuscular junction. Depolarization of motor nerve terminals causes release of ACh, which binds to and activates nicotinic cholinergic receptors on muscle endplates. The electrical events at the postsynaptic membrane are miniature endplate potentials (MEPP) and endplate potentials (EPP). MEPPs are due to spontaneous release of ACh. The release is random, at an average frequency of 1 per second. The amplitude of a MEPP is approximately 0.5 mV and is believed to result from the release of a single vesicle of ACh, also referred to as a "quanta" of neurotransmitter. EPPs are 15-40 mV in amplitude and are caused by the synchronous release of ACh from more than 100 vesicles in the nerve terminal. Activation of the nicotinic receptors causes an increase in the permeability of Na^{+} and K^{+}. The resulting increase in gK and gNa depolarizes the membrane to the algebraic sum of the E_K and E_{Na} (Table 1-1, p. 3). Thus the equilibrium potential for the EPP (E_{EPP}) is -15 mV.

Depolarization of the endplate causes a local circuit current to flow from the adjacent electrically excitable membrane. When the membrane potential of this region is depolarized to threshold, an action potential is generated. The action potential propagates in both directions along the muscle fiber, causing it to contract. Once ACh activates nicotinic receptors it dissociates from them and is metabolized by acetylcholinesterase. Approximately half of the choline is taken back up into the presynaptic nerve terminal and used to synthesize new ACh.

Under normal conditions neuromuscular transmission is 100% efficient, since motor neuron action potentials always result in muscle action potentials. Under pathological conditions EPPs may be too small to generate action potentials. Two such conditions are **myasthenia gravis,** where there is a decrease in the number of neuromuscular nicotinic ACh receptors, and in **myasthenic syndrome,** where there is a depression of evoked neurotransmitter release sites. The pharmacology of neuromuscular transmission should be reviewed for the effects of agonists and antagonists of ACh and cholinesterase inhibitors.

Neuronal Synapses

The neuromuscular junction is a simple synapse with a single neurotransmitter, ACh. Activation of the muscle action potential is normally an all-or-none process. Neurotransmission between neurons is much more complex with many neurotransmitters and receptors, and many synapses on each neuron. Activation of receptors at neuronal synapses produces changes in gNa, gK or gCl. This leads to small depolarizations or hyperpolarizations of postsynaptic membranes. Postsynaptic potentials are graded potentials with amplitudes proportional to the strength of the stimulus. Postsynaptic neurons summate postsynaptic potentials spatially and temporally and "integrate" the total excitatory and inhibitory current flow at the **axon hillock** or **initial segment.** An action potential is generated when summated postsynaptic potentials bring the axon hillock region to threshold via local circuit currents.

Excitatory synapses produce **excitatory postsynaptic potentials** (EPSPs). EPSPs are 1-10 mV depolarizing potentials of several msec duration. They produce an inwardly-directed current flow through channels in the postsynaptic membrane. The neurotransmitter activates receptors to increase gNa and gK. The EPSP equilibrium potential is close to the algebraic sum of the E_{Na} and E_K, which is near 0 mV. (Typical excitatory transmitters are glutamate and aspartate.)

Inhibitory synapses produce **inhibitory postsynaptic potentials** (IPSPs). IPSPs are 1-5 mV hyperpolarizing potentials of several msec duration (often tens of msec). They are produced by outwardly-directed current. Inhibitory neurotransmitter activate receptors that increase gCl or gK. When the resting potential is close to E_{Cl}, then an IPSP will not be observed with an increase in gCl . The increased gCl effectively "clamps" the membrane potential at the E_{Cl} and inhibits depolarization. (Examples of inhibitory transmitters are glycine and GABA).

Presynaptic Inhibition. Synaptic transmission can also be inhibited by releasing less excitatory neurotransmitter. Presynaptic inhibition occurs at axo-axonic synapses and is accomplished by reducing the amount of neurotransmitter released from the presynaptic terminal. This inhibition is due to the reduction of the influx of Ca^{2+} that occurs when the nerve terminal is depolarized by an action potential. Neurotransmitter binding to receptors on the presynaptic nerve terminal depolarizes the nerve terminal and brings the membrane potential closer to threshold. This results in less neurotransmitter being released from the presynaptic terminal. Presynaptic inhibition could also be called "disfacilitation", since it is due to less excitation.

The structural and electrical elements of presynaptic inhibition are diagramed in Fig. 1-2. The axon of Neuron #3 forms an axo-axonic synapse with Neuron #1. When Neuron #3 is not active, a normal action potential is recorded in Neuron #1 (upper left of right side diagram), and an EPSP with normal amplitude is generated (lower left). If Neuron #3 is active, the terminal action potential is reduced in amplitude (upper right). Consequently, less neurotransmitter is released, and an EPSP with a decreased amplitude is generated in Neuron #2 (lower right).

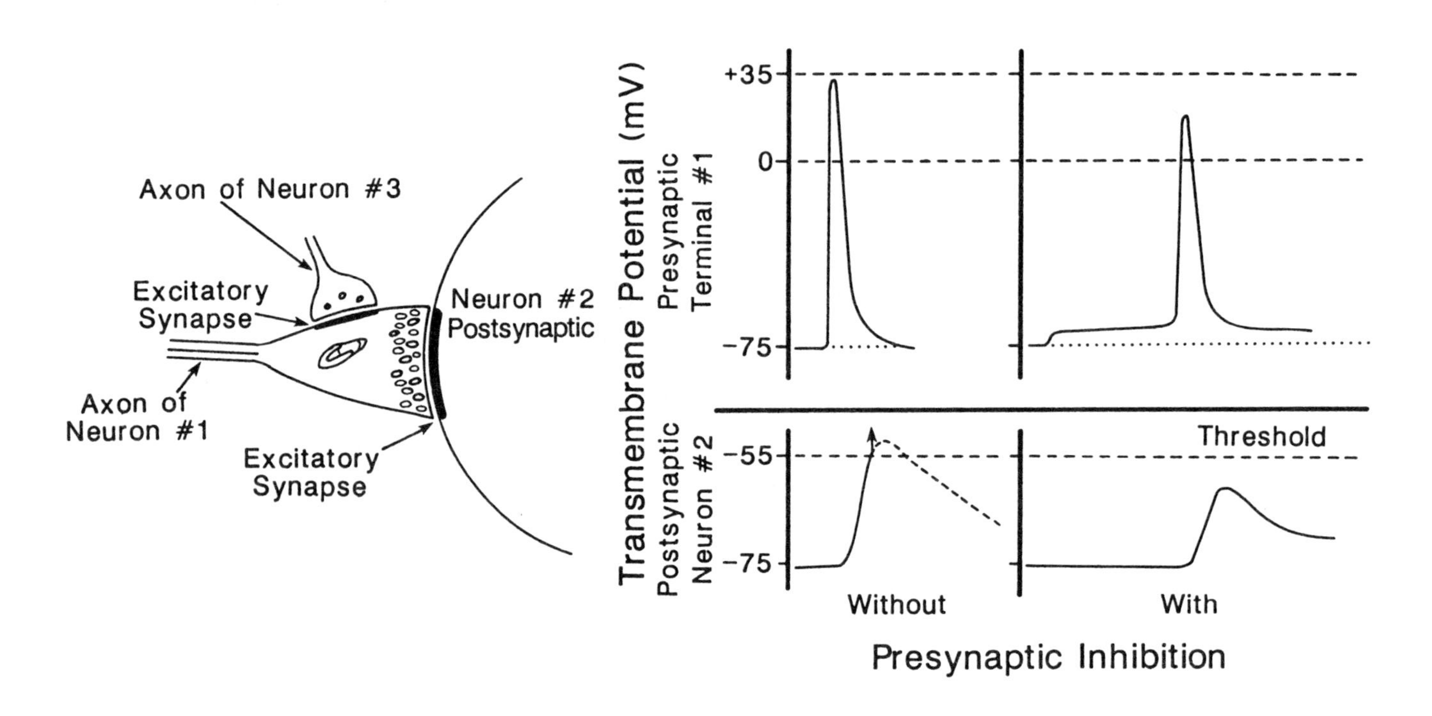

Figure 1-2. Anatomy and electrophysiology of presynaptic inhibition.

Regardless of the mechanism of excitation or inhibition, the postsynaptic neuron is morphologically arranged to algebraically sum all synaptic potentials. The extensive dendritic membrane of most neurons allows EPSPs and IPSPs to spatially **summate**. Because of the relatively long duration of synaptic potentials, the EPSPs and IPSPs may overlap and temporally summate as well. There must be extensive summation of EPSPs to fire neurons.

Electrical Synapses. While most neuronal communication is by chemical transmission, in rare instances electrical transmission can occur. This is accomplished through gap junctions and is similar to the electrical conduction across the myocardium via intercalated discs.

Neurotransmitters

A variety of chemicals are secreted as neurotransmitters. Acetylcholine is an excitatory transmitter in the autonomic and central nervous systems (CNS). Norepinephrine is used for postganglionic excitation in the sympathetic nervous system and for excitation in the CNS. Of its related amines, epinephrine is excitatory, and dopamine is inhibitory. Glutamate is the major excitatory transmitter in the CNS. GABA is the major inhibitory transmitter in the brain, while glycine is the major inhibitory transmitter in the spinal cord. These neurotransmitters are synthesized in the nerve terminals where they are released. On the other hand, peptide neurotransmitters, such as endorphins, substance P and bradykinin, are synthesized in cell bodies and transported along axons to nerve terminals. This mechanism is also true for the hormones oxytocin and vasopressin (ADH). Some special functions of neurotransmitters or the effect of their loss from the brain are shown in Table 1-2.

Table 1-2. Prime locations of transmitters in various brain regions and their functions.

Transmitter	Locus	Function
ACh	Basal nucleus of Meynert and diffuse	Loss associated with Alzheimer's disease
ACh and GABA	Neostriatum	Loss gives Huntington's disease
Dopamine	Substantia nigra	Loss gives Parkinson's disease
Norepinephrine	Ventromedial nucleus of hypothalamus	Satiety
Norepinephrine	Locus ceruleus	(highest concentration)
Serotonin	Raphe nucleus	Facilitates motor activity
Melatonin	Suprachiasmatic nucleus	Circadian rhythms
Endorphins	Anterior pituitary	Pleasure feelings

The structure of melatonin is similar to that of serotonin. Melatonin is secreted at night with longer pulses in the winter when days are shorter. Patients with excessive secretion have seasonal affective disorder (SAD), which is treated by exposure to bright lights during the day.

Review Questions

12. Miniature endplate potentials recorded at mammalian neuromuscular junctions

    A. represent the postsynaptic response to the release of one molecule of acetylcholine
    B. are a response to stimulation of the motor axon
    C. are propagated responses
    D. are associated with increases of Na^+ and K^+ conductances of endplate membranes
    E. have an equilibrium potential of about +30 millivolts

13. Anticholinesterase inhibitors, such as neostigmine, may relieve the muscle weakness of a patient with myasthenia gravis, because they

    A. prevent the release of acetylcholine
    B. prevent acetylcholine from depolarizing the muscle membrane
    C. prevent the rapid enzymatic degradation of acetylcholine
    D. inhibit the synthesis of acetylcholine
    E. inhibit the rapid synthesis of cholinesterase

14. The neurotransmitter at an excitatory synapse produces an EPSP of greatest amplitude when the membrane potential of the postsynaptic membrane is

    A. -80 mV
    B. -60 mV
    C. -40 mV
    D. 0 mV
    E. +10 mV

15. At the presynaptic nerve terminal calcium

    A. is found in high concentrations within the nerve terminal
    B. permits fusion of synaptic vesicles with the presynaptic nerve terminal membrane
    C. conductance decreases across the presynaptic nerve terminal with membrane depolarization
    D. inhibits release of neurotransmitter
    E. increases sodium conductance

16. Synaptic inhibition may be caused by

    A. increase in gCl at the postsynaptic membrane
    B. increase in gCa at the postsynaptic membrane
    C. decrease in gK at the postsynaptic membrane
    D. increase in gNa at the postsynaptic membrane
    E. increase in gCa at the presynaptic membrane

SKELETAL MUSCLE

Action Potentials

Action potentials in skeletal muscle are generated by the same changes of ionic conductances as those in neurons. They have a larger overall amplitude, since the resting potential of muscle is typically -90 mV, whereas resting potential of neurons is typically -70 mV. Skeletal muscle action potentials also repolarize in two phases, with a slower late phase after repolarizing to about -60 mV.

Excitation-Contraction Coupling

Depolarization of the muscle membrane triggers an action potential, which passes through transverse tubules (**T-tubule**) formed by invaginations of the muscle membrane. This electrical signal travels from t-tubules into the **lateral sacs** of the sarcoplasmic reticulum, initiating release of Ca^{2+} from the lateral sacs into the cytoplasm, thus causing muscle contraction. Specialized **regulatory proteins** and Ca^{2+} control the interaction between **actin** (thin filaments) and **myosin** (thick filaments). In the absence of Ca^{2+} (the resting state) the rod-like protein known as **tropomyosin** sits in the F-actin grove, thus masking the myosin binding site on actin. **Troponin** is the other major regulatory protein; it consists of three major polypeptides, troponin-T, troponin-I, and troponin-C. Troponin-T keeps the tropomyosin in the grove of F-actin. Troponin-I inhibits the interaction between actin and myosin, and troponin-C binds Ca^{2+}. This binding of Ca^{2+} causes conformational changes in troponin-C and produces cooperative changes in troponin-I and troponin-T (Fig. 1-3).

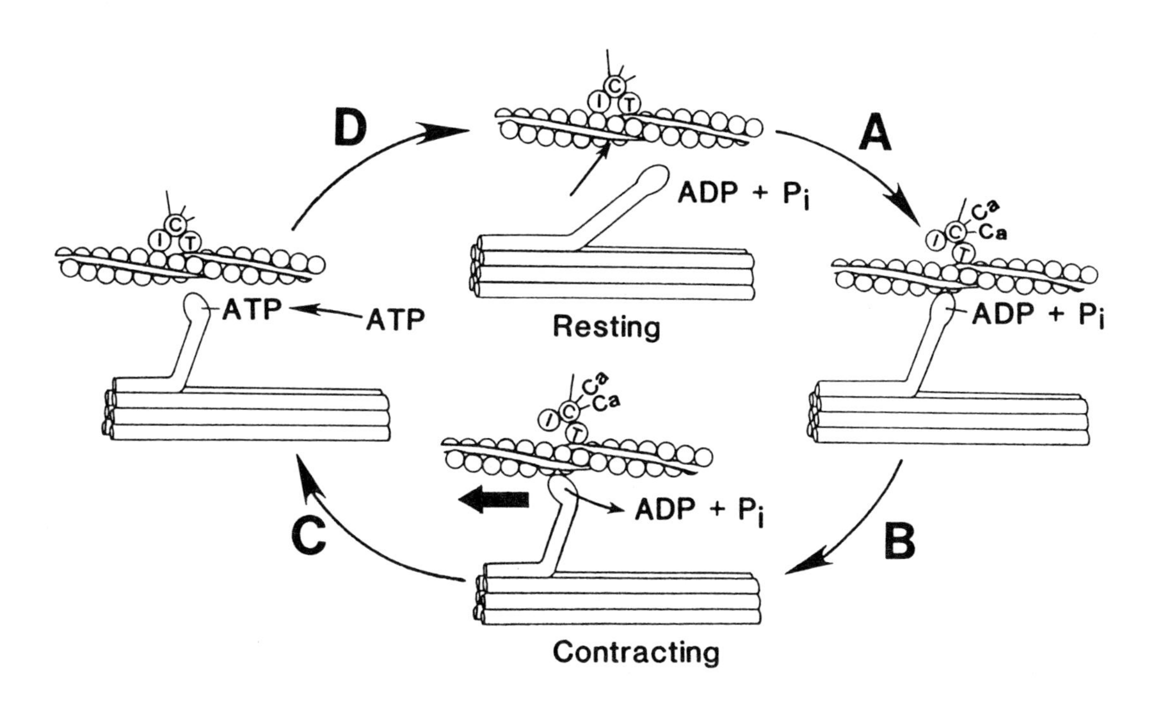

Figure 1-3. Molecular mechanism of muscle contraction.

This allows tropomyosin to move away from its blocking position, thus exposing the myosin binding site on the actin (Step A in Fig. 1-3). Then the myosin head with its ATP-hydrolyzed products binds to the actin, and the hydrolyzed products, ADP+P_i, are released into the cell (Step B in Fig. 1-3). Simultaneously, the myosin head changes its conformation and pulls against the actin filaments. Then a new ATP molecule binds to the myosin head, causing its detachment from the actin filaments (Step C in Fig. 1-3). Finally, the myosin head hydrolyzes the ATP molecule into ADP+P_i (Step D in Fig. 1-3). Calcium ions remain bound during steps C and D, and this cycle continues many times during a muscle twitch to move the actin and myosin filaments past one another. The contraction ends when reuptake of Ca^{2+} into the sarcoplasmic reticulum in an ATP-dependent process reduces the Ca^{2+} concentration in the cytoplasm to as low as 10^{-8}M. Then Ca^{2+} is removed from troponin as shown in Step C of Fig. 1-3.

Mechanics

The response of a muscle fiber to a single action potential is called a **twitch.** One motor neuron and the many muscle fibers that it innervates is a **motor unit.** The strength of muscle contraction is dependent upon the number of motor units to a muscle that are activated. If a second stimulus arrives before the muscle is completely relaxed from an initial stimulus, then the subsequent contraction adds to the first contraction. Stimuli applied rapidly produce a **tetanus,** a summation of twitches. The resulting tension is 3- to 4-fold greater than the tension produced by single twitch. Tetanic tension is higher than twitch tension, because higher concentrations of intracellular calcium are maintained during tetanic contraction rather than decaying immediately as cytosolic Ca^{2+} is pumped back into the sarcoplasmic reticulum. In addition, the elasticity of muscle is not fully stretched during the contractile phase of a twitch. Partial tetani are used during muscle contractions in man, with activation at frequencies of 5-50/sec. Since a 10-fold increase of frequency produces less than a 4-fold increase of tension, gradation of muscle contraction is primarily by recruitment of additional motor units.

Muscle contractions can be expressed on the length-tension relationship (Fig. 1-4).

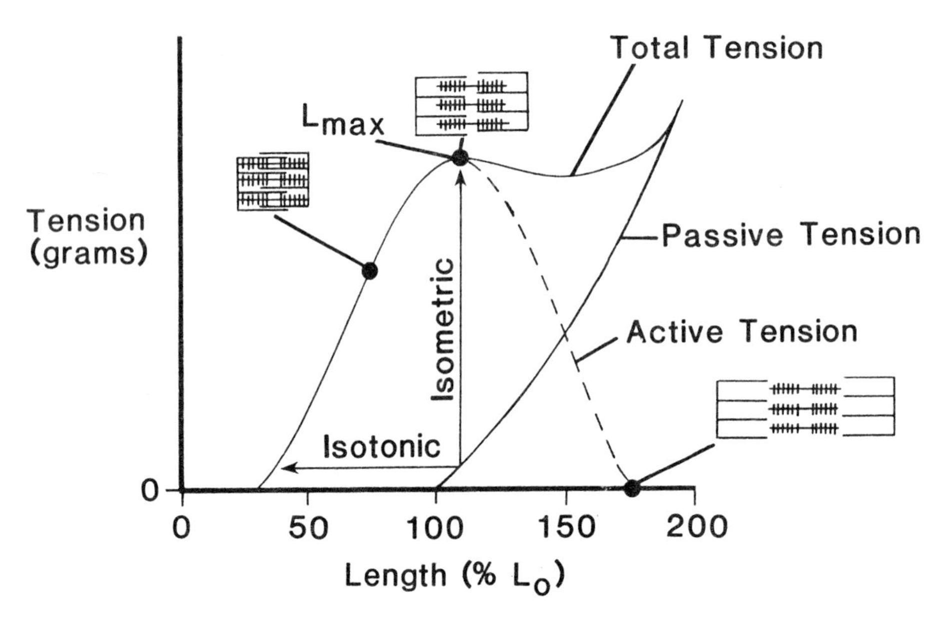

Figure 1-4. Length-tension relationship of skeletal muscle.

Passive tension (preload) is developed by stretching a muscle with various weights, showing its elastic nature. The elasticity of the muscle is due to the sarcolemma, the sarcoplasmic reticulum, and connective tissue between myofibrils. The force developed by the muscle **(active tension)** varies with the initial length of the muscle (Fig. 1-4). The total tension developed by the muscle is the sum of active tension and passive tension. Active tension is due the interaction between crossbridges of myosin from thick filaments to actin in thin filaments. During **isometric** contractions external muscle length is held constant, and developed tension is measured. On the length-tension curve an isometric contraction is a vertical line (Fig. 1-4). During **isotonic** contractions tension is constant, and the muscle is allowed to shorten carrying the load. This is work (weight times distance). On the length-tension curve an isotonic contraction is a horizontal line (Fig. 1-4). During **afterloaded** contractions the muscle contracts first isometrically (no change in length), producing sufficient tension to hold the load, and then isotonically, moving the load through a distance. Almost all movements by humans are afterloaded.

The sliding filament theory of muscle contraction explains muscle mechanics by the varying overlap of thick and thin filaments. The greatest development of active tension is just above L_0 (Fig. 1-4), where there is maximum interaction of myosin crossbridges with actin in thin filaments. At shorter lengths than L_0 (left side of Fig. 1-4), the decline in tension is due to overlap of thin filaments from adjacent sarcomeres, causing interference with the crossbridges. At longer lengths than L_0, tension declines because less cross bridges are interacting with the thin filaments.

Review Questions

17. The total amount of tension in the muscle is determined by the

    A. amount of sarcoplasmic reticular membranes
    B. distance between myosin and actin filaments
    C. amount of ATP available
    D. number of t-tubules
    E. number of myosin cross bridges interacting with actin

18. Tetanic muscle tension is greater than the twitch tension, because of

    A. intracellular calcium concentration declining during a tetanus
    B. smaller turnover of cross bridge cycles taking place
    C. elevated levels of cytoplasmic Ca^{2+}
    D. overlap of thin filaments between adjacent sarcomeres, thus interfering with cross bridge cycling
    E. blocking of myosin binding sites on actin by tropomyosin

19. Immediately after death, muscles undergo rigor mortis because

    A. most of the Ca^{2+} is pumped back into the sarcoplasmic reticulum
    B. the crossbridges are interacting with thin filaments
    C. myosin and actin are dissociated
    D. myosin with its hydrolyzed products is permanently attached to thin filaments
    E. binding of new ATP molecules to myosin heads

20. In Figure 1-4 the tension development at 1.75 L_0 is due to

    A. maximum overlap of thick and thin filaments
    B. the overlap of thin filaments from adjacent sarcomeres, thus interfering with crossbridge interactions
    C. no overlap of thick and thin filaments with the fiber stretched to its maximum limits
    D. pumping Ca^{2+} back into the sarcoplasmic reticulum
    E. Ca^{2+}-induced Ca^{2+} release from the lateral sacs

CARDIAC MUSCLE

Action Potentials

Cardiac muscle action potentials have a fast depolarization similar to skeletal muscle, but they remain depolarized for 200-400 msec. The cardiac action potential consists of five stages (Fig. 1-5A).

Stage 0:	Upstroke of depolarization is produced by regenerative, depolarization-triggered increase in gNa. The increased gNa is time dependent, and inactivation occurs in a few msec as in skeletal muscle.
Stage 1:	Early slight repolarization resulting from brief increase in gCl.
Stage 2:	Plateau of depolarization maintained by both 1) increased gCa from its resting low value and 2) decreased gK from its resting high value.
Stage 3:	Repolarization caused by 1) gCa returning to its resting low value and 2) gK returning to its resting high value.
Stage 4:	Resting potential in diastole is maintained by high gK.

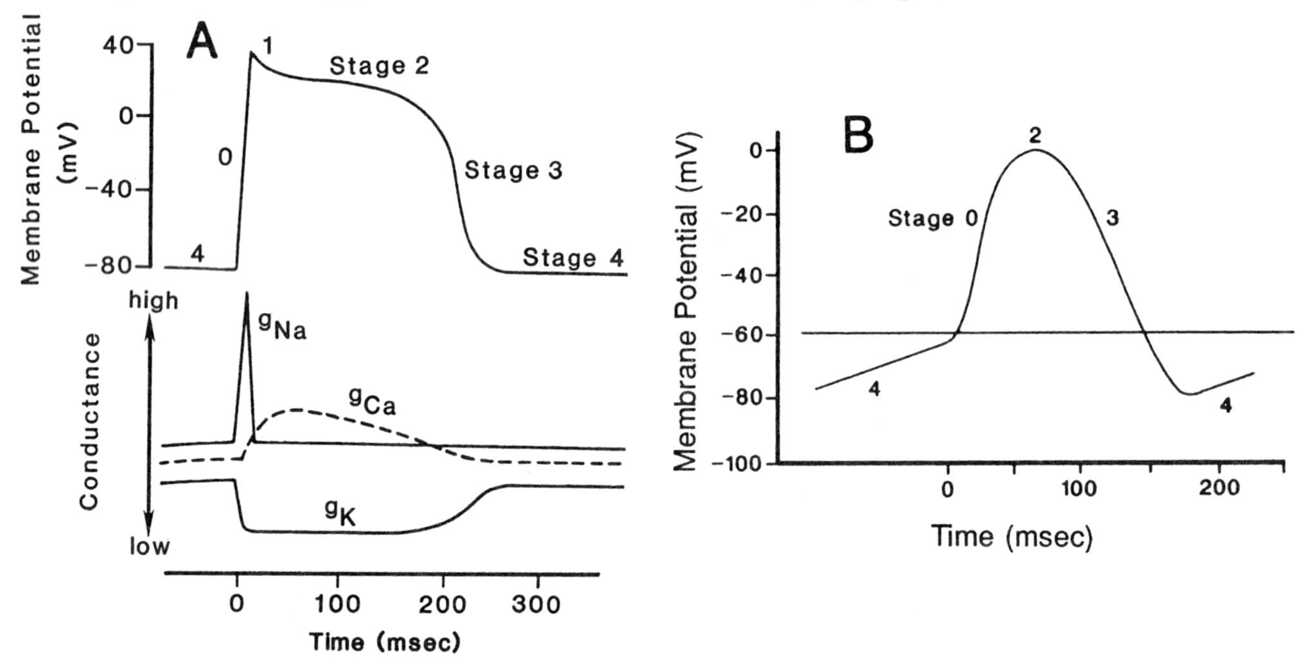

Figure 1-5. Cardiac action potentials. A - Ionic changes and stages of muscle action potentials, B - Stages of pacemaker potentials in the conducting system.

Cardiac tissue shows **automaticity** (spontaneous activity) and **rhythmicity** (the ability to beat). Stage 4 depolarization of cardiac pacemaker cells (eg, SA node) results from 1) spontaneously increasing gCa, causing depolarization and 2) possibly a spontaneous decrease in gK (Fig. 1-4B). Stage 0 triggered depolarization is primarily due to increased gCa. Agents that decrease the rate of depolarization, such as acetylcholine, increase gK hyperpolarization in Stage 4 and slow Stage 0. Agents that increase the rate of spontaneous depolarization, such as **norepinephrine (NE)**, increase gCa or gNa. Movements of Ca^{2+} slow with aging, which contributes to prolonged phases within the action potential and in coupling to contraction.

Excitation-Contraction Coupling

In cardiac muscle excitation-contraction coupling requires extracellular calcium ions, which induce release of Ca^{2+} stored in lateral sacs of the sarcoplasmic reticulum, thereby increasing free Ca^{2+} concentrations inside the cell. Thus in cardiac muscle excitation-contraction coupling requires both extracellular and sarcoplasmic reticular Ca^{2+}. The increase in cytoplasmic Ca^{2+} binds to troponin-C, thereby producing cardiac contraction. Contraction is terminated by both reuptake of Ca^{2+} into the sarcoplasmic reticulum and extrusion of Ca^{2+} into the extracellular space.

Mechanics

The contractile protein assembly and the molecular mechanism of contraction in myocardium is identical with that of skeletal muscle. Cardiac muscle cannot be tetanized, due to the long duration of cardiac action potentials. The mechanical response of the cardiac muscle is completed within the refractory period. The heart beat is the almost synchronized sum of all cardiac twitches. Myofibrillar ATPase activity declines with age.

The force of cardiac contraction can be modified in two ways. A change in **preload** or initial starting length causes the force of the next beat to be greater. A change in **contractility** also changes force development. Contractility is the altered ability to develop force at a given length. That is, if force of contraction increases while preload remains constant, then contractility has increased.

Review Questions

21. In cardiac pacemaker cells of the sino-atrial node, Stage 4 spontaneous depolarization

    A. slows with elevated plasma potassium (hyperkalemia)
    B. results primarily from decreasing potassium conductance, gK
    C. results primarily from decreasing sodium conductance, gNa
    D. speeds up in the presence of acetylcholine
    E. depends upon a slowly decreasing intracellular calcium concentration

22. The discharge rate of cardiac pacemaker cells is primarily controlled by the

    A. duration of their refractory period
    B. threshold for "slow" channel activation
    C. rate of slow diastolic depolarization
    D. magnitude of the transmembrane potential
    E. temperature of blood in the right atrium

23. Activity of the vagus nerve at the sino-atrial node results in

   A. increased gCa
   B. increased gK
   C. increased rate of spontaneous depolarization
   D. increased threshold for a propagated action potential
   E. decreased extracellular concentration of K^+

SMOOTH MUSCLE

It is difficult to generalize about the activity of smooth muscles, since there are so many variations. Some smooth muscles contract in response to action potentials, others in response to graded changes in membrane potential, while some contract in the absence of changes in membrane potential. The major source of Ca^{2+} for contraction of some smooth muscle cells is the extracellular fluid, while for others it is intracellular stores. Contractions of some smooth muscles are relatively weak, while others can generate forces equal to, or greater than, those generated by skeletal muscle.

Classification

Smooth muscle tissues can be separated into two broad classifications; 1) **multiunit** smooth muscle and 2) **single-unit** or **unitary** smooth muscle. In multiunit tissues there is little communication among individual cells, and each smooth muscle cell may operate relatively independently from other cells of the tissue. Examples of multiunit tissues are the iris and piloerector muscles. In single-unit tissues the individual smooth muscle cells communicate via low resistance pathways **(gap junctions)** between cells. Thus electrical events occurring in one cell can be passed on to adjacent cells, and cells of these tissues contract in a coordinated fashion. Examples of single-unit tissues are smooth muscles of the gastrointestinal tract, the uterus and smaller blood vessels.

Excitation-Contraction Coupling

When at rest (not contracting) smooth muscle cells in some tissues have relatively stable membrane potentials, while muscle cells in other tissues exhibit rhythmic oscillations in membrane potential **(slow wave depolarizations)**. There are two mechanisms for excitation-contraction coupling in smooth muscle: electromechanical and pharmacomechanical coupling. In **electromechanical coupling** (sometimes referred to as e-c coupling), excitation of the cell results in partial depolarization of the membrane and an increase in cytosolic Ca^{2+}. Depending upon the tissue, the depolarization may take the form of an action potential or a graded change in membrane potential. In **pharmacomechanical coupling** the cell is excited to contract, and cytosolic Ca^{2+} increases, in the absence of a significant change in membrane potential.

Regardless of the mechanism of excitation, an increase in cytosolic Ca^{2+} is required to activate the contractile process. There are two potential sources of Ca^{2+}: the extracellular fluid and the sarcoplasmic reticulum of the smooth muscle. Extracellular Ca^{2+} can enter the cell through **voltage-sensitive channels** (electromechanical coupling) and/or **receptor-operated channels** (pharmacomechanical coupling).

Several mechanisms can cause release of Ca^{2+} from the sarcoplasmic reticulum. The relative contribution of Ca^{2+} from extracellular fluid and sarcoplasmic reticulum in raising cytosolic Ca^{2+} varies with different smooth muscle tissues and different stimuli.

Contraction

Smooth muscles contain the same contractile proteins as skeletal muscles, but the molar ratio of actin to myosin is much higher in smooth muscle (30:1) compared to skeletal muscle (3:1). Smooth muscle does not contain troponin, and its actomyosin ATPase activity is significantly lower than in skeletal muscle.

The mechanism for initiating contraction in smooth muscle involves calmodulin and myosin light chain kinase. Increased cytoplasmic Ca^{2+} binds to **calmodulin**, and the Ca^{2+}-calmodulin complex activates **myosin light chain kinase** (MLCK). MLCK phosphorylates myosin light chains, which then interact with actin in thin filaments. Increased actomyosin ATPase activity accompanies contraction.

Smooth muscle contractile activity is determined by the relative activity of the two enzymes, myosin light chain kinase which is regulated by Ca^{2+} and **myosin light chain phosphatase** (MLCP). When Ca^{2+} concentrations are high, myosin is phosphorylated by MLCK and contraction occurs. When cytosolic Ca^{2+} concentrations are reduced to intermediate levels, myosin is dephosphorylated by MLCP. The dephospho-form of myosin cross bridges may remain attached to actin in thin filaments to maintain force. This is known as the **latch-state** and is important for tissues that maintain active tone (eg, blood vessels and sphincters). Relaxation occurs when cytosolic Ca^{2+} concentrations decline to resting levels. Such decline is dependent upon extrusion of Ca^{2+} from the cytoplasm into the extracellular space and/or into the sarcoplasmic reticulum.

Review Questions

24. Electromechanical coupling

    A. does not occur in single-unit smooth muscle
    B. does not require elevated cytosolic Ca^{2+} to cause contraction
    C. activates receptor-operated channels
    D. activates potential-dependent channels
    E. does not result in phosphorylation of myosin light chains

25. Contraction of smooth muscle begins when

    A. Ca^{2+} binds to troponin-C
    B. Ca^{2+} binds to the cytosolic protein calmodulin
    C. myosin light chains are phosphorylated
    D. myosin is dephosphorylated by phosphatases
    E. troponin-I is phosphorylated

AUTONOMIC NERVOUS SYSTEM

Peripheral Control

The peripheral nervous system is composed of the **autonomic nervous system (ANS)** and the somatic nervous system. The ANS innervates smooth muscle, cardiac muscle, and glands. The somatic nervous system innervates skeletal muscle. The ANS originally was defined by its efferent innervation or motor control of the viscera. However, sensory or afferent innervation from visceral organs to the central nervous system is critical for central nervous system processing of autonomic motor outflow. Anatomically the ANS can be separated into two major divisions; the **parasympathetic nervous system** and the **sympathetic nervous system.**

In both divisions two neurons are connected in series, and these neurons carry efferent information from the central nervous system to the effector or end-organ. As shown in Fig., the first neuron in the series, the **preganglionic neuron** (labeled "1"), originates in the central nervous system, and subsequently makes synaptic contact on the cell body of the **postganglionic neuron** (labeled "2") located in the peripheral autonomic ganglia. The postganglionic neuron directly innervates the effector organ. In comparison, the somatic nervous system is composed of a single neuron (labeled "3" in Fig. 1-6) which originates in the spinal cord and whose axon innervates the effector organ (ie, skeletal muscle). The parasympathetic system is also called the **craniosacral division** of the ANS. The term "cranio-" describes parasympathetic nervous outflow from preganglionic cell bodies located in the midbrain and medulla, which gives rise to cranial nerves 3, 7, 9 and 10. The term "-sacral" refers to preganglionic cell bodies in spinal cord segments S_2 to S_4 (not drawn in Fig. 1-6). Each parasympathetic preganglionic neuron (labeled "P" in Fig. 1-6) has a long axon that innervates postganglionic neurons located near or on the end-organ. Therefore, postganglionic parasympathetic neurons extend only a very

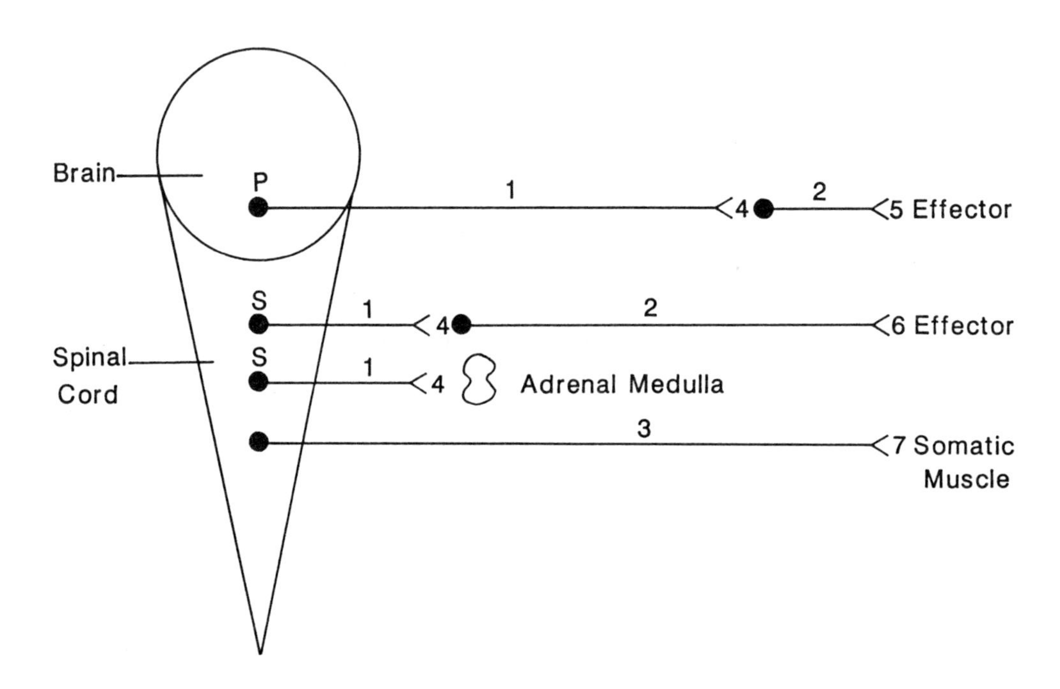

Figure 1-6. Patterns of autonomic efferent and somatic neurons.

short distance. This arrangement allows parasympathetic control to exert relatively discrete control on peripheral organ function. The sympathetic nervous system is also referred to as the **thoracolumbar division** of the ANS. The term "thoracolumbar" describes the location of sympathetic preganglionic neuronal cell bodies in spinal cord segments T1-L3, specifically in the intermediolateral cell column. Sympathetic preganglionic neurons have relatively short axons which terminate on neurons located in the paravertebral ganglia (sympathetic chain) or the more distal prevertebral ganglia such as the superior mesenteric ganglia. Sympathetic preganglionic neurons also innervate the adrenal medulla. Sympathetic postganglionic axons are relatively long and directly innervate end-organs. The notable exception for direct sympathetic innervation of end-organs is the GI tract, where sympathetic neurons innervate enteric nerve plexuses. Historically, the sympathetic outflow was believed to be widespread or generalized. However, recent evidence shows that the sympathetic nervous system can produce discrete reactions to appropriate stimuli.

Physiology

The two divisions of the ANS are most active during different conditions. The parasympathetic division is involved in the maintenance, conservation, and protection of bodily resources and energy stores **(anabolic action)**. Parasympathetic neurons excite end-organs responsible for absorbing food (gastrointestinal tract) and reduce activity of organs that expend energy (eg, heart rate is decreased). The sympathetic division is involved in expenditure of bodily resources or energy **(catabolic action)**. It can act as a whole to promote rapid adaptation to or preparation for stress. Sympathetic activation supports the "fight-or-flight" response in reaction to emergency situations, and it includes release of epinephrine from the adrenal medulla. In many organs stimulation of the parasympathetic and sympathetic nervous systems produce opposite responses. For example, parasympathetic stimulation decreases heart rate while sympathetic stimulation increases heart rate. Antagonistic effects due to simultaneous activation of both systems do not occur under physiological circumstances; rather the two divisions work in concert, a process referred to as **functional synergism**. For example, to increase heart rate, sympathetic outflow to the heart increases, while the inhibitory influence of parasympathetic outflow on heart rate is withdrawn. Another important concept in autonomic physiology is **tone**. Tone describes the resting or basal outflow from either the sympathetic or parasympathetic nervous systems. Tonic outflow provides the ANS with the ability to produce a response by reducing background neural activity to an organ. In the case of heart rate control, withdrawal of background parasympathetic inhibitory influence allows heart rate to increase. An important implication of autonomic tone is that dual innervation of an organ or organ system by both sympathetic and parasympathetic systems is not required. For example, control of vascular smooth muscle is maintained by sympathetic outflow; therefore, a reduction in tonic sympathetic outflow to vascular smooth muscle results in vasodilation. An example of parasympathetic tone is tonic constriction of the sphincter muscle of the iris. The possibility also exists that an organ or tissue may be innervated without receiving tonic innervation. For instance, sympathetic tone is absent under resting conditions but can be increased when needed to influence sweating and lipolysis. Specific end-organ responses to parasympathetic and sympathetic excitation should be reviewed in a textbook.

Neurotransmission and Receptors

All ANS neurons leaving the central nervous system (ie, preganglionic sympathetic and parasympathetic neurons) secrete ACh as their neurotransmitter. Thus, the synapses labeled "4" in Fig.

1-6 are **cholinergic**, ie, they secrete ACh. Acetylcholine released by preganglionic neurons excites **nicotinic receptors** on postganglionic neurons. Somatic motor neurons also release ACh, which excites nicotinic receptors on skeletal muscle (neuroeffector junction "7"), but they are different from ganglionic nicotinic receptors. Neuromuscular blocking drugs, such as curare, block nicotinic receptors at the neuromuscular junction but do not affect ganglionic nicotinic receptors. ACh is also the neurotransmitter for postganglionic parasympathetic neurons (neuroeffector junction labeled "5" in Fig. 1-6). The cholinergic receptors located on effector organs are termed **muscarinic** receptors; they are blocked by atropine.

Postganglionic sympathetic neurons secrete the neurotransmitter **NE (noradrenaline)** and are classified as adrenergic neurons (neuroeffector junction "6" in Fig. 1-6). NE excites two types of receptors on effectors; **alpha (α)-receptors** and **beta (β)-receptors**. Alpha-adrenergic receptors are subclassified into $alpha_1$ and $alpha_2$ subtypes. $Alpha_1$ receptors are located on target organs and tissues and are primarily responsible for mediating the actions of sympathetic release of norepinephrine. $Alpha_2$ receptors are located presynaptically on sympathetic nerve terminals and postsynaptically. Activation of presynaptic $alpha_2$ receptors produces a feedback inhibition of norepinephrine release. Postsynpatic $alpha_2$ receptors mediate responses similar to $alpha_1$ receptors. Beta-adrenergic receptors are subclassified into $beta_1$ and $beta_2$ receptors. Activation of $beta_1$ receptors produce excitatory effects on the heart, such as increased heart rate, conduction velocity and contractility. $Beta_2$ receptors mediate the inhibitory effects of beta-receptors, as well as dilator actions on vascular and bronchial smooth muscle. The adrenal medulla, which is analogous to postganglionic sympathetic neurons, secretes both **epinephrine (adrenaline)** and norepinephrine into the blood (80% epinephrine and 20% norepinephrine). Some postganglionic sympathetic terminals secrete ACh which activates muscarinic receptors on eccrine sweat glands (conventional temperature regulation).

Review Questions

For Questions 26-29. For each manipulation listed in the questions below, MATCH the most likely combination of effects shown in each row of the table.

	Lipolysis	Heart Rate	Vascular Muscle Tone	Sweat Gland Secretion	Tone of Circular Muscle of Iris
A.	0	↓	0	0	↑
B.	↑	↑	↑	↑	0
C.	0	0	↑	↑	↑
D.	0	↑	↓	0	↓
E.	↑	↑	↑	0	0
F.	0	↑	0	↓	↓

Symbols ↑ = increase; ↓ = decrese; 0 = no change

26. Activation of sympathetic nervous system

27. Administration of a large dose of epinephrine to a patient

28. Activation of parasympathetic nervous system

29. Administration of parasympatholytic agent (or muscarinic antagonist) such as atropine

30. Stimulation of β_1-adrenergic receptors produces which of the following effects?

    A. Vasoconstriction
    B. Bronchial smooth muscle constriction
    C. Increased sweating
    D. Contraction of the urinary bladder
    E. Increased myocardial contractility

31. In the sympathetic division of the autonomic nervous system

    A. the cell bodies of preganglionic neurons are found in the brainstem
    B. postganglionic neurons release epinephrine
    C. preganglionic neurons secrete ACh to excite muscarinic receptors
    D. activity is highest under stressful conditions
    E. preganglionic neurons are long and form synapses on or near the target organ

ANSWERS TO ELECTROPHYSIOLOGY QUESTIONS

1. A. The Na^+, K^+ pump uses ATP (A). Glucose and amino acids are transported into cells by secondary active transport, using the Na^+ concentration gradient as the source of energy. Filtration depends upon pressure generated by the heart (E), as does the balance of osmotic and pressure forces across cell membranes (D).

2. B. With the Nernst equation a 100X concentration ratio gives $61 \times \log_{10}(100) = 122$ mV with the 0.1 M side positive to oppose the concentration gradient.

3. D. Chloride concentration is equal on both sides (110 mM). So, net negativity on the left side results from sodium ions moving down their concentration gradient to the right.

4. C. E_K will be close to zero, since its concentration will be similar on both sides of the membrane, and permeability to K^+ will still be high. Poisoning the pump will cause the membrane potential to decrease gradually, as K^+ leaks out and Na^+ leaks in (A). The other procedures will have minor effects on resting potential.

5. B. According to the chord conductance equation, any increase in gNa relative to gK will depolarize the membrane towards E_{Na}. The depolarization continues as long as the high gNa is maintained by the toxin.

6. E. This is the absolute refractory period. gNa is still greater than gK (A), and transmembrane potential reverses sign at the peak of the action potential (B). Driving force (E_K-RP) for K^+ is at its highest value (D), and for Na^+ is at its lowest value (E_{Na}-RP) (C).

7. E. E_{Na} will decrease 19 mV. as a function of reduced $[Na^+]_o$ (A). This reduces overshoot, since E_{Na} is closer to 0 mV (A), and slows the rate of action potential rise because of decreased driving force on Na^+ (C). The membrane potential would increase about 1 mV. with the decrease of E_{Na} (E).

8. A. gK is higher than at rest (B). Threshold is higher (C). E_K and E_{Na} are unchanged (D, E).

9. D. Conduction velocity is significantly increased in myelinated fibers compared to unmyelinated fibers of equal axon radius (A). Myelination has no effect on refractory periods of action potentials (B), but it does decrease the energy expenditure necessary for membrane recovery by the Na^+, K^+ pump (C). C-fibers are unmyelinated by definition (E).

10. C. Procaine blocks voltage-sensitive membrane channels for Na^+ (A), but it does not affect gK or K^+ efflux (B) or the Na^+, K^+ pump (D). It has no effect on resting membrane potential (E). For a given length of axon smaller fibers have less Na^+ conductance channels to block, so a greater fraction of gNa channels in smaller axons are inactivated.

11. D. Conduction velocity (CV) equals conduction distance (CD) divided by conduction time (CT) between stimulation sites. The hypothenar muscle is the recording site while the stimulation site is varied. So, CV = CD/CT. Rearranging for conduction time: CT = CD/CV. Then

$$CT = \frac{24 \text{ cm}}{60 \text{ m/ms}} = \frac{240 \text{ mm}}{60 \text{ mm/msec}} = 4 \text{ msec}$$

12. D. MEPPs are local, decremental responses (C) from spontaneous release (B) a single synaptic vesicle which contains thousands of molecules of ACh (A). $E_{MEPP} = E_{EPP}$ is -15 mV (E).

13. C. Answers A and D imply a presynaptic action of neostigmine that would make myasthenia gravis worse. ACh-induced depolarization of the postsynaptic membrane would be augmented (B). Cholinesterase inhibitors prevent the action of the enzyme that metabolizes ACh, but synthesis is normal (E).

14. A. Although both gK and gNa increase during an EPSP, Na current predominates because of the small driving force on K^+. When the membrane potential is hyperpolarized to -80 mV, the driving force on Na^+ is maximal while that on K^+ is minimal. Then greater current flows, and EPSP amplitude is increased.

15. B. Depolarization of the nerve terminal by an invading action potential increases gCa (C), and Ca^{2+} diffuses into the nerve terminal along its concentration gradient (A). The increase in cytosolic calcium concentration causes synaptic vesicles to fuse to the presynaptic membrane and release neurotransmitter into the synaptic cleft (D).

16. A. An increase in chloride conductance hyperpolarizes and/or clamps the membrane potential and inhibits depolarization of the postsynaptic membrane by an EPSP. An increase in gCa or in gNa or a decrease in gK would depolarize the postsynaptic membrane (B, C, D). An increase in gCa at the presynaptic membrane would cause release of neurotransmitter (E).

17. E. Sarcoplasmic reticular membranes are involved in the relaxation process by removing Ca^{2+} (A). The distance between the filaments does not vary (B). ATP by itself has no effect on the generation of tension (C). T-tubules are responsible for the propagation of action potentials but not the generation of tension (D).

18. C. Intracellular Ca^{2+} concentration is maintained by repetitive stimuli (A). The development of tension is directly proportional to the cross bridge turnover (B). Overlap of thin filaments would cause less cross bridge interaction, producing less tension (D). Blocking of myosin binding sites would result in no interaction between filaments and thus produce no tension (E).

19. D. Uptake of Ca^{2+} from the cytoplasm into the sarcoplasmic reticulum causes muscles to relax (A). Crossbridge interaction with actin results in muscle contraction (B). Myosin and actin dissociation causes muscle to relax (C). Binding of new ATP molecules during the contraction cycle causes the dissociation of myosin and actin (E).

20. C. Maximum cross bridge recycling and active tension development occurs at L_0 (A). Overlap of thin filaments occurs at 0.5 L_0 or less (B). Uptake of Ca^{2+} from the cytoplasm causes muscle to relax, but this is unrelated to sarcomere length (D). Ca^{2+}-induced Ca^{2+} release from lateral sacs causes increased cytoplasmic Ca^{2+} and therefore increased muscle contraction (E).

21. B. Hyperkalemia depolarizes the membrane, bringing Stage 4 closer to AP threshold but not continuing to depolarize (A). Decreased gNa, acetylcholine, and increased gCl slow or inhibit Stage 4 depolarization (C, D). Decreasing gCa would oppose depolarization (E).

22. C. The key is "primarily" and suggests C. SA node pacemakers do not have true refractory periods. Prolonged excitation or artificial driving of the SA node leads to "overdrive suppression" and quiescence (an ectopic focus may continue to fire at the high rate) (A). Slow (gCa) channel activation threshold does not change normally (B). D does not change normally except in the presence of ACh. A decreased blood temperature would slow discharge rate; this is not normal nor a "primary" control.

23. B. The vagal axon endings release ACh, which increases gK and hyperpolarizes the pacemaker cell, decreasing the slope of depolarization (C) and pacemaker rate. Threshold remains constant (D), ACh does not effect gCa (A), and increased $[K^+]_o$ is not sufficient to alter any pacemaker events (E).

24. D. Electromechanical coupling results in depolarization which activates potential-dependent (D) but not receptor-operated (C) channels. Electromechanical coupling occurs in single-unit smooth muscle (A), results in elevation of cytosolic Ca^{2+} (B) and phosphorylation of myosin light chains (E).

25. C. Troponin is absent in smooth muscle (A, E). Binding of Ca^{2+} to calmodulin alone has no effect (B). Dephosphorylation of myosin by phosphatases causes relaxation of smooth muscle (D).

26. B. Activation of the sympathetic nervous system increases energy utilization, so increased lipolysis is promoted. Heart rate, vascular smooth muscle tone, and sweat gland secretion are all increased. Remember sweat gland secretion is mediated through sympathetic postganglionic nerves which release ACh. The sympathetic nervous system does not innervate the circular muscle of the iris, but it can affect pupillary diameter through effects on radial muscles of the iris.

27. E. The actions of epinephrine are very similar to that produced by stimulation of the sympathetic nervous system. At low concentrations epinephrine will activate β_2-adrenoreceptors predominantly found in the skeletal muscle vasculature, while at higher doses epinephrine will activate α-adrenergic receptors throughout the entire vasculature, causing vasoconstriction and increased peripheral resistance. In contrast to activation of the sympathetic nervous system, epinephrine will not stimulate eccrine sweat glands, which are stimulated by sympathetic release of acetylcholine on to muscarinic receptors.

28. A. Parasympathetic nervous system innervation does not influence lipolysis, vascular tone or sweat gland secretions. Parasympathetic activation reduces heart rate, and the circular muscle of the iris is stimulated to produce pupillary constriction (ie, miosis).

29. F. The tachycardia and decrease in circular muscle tone of the iris result from blockade of tonic parasympathetic innervation to these tissues. Lack of effects of atropine on lipolysis and vascular muscle tone indicate paucity of parasympathetic tone to these systems. Eccrine sweat gland secretion is blocked by atropine, since the sympathetic nervous system controls these glands through release of acetylcholine which activates muscarinic receptors.

30. E. $Beta_1$ adrenergic receptors mediate the excitatory effects of the sympathetic nervous system on the myocardium (E). Vasoconstriction is mediated by alpha-adenergic receptors (A). Bronchial smooth muscle constriction, increased sweating, and contraction of the urinary bladder are produced through activation of cholinergic muscarinic receptors (B, C. D).

31. D. Sympathetic preganglionic cell bodies are found in spinal cord at levels T1-L3 (A). Sympathetic postganglionic neurons release norepinephrine (B). These neurons release ACh to excite nicotinic receptors on postganglionic cell bodies (C). The parasympathetic division has long preganglionics and short postganglionics (D).

NEUROPHYSIOLOGY

SENSORY RECEPTORS

Sensory systems process information that causes either a sensation or a perception of an event within the body or of the outside world. In **sensory transduction** a sensory receptor converts one form of energy into an electrical signal, eg, a neuronal action potential. Each receptor is most sensitive to one particular form of energy; this is called its **adequate stimulus.** Sensory receptors can be specialized endings of sensory (or afferent) neurons, such as mechanoreceptors like Pacinian corpuscles, or separate cells adjacent to sensory neurons, such as vestibular hair cells or photoreceptors. Separate cells produce **receptor potentials** that release chemical transmitter onto an adjacent nerve terminal of afferent neurons. **Generator potentials** occur in the specialized or unspecialized ending of the primary afferent neuron itself. Both types of graded potentials trigger action potentials in an adjacent nerve membrane via local circuit currents. Sensory information is transmitted by action potentials over long distances to the central nervous system.

The sensory receptor transduction process involves the following sequence:

1. A stimulus opens Na^+ channels in the receptor membrane.
2. Increased Na^+ conductance depolarizes the membrane to produce a generator potential (or receptor potential) that is decrementing and non-propagating.
   a. The amplitude of the generator potential is a function of the stimulus intensity.
   b. Generator potentials can undergo spatial (stimuli at two or more stimulus sites) and temporal (repeated stimuli at the same site) summation.
3. Initiation of action potentials at the first node of Ranvier (or adjacent electrically-excitable membrane in unmyelinated axons).
   a. Action potential frequency is a function of receptor type and stimulus intensity.
   b. Larger graded potentials cause more action potentials per second.

A **sensory unit** is the area innervated by one afferent axon. Ability to discriminate between two adjacent stimuli in sensory space (eg, two-point discrimination in somesthesia) improves with smaller sensory units and less overlap. Receptors may adapt to the presence of a stimulus of constant intensity; **adaptation** probably involves decreased Na^+ conductance (g_{Na}). Adaptation causes a decrease of generator potential amplitude and therefore a reduction in action potential frequency. **Rapidly adapting** or **phasic receptors** have a generator potential that decays rapidly even when the stimulus intensity remains constant. Consequently, they signal **onset, offset,** or **change** in stimulus intensity, but they cannot individually code steady-state intensity. Examples of rapidly adapting sensory receptors are **velocity detectors** (touch receptors) which respond to a constant change in skin or hair displacement, and **acceleration detectors** (vibration receptors) which respond only when rate of displacement is changed. Rapidly adapting receptors code intensity of stimuli by **recruitment** of both additional sensory units and higher threshold receptors from within a given sensory unit.

Slowly adapting or **tonic receptors** maintain a generator potential of relatively constant amplitude in the face of constant stimulus intensity. Generator potentials of slowly adapting receptors often have a **dynamic phase** preceding the **static phase** that may signal stimulus **onset, velocity** of stimulus change and **direction** of change. The static phase of these receptors provides information to the central nervous system about the steady state level of a stimulus. The **frequency** of action potentials produced in the static phase codes stimulus intensity. Examples of slowly adapting receptors include the muscle spindle, carotid sinus pressure receptors, and pain receptors. Mathematical functions that relate action potential frequency to stimulus intensity generally resemble the **Power Law**, $F = kI^x$, where F is action potential frequency, k is a constant of proportionality, I is stimulus intensity, and x is an exponent specific to the receptor or sensory modality. Special cases of the Power Law include linear and logarithmic functions. Exponents (x) for **subjective** judgments of stimulus intensity may vary from 0.3 for relatively innocuous stimuli to 3 for painful stimuli. Some physiological systems (eg, heat receptors in the hypothalamus sensing extracellular or blood temperature) may operate at powers greater than 30. Receptors can code wide variations in stimulus intensity because of their responses according to the Power Law.

Central sensory pathways show the following important characteristics:

- **Spontaneous activity** allows changes of stimulus direction to be coded by either increases or decreases of action potential frequency above and below the spontaneous rate.
- **Surround or lateral inhibition** mediated by neural connections between sensory units enhances contrast at stimulus borders.
- **Descending control** of sensory pathways allows higher CNS centers to modify sensory thresholds at lower relay stations as a function of stimulus priority and attention. This is well known in suppression of transmission of pain sensations.
- The **law of projection** states that the sensation produced by stimulation anywhere along a sensory pathway is (perceived as) equivalent to stimulation of the receptor.
- The **neural code** for sensory information may be contained in action potential frequency, in the combination of sensory fibers firing simultaneously, or in patterns of variations in intervals between action potentials.

Review Questions

1. In the skin sensory units are the

   A. receptors that serve a single dermatome
   B. area served by a given receptor
   C. receptors that serve a single sensation
   D. afferent axons of a single nerve trunk
   E. area innervated by a single afferent axon

2. The generator potential of slowly adapting sensory receptors is

   A. "all-or-nothing" in response
   B. graded according to stimulus intensity
   C. independent of stimulus duration
   D. due to increased gK
   E. due to increased gCl

3. Spontaneous neural activity recorded from the ulnar nerve

   A. is found in most sensory fibers
   B. does not occur in sensory fibers
   C. is a sign of pathology
   D. prevents transmission of normal signals
   E. allows transmission of more information, by both increases and decreases of frequency

SOMESTHETIC SYSTEM

Primary somesthetic modalities include cutaneous (vibration, touch, pressure, temperature, pain), proprioceptive (muscle length, tension, stretch velocity), and kinesthetic (joint position, joint movement) senses. Proprioception is covered later in the Spinal Motor Control section (p. 47). Sensory fiber types and functions are shown in Table 2-1.

Table 2-l. Summary of **afferent** fiber types and their functions.

Group	Diameter (microns)	Conduction velocity (M/sec)	Subgroup	Function Afferents come from:
I (Aα)*	12-20	72-120	Ia Ib	Muscle spindle primary endings Golgi tendon organs
II (Aβ)	6-12	36-72	Muscle Skin	Muscle spindle secondary endings Pacinian corpuscles, touch receptors
III (Aδ)	1-6	6-36	Muscle Skin	Pressure-pain endings Touch, temperature, and fast pain receptors
IV (C)	1	0.5-2	Muscle Skin	Pain receptors Touch, temperature, and slow pain receptors

*Older nomenclature derived from motor fibers

Cutaneous receptors include mechanoreceptors, thermoreceptors and nociceptors. Mechanoreceptors can be further classified as position (tonic), velocity (phasic) and acceleration (change in velocity, phasic) detectors. The **receptive field** for any afferent neuron is the region of the skin where stimuli excite the neuron. The size of the receptive fields varies over the body surface. Two-point discrimination is the minimal distance that two touch stimuli must be separated to be recognized as separate. It improves with greater receptor density and more cortical representation. It is best on a finger tip or lip (2 mm) and worst on the shoulder, back, and calf (50 mm).

Proprioception is used to perceive posture and movement. Various sensitive mechanosensors, including those in joints, muscle spindles, tendons and the skin, work in combination to provide conscious perception of joint positions and movements. Both direction and velocity of movement is perceived when the position of a joint is changed. The threshold amplitude for perception of joint movements depends upon angular velocity. **Kinesthesia** is that part of position sense which allows conscious recognition of the rates of movement of different parts of the body.

Afferent somesthetic information is conveyed to the thalamus and cerebral cortex by both the **dorsal column-medial lemniscal** and the **anterolateral (spinothalamic tract) systems** (Fig. 2-5 on p. 51 shows spinal cord locations). In both systems each side of the body is connected with the contralateral half of the brain. The dorsal-column system crosses over in the medulla, and the anterolateral system crosses over in the spinal cord. Characteristics of the dorsal column system include 1) preservation of modality specificity, 2) precise mapping of body surface carried through all relays and on to the neocortical surface, 3) high synaptic security, and 4) transmission via Group I and Group II afferents carrying information from receptors associated with these afferents (Table 2-1). Preservation of modality specificity and precise localization of the stimulus is enhanced by neurons that are excited when stimuli are applied to the center of their receptive field, while neurons in the surrounding area are inhibited, a process called **lateral inhibition**. Characteristics of the anterolateral system include the following: 1) ascending pathways directed to the brainstem, medial thalamus, and lateral thalamus, 2) imprecise mapping of body surface, 3) cross-modality (eg, cutaneous & muscle) convergence, 4) low synaptic security, and 5) transmission via Group III and Group IV afferents and information from their receptors (Table 2-1).

Afferent information to the **cerebral cortex** is projected to primary **Somatic Sensory Areas I and II**. Somatic Sensory Area I consists of Brodmann areas 1, 2 and 3 on the postcentral gyrus. Body surface is mapped onto the cortical surface in a strict, topographical fashion to form a sensory **homunculus**. Somatic Sensory Area II adjoins the lower posterior portion of Area I; it receives information from **both** sides of the body, has low synaptic security, no joint afferents, and is particularly sensitive to direction of stimulus movement on the body surface. Fine tactile information is entirely conveyed to the contralateral postcentral gyrus. The parietal cortex directly posterior to the postcentral gyrus contains the secondary somesthetic area that mediates stereognosis. More posteriorly, parietal cortex mediates complex somesthetic perceptions (eg, spatial orientation) and forms associations between the somesthetic, auditory, and visual systems.

Review Questions

4. An operation is performed to expose the parietal cortex. A recording electrode is used to determine the characteristics of cells in the superficial cortex. Somatic sensory Area I is characterized by

   A. bilateral representation of body surface
   B. precise topographic representation of the contralateral body surface
   C. absence of information concerning joints
   D. receiving information only from the dorsal column medial lemniscal system
   E. integrating visual and somatic input

5. In a neurologic exam a wisp of cotton is dragged over the hairy skin to activate receptors that

   A. are slowly adapting receptors
   B. respond to noxious heat
   C. discharge at a constant rate as skin is indented at a constant rate
   D. signal static displacement of the stimulus
   E. respond when rate of displacement is changed

6. A patient is asked to close their eyes while you move the elbow joint to a new position. A normal person can identify the new position, because information concerning joint position is coded by

   A. rapidly adapting receptors
   B. slowly adapting receptors
   C. dynamic phase discharge of joint receptors
   D. receptors that respond to joint movements
   E. receptors that respond to noxious stimuli

7. Optimal discrimination between closely-spaced tactile stimuli, as in "two-point discrimination," is found in skin regions where the

   A. receptor thresholds are lowest
   B. sensory axons leaving the region have uniform conduction velocities
   C. receptive field sizes are largest
   D. receptive field sizes are smallest
   E. density of receptors is lowest

PAIN

PAIN is an unpleasant sensory and emotional experience associated with actual or potential tissue damage or perception of such damage. **ACUTE pain** is a complex constellation of unpleasant sensory, perceptual, and emotional experiences; these are associated with certain autonomic, psychologic, emotional, and behavioral responses. Many acute diseases or injuries often heal in 2 to 4 weeks, then acute pain subsides. **CHRONIC pain** continues a month beyond the usual course of recovery from an acute disease. Chronic pathological processes occur that cause continued pain or the pain recurs at intervals for months or years.

Pain is a subjective experience. Pain is **THE** presenting sign for medical treatment; it serves a protective function by signaling tissue injury or processes which are capable of damaging the tissue. **Fast pain** is conducted in myelinated Group III fibers; it is a pricking, sharp pain and usually cutaneous in origin. **Slow pain** is conducted in unmyelinated Group IV fibers; it is a burning, intolerable pain and may be of cutaneous, muscle, joint, or visceral origin. Both types of pain are mediated by specific receptors called **nociceptors**. Nociceptors are tonic, non-adapting free nerve endings that are sensitive to noxious (painful) stimuli or to stimuli that become noxious if prolonged. These may be mechano-, thermo, or chemosensitive (to metabolites released in injured tissue) or "polymodal" with responses to all of these three stimuli. **Analgesia** is complete loss of pain. Both slow and fast pain may be blocked by local anesthetics (eg, procaine); higher concentrations are needed to block fast pain, because it is carried in larger fibers. **Primary hyperalgesia** is an increased response to a stimulus that is normally painful; it occurs several minutes after the onset of intense vasodilation at the injury site and development of edema at the same site. **Secondary hyperalgesia** develops in the unstimulated, undamaged region surrounding the zone of primary hyperalgesia and may depend on activity in unmyelinated primary afferent fibers.

Visceral and **muscle** pain are poorly localized and cause diffuse, aching, suffering sensations. This pain results from chemical and mechanical stimuli (cramps, spasms, over-distension of hollow viscera). Ischemia may lead to production of acid by-products and kinins (bradykinin) that stimulate nociceptors. This may occur during cramps in skeletal muscles and occlusion of coronary arteries (angina pectoris). **Referred pain** occurs because somatic and visceral afferent fibers converge onto the same cell of the spinothalamic tract. Referred pain originates from the viscera but is perceived as if coming from an overlying or nearby somatic structure, possibly due to life-long experiences of somatic stimuli. It is also sensed from more proximal somatic fields and is similar to deep muscle pain, since the same spinal neurons respond to both types of pain. **Parietal pain** is produced by noxious stimulation of somatic afferent fibers innervating the inner surface of the body wall.

Pain produces autonomic reflexes, including cardiac acceleration, peripheral vasoconstriction, pupillary dilation, and sweating. It also produces somatic reflexes, including flexor reflexes and sustained muscle contractions that cause **splinting pain**, which results from maintained muscle contraction. Impulses arising from nociceptors in deep somatic or visceral tissues elicit reflex contractions of adjacent (but sometimes distant) skeletal muscles.

The **spinothalamic tract** originates primarily from cells located in laminae I and V of the spinal cord gray matter and ascends to the thalamus, especially to the ventral posterior lateral (lateral) nucleus, posterior nucleus and intralaminar nuclei (medial) (Fig. 2-1).

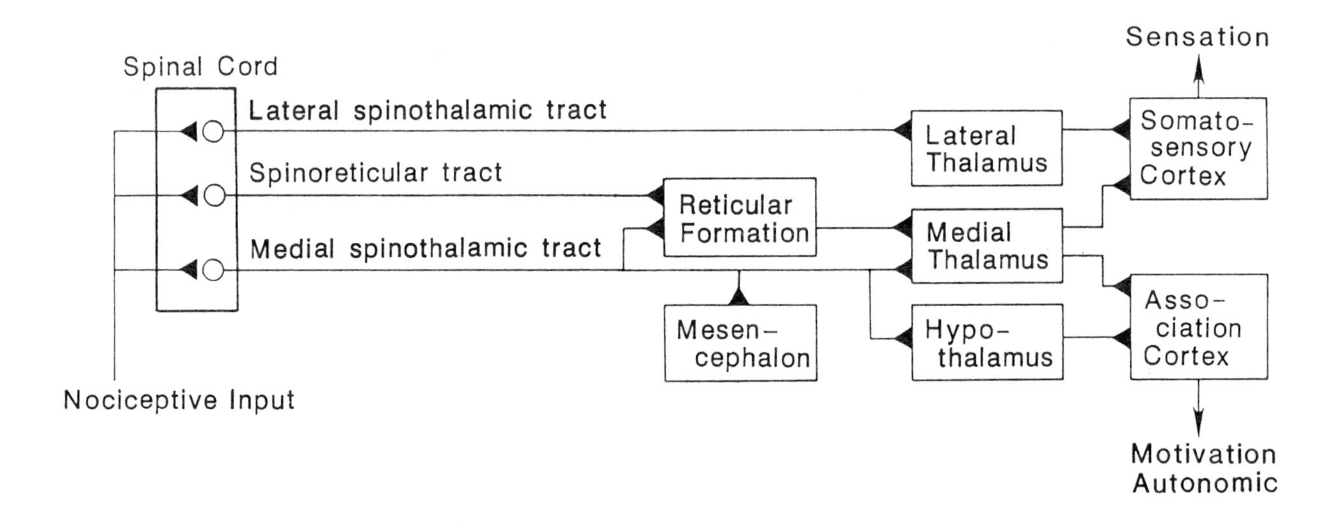

Figure 2-1. Ascending pathways for pain transmission

The **lateral spinothalamic** tract transmits sensory-discriminative aspects of pain, that is, processes that are important for the localization and identification of noxious stimuli. The paramedial ascending system includes the **medial spinothalamic** tract and projections from the spinal cord to the brainstem reticular formation, midbrain and hypothalamus. Nociceptive information transmitted by this system is usually associated with powerful motivational behavior, anxiety, and unpleasant effects that trigger a desire to escape. This system also activates supraspinal reflex responses involving respiratory, circulatory and endocrine functions. The **trigeminothalamic** system transmits information arising from the face and oral and nasal cavities. Noxious information is carried in afferent fibers of the trigeminal nerve, synapses in the nucleus interpolaris and nucleus caudalis of the spinal nucleus of Nerve V. It then projects to the ventral posterior medial, posterior and intralaminar nuclei of the thalamus. Innocuous information synapses in the Nerve V nucleus oralis and then projects to the ventral posterior medial nucleus.

Headache. Intracranial headaches produce deep, somatic pain originating from supporting tissues of the brain, not the brain parenchyma itself. They may be due to hypertension, histamine release, inflammation, meningitis, or changes in intracranial pressure. **Extracranial headaches** may be due to "eye-strain", sinus congestion, or muscle tension. **Migraine headaches** are usually unilateral and of fast onset, often preceded by sensory disturbances, such as nausea, vomiting, photophobia and visual aura. They are associated with increased pulsation and vasodilation of cranial arteries, leading to edema and production of pain by the release of peptides (such as bradykinin or Substance P). **Cluster headaches** have a slower onset without aura and are thought to result from a localized increase of cerebrovascular permeability and edema. A genetic susceptibility has been suggested for both migraine and cluster headaches.

Treatment of pain. Intractable or chronic pain may be interrupted surgically at successively higher levels of the neuraxis by anterior cordotomy, medullary tractotomy, thalamotomy, or frontal lobotomy. Spinal cord stimulation, transcutaneous electrical nerve stimulation (TENS) and acupuncture excite larger afferent fibers that may "close the (spinal) gate" for relay of noxious stimuli conducted in smaller afferent fibers **(Gate control hypothesis). Endogenous opioid peptides** (enkephalins, beta-endorphins, dynorphins) may activate opiate receptors that suppress pain transmission at the level of the spinal cord, especially of spinothalamic tract cells. Release of peptides can be produced by activation of a common descending pathway that originates from the periaquaductal gray of the midbrain, synapses in the nucleus raphe magnus, descends to the gray matter of the spinal cord and makes either presynaptic or postsynaptic contact onto interneurons or spinothalamic tract cells. **Serotonin** is the transmitter for the relay cells of the nucleus raphe magnus.

Thermoregulation

Temperature sensation is signaled by thermoreceptors located on free nerve endings of small myelinated (A-delta) and unmyelinated (C) fibers. Each thermoreceptor has a discrete receptive field to encode warm and cold sensations. Warm thermoreceptors begin to respond at a skin temperature of 30° C, reach maximal discharge at about 45° C, and cease activity at about 47° C. Cold thermoreceptors begin to respond at skin temperature of about 15° C, reach maximal response at about 25° C, and cease discharging at about 42° C. Maximal discharge rate for these receptors is lower than for warm receptors. Both warm and cold thermoreceptors have an early, rapidly adapting phase followed by a tonic phase. Cold thermoreceptors begin to discharge again at skin temperatures above 45° C. This temperature is also the threshold for activating pain fibers. Thus, when the skin is heated above 45^{0} C, a feeling of coolness accompanies pain sensation; this is called **paradoxical cold.**

Review Questions

8. The ascending pain system

   A. is essential for stereognosis
   B. is excited by noxious heat and noxious cold
   C. releases enkephalins to decrease pain
   D. is arranged by fiber diameter in the ventral posterior lateral thalamic nucleus
   E. transmits only somtic sensations

9. The leakage of the acidic contents from the duodenum evokes parietal pain that is caused by

   A. excitation of visceral nociceptors
   B. excitation of Pacinian corpuscles
   C. sustained contraction of muscle overlying the diseased visceral organ
   D. irritation of the inner surfaces of the body wall
   E. convergence of visceral and somatic afferent fibers

10. A pin prick to the skin evokes information that

   A. is transmitted only to the reticular formation
   B. has a long onset latency
   C. is blocked by stimulating Group IV fibers
   D. is carried primarily by Group III fibers
   E. is transmitted via the dorsal column-medial lemniscal pathways

11. The sensation of pain can be

   A. initiated at receptors with a lower threshold for pain than for any other modalities
   B. produced by excessive stimulation of a variety of sensory terminals
   C. enhanced by direct electrical stimulation of the posterior columns
   D. blocked by destruction of all Group IV fibers

12. Pain relief in the left leg is experienced after anterolateral cordotomy, because it interrupts the

   A. right dorsal column
   B. right lateral spinothalamic tract
   C. left lateral spinothalamic tract
   D. right lateral corticospinal tract
   E. left lateral corticospinal tract

VISION

Physiological Optics

Light waves are bent or refracted as they travel at an acute angle through the interface of two transparent substances with different densities. **Refractive index** is the ratio of the velocities of light in the two substances; for air to glass or water the index is 1.4. **Convex (+)** lenses converge light rays; **concave (-)** lenses diverge light rays. The refractive power of a lens is calculated in diopters (D = Refractive power = 1/focal length). A +1D convex lens converges parallel light rays to a focal point 1 m (the focal length) behind the lens; a +5D lens converges at 200 mm from the lens (1 m/5D = 200 mm). A -1D concave lens has a "virtual" focal length of 1 m in front of the lens at a point where the diverging rays would intersect if extended backwards (ie, -1 m; focal length of a -2D lens is -50 cm).

The total converging power of the resting eye is about +59 D, sufficient to produce a focal point 17 mm behind the center of the lens system. Most of this convergence (+49 D) occurs at the interface of the cornea and air, because that is where the refractive index is greatest. At the inside of the cornea there is divergence of -6 D, and the rest of the convergence is controlled by the lens (+16 D), giving a total of +59 D. Nearer objects are focused on the retina by **accommodation** or increasing the curvature of the lens. Maximum accommodation occurs in children (+13 D) and allows the eye to focus an object on the retina that is a minimum of 8 cm in front of the eye, the **near point** (Fig. 2-2). Lens curvature is controlled by the **ciliary muscles**. Muscle contraction counteracts the continuous stretch on the lens by supporting ligaments, allowing the lens to bulge more spherically for near vision. With aging the lens gradually loses its elasticity until only about +1 D. of accommodation is possible. The resultant inability to focus closely is called **presbyopia** ("elder vision"); the near point moves to as much as 1 m from the

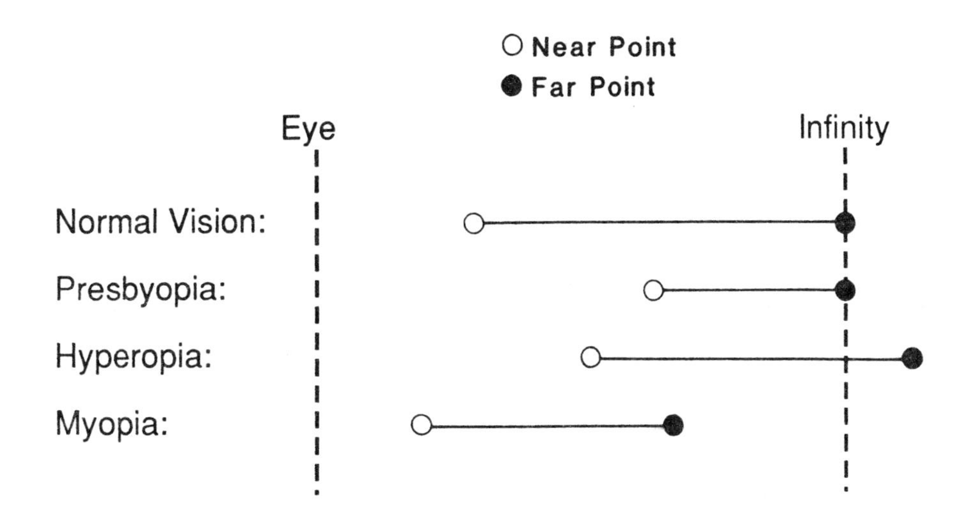

Figure 2-2. Relative location of near point and far point for various defects of vision.

eye (Fig. 2-2). Distant vision is still maintained, so the far point remains at infinity (Fig. 2-2). In the **near response** the eyes adjust in three ways to close objects; convergence, accommodation, and pupillary constriction (which increases depth of field) to focus an image on the fovea.

Normal vision is **emmetropia.** 20/20 vision is defined as the ability to resolve symbols at a distance of 20 ft that a "normal" person can resolve at 20 ft. 20/200 means resolving at 20 feet a stimulus that should normally be resolved at 200 ft. Normal acuity corresponds to a resolution of 1 minute of arc (1/60th of a degree), a resolution of 5 microns on the retinal surface.

Optical defects arise when the eyeball is too short or too long, or the cornea or lens has too much or too little curvature. In **hyperopia** (hypermetropia) near images focus behind the retina, producing **far-sightedness.** The near point and far point are farther from the eye than normal (Fig. 2-2). Hyperopia is corrected by adding convergence to the eye with a convex lens. In **myopia** far images focus in front of the retina, making the eye **near-sighted.** Both near point and far point are closer to the eye than normal (Fig. 2-2); it is corrected by adding a diverging lens. Surgical correction involves procedures that flatten the cornea. Myopia may be caused by continuous, slight accommodation. With aging the myopic person does not need to correct for presbyopia as soon as an emmetrope, since their near point has farther to move from the eye before reading is impaired. In **astigmatism** the refractive power of the eye varies in different planes parallel to the optical axis (ie, an asymmetric lens); it is corrected with cylindrical lenses.

Pupillary aperture varies to control light reaching the retina over a 30-fold range. This variation of intensity is minimal compared to the million-fold range of retinal receptor light adaptation. The pupillary reflex is an efferent motor control system mediated in the brain stem. Under normal conditions changes in parasympathetic tone contract circular sphincter muscles to **constrict** the pupil (miosis) or relax them to **dilate** the pupil (mydriasis). Atropine and related alkaloids block tonic neuromuscular transmission by ACh at parasympathetic endings, thereby dilating the pupil. Sympathetic activation (as in fright) contracts the radial fibers to dilate the pupil. "Dilation" of the eyes with homatropine for examination is primarily to paralyze accommodation and only secondarily to view the retina.

Absorption of Light by the Retina.

Light travels through several layers of the retina to reach rod and cone receptors. Photons react with photopigments to **hyperpolarize** the receptors by **decreasing** gNa. Transduction in other receptors always increases gNa. Resting gNa is **high** in retinal receptors at great metabolic cost. In **rods** the photopigment **rhodopsin** is "bleached" by light and spontaneously regenerates in the dark. Vision mediated by rods only is called **scotopic** vision. With increasing illumination **light adaptation** occurs as the rods become less sensitive to light with greater bleaching. Rods respond over a range of luminance of 10^4 and are totally inactivated at moderate light intensities. Reducing illumination allows increasing **dark adaptation,** and rods become increasingly sensitive. Similar photochemical processes occur in cones with responses over a range of 10^3. Vision mediated entirely by cones is called **photopic** vision. Wavelength sensitivity of the cones is more restricted, with ranges of about two-thirds of the visible spectrum for the red and green cones and the lower one-half of the spectrum for blue cones. Blue cone sensitivity is only about 10% of the absolute sensitivity of the red and green cones, because of partial absorption of shorter wavelengths by the pre-retinal optical system.

Visual afterimages result from bleaching of receptor pigments. A **negative afterimage** of complementary color is seen while gazing at a white background after intense color stimulation of the retina. The retina is less sensitive to the original color and therefore more sensitive to the remaining colors in white. A **positive afterimage** of the same color is seen on a dark background due to after-discharge of ganglion cells. **Color blindness** is the congenital absence of one or more color-type cones and is a recessive X-linked characteristic, occurring in 2% of males. Red-green blindness indicates lack of either red or green sensitive cones and leads to confusion of these hues; the lack of blue sensitive cones is rare. The more common **color weakness** (6% of males) is a decreased sensitivity of one or more cone systems. A **protonope** has low sensitivity to red with normal green vision; a **deuternope** has low sensitivity to green with normal red vision.

Four factors contribute to the increased spatial visual acuity of the **fovea**; 1) its high concentration of cones (100% cones at center), 2) close packing of thin cones, 3) thinning of overlying retinal layers to minimize light scattering by displacement of bipolar cells, ganglion cells and their axons radially away from the fovea, and 4) absence of blood vessels. The **blind spot** is the absence of receptors where the optic nerve and blood vessels penetrate the eyeball. The blind spot is not noticed, because visual perceptual processes automatically "fill in" the area. Cone density decreases radially from the fovea, while rod density increases. Therefore, dim objects, such as faint stars, are detected more easily when their image is extrafoveal. Loss of foveal vision and ability to read occurs in the elderly with macular degeneration. Loss of rods (extrafoveal vision) occurs in patients with retinitis pigmentosa; they have "tunnel" vision.

Neural Processing in the Retina

Receptors, bipolar and horizontal cells produce only local potentials; ganglion (and amacrine) cells also produce action potentials. Hyperpolarization of rods or cones by light decreases their continuous release of inhibitory or excitatory transmitter onto bipolar cells. If the transmitter is inhibitory to bipolar cells, then they will be **disinhibited** in response to light and secrete more excitatory transmitter onto ganglion cells, causing facilitation and firing. This is characteristic of light shined in the center of the receptive field of an "on-center, off-surround" (light center) ganglion cell. Other bipolar cells will respond to the same transmitter as excitatory. Consequently, they will be less facilitated in response to light and secrete less excitatory transmitter onto ganglion cells. Such ganglion cells have "off-center, on-surround" (dark center) receptive fields and tend to fire a burst of action potentials when light is turned off.

Lateral connections between receptors, bipolar and ganglion cells via horizontal and amacrine cells mediate enhancement of stimulus contrast by **lateral inhibition**. When adjacent receptors are illuminated at different intensities, the less intensely activated (illuminated) pathways will inhibit their neighbors less, and the more intensely activated (illuminated) pathways will inhibit their neighbors more. Consequently, the difference in apparent illumination will be enhanced at the border of the two light intensities. The result of lateral inhibition is shown in Figure 2-3. Lateral inhibition is extended radially as **surround inhibition** for a spot of light against a dark background (or dark spot against a light background). This creates circular receptive fields of ganglion cells. Illuminating both center and surround of a receptive field simultaneously results in only a slight increase in spike frequency. There are two major types of ganglion cells; 1) X cells are color and pattern sensitive with small color sensitive center-surround fields (eg, red ON-center, green OFF-center), and 2) Y cells are motion (transient) sensitive with large receptive fields without color sensitivity.

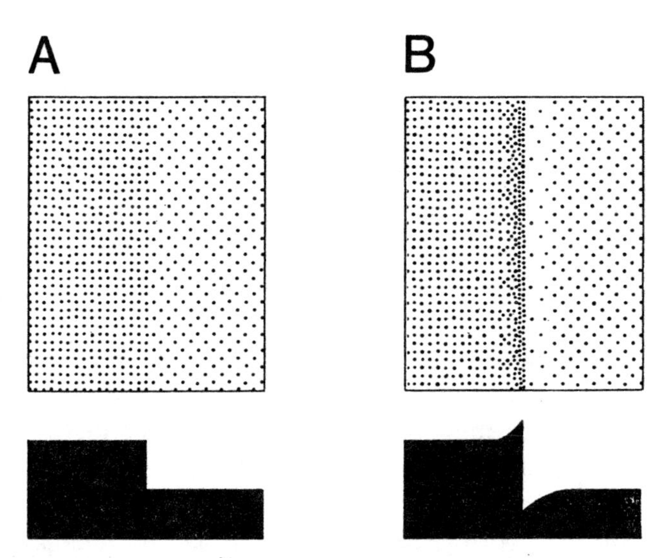

Figure 2-3. The perceived enhancement of differences in intensity at an edge (as in B) from different light intensities (as in A) is mediated by lateral inhibition.

Central Processing of Visual Information

Cells of the **lateral geniculate body** relay visual information from ganglion cell terminals not only primarily to the striate cortex (Brodmann Area 17), but also to secondary visual areas in the cortex and visual reflex mediating areas in the brain stem. The excitability of lateral geniculate neurons can be modified by efferent control pathways from the cortex, reticular formation, and superior colliculus. These neurons are particularly sensitive to foveal receptors, exhibit surround inhibition like the retina, and maintain topographic representation of the retinal field. Retinal afferents to each lateral geniculate body come from both right and left visual fields, but the right and left inputs are segregated in a series of six layers, so there are few neurons with binocular input. Individual lateral geniculate neurons may have "X-like" or "Y-like" characteristics.

Visual information is transformed to visual perceptions in the **cerebral cortex**. Although **foveal** receptors occupy less than 1% of total retinal area, they project to about 50% of the primary striate cortex. Cortical neurons have various kinds of receptive fields, the more complicated occurring with greater frequency in the secondary visual Areas 18 and 19. **Simple** cells have elliptically-shaped receptive fields. An excitatory region may lie along side or within the long axes of the ellipsoid parallel to an inhibitory region. Maximal activation occurs with **stationary** stimuli at the proper angle or **orientation. Complex** cells respond best to an edge stimulus in the receptive field with the edge at the preferred orientation, usually perpendicular to the long axis of the receptive field. **Moving** stimuli also have a preferred direction for optimal activation. **Hypercomplex** cells have a central excitatory field flanked by one or two inhibitory fields. They respond best to bars of optimum **length** moving in one direction along a preferred orientation. This hierarchy may develop in a cascade of inputs from simple to more complex cells.

The visual cortex has **topographic** representation with contralateral projection of the visual fields. There is also columnar organization. Neurons in adjacent columns respond to different stimulus characteristics, such as orientation, movement, direction, intensity, and to left or to right eyes. If an infant has impaired vision in one eye, neurons in the columns for that eye do not develop normally. Vision can be lost permanently, so treatment should begin at an early age.

Most cortical neurons are **binocular**, receiving input from both retinas. Varying degrees of discrepancy between the two retinal loci is a basis for binocular vision, and specific cells respond to only precise degrees of discrepancy. Since true "binocular" vision is lost beyond 5 m, **depth perception** depends first on relative size, second on relative movement, and only third on the slightly different images in the two retinas. Higher level visual information processing occurs in inferotemporal cortex. Integration of color and pattern information occurs here with mediation of visual learning and memory. Visual afferent projections to the superior colliculus overlap somesthetic and auditory inputs from the same point in "sensory space" to facilitate body orientation to stimuli.

Eye Movements

Eye movement is smoothly controlled by six fast, striated ocular muscles via cranial nerves III, IV, and VI. Precise control is partly due to motor units with low innervation ratios, one motoneuron to as few as ten muscle cells. **Voluntary** eye movements are organized in Brodmann Area 8, the "frontal eye field", while most other movements are controlled from the brain stem. There are five important eye movements. The first three are **conjugate**; the eyes move as if yoked together.

- **Saccades** - rapid flicks, fast movements to shift the gaze.
- Smooth **pursuit** movements - tracking or following of visual stimuli.
- **Vestibular** movements - compensation for change in head orientation with body movements.
- **Vergence** movements - disjunctive movements of convergence or divergence, for focusing on near or far objects.
- **Microtremors** (microsaccades) - constant tiny involuntary movements during fixation to shift the target image to unadapted receptors. If the target image were not shifted, the target would fade by receptor bleaching.

Review Questions

13. If the near point is farther from the eye than normal, then

    A. correction to normal vision requires a cylindrical lens
    B. correction to normal vision requires a negative (-) lens
    C. patient has emmetropia
    D. patient has hyperopia
    E. patient has myopia

14. Contraction of ciliary muscles

    A. decreases the anterior-posterior dimension of the eyeball
    B. causes the crystalline lens to flatten by increasing tension on its attachments
    C. constricts the pupil
    D. relieves tension on the attachment of the lens, allowing it to bulge by elastic forces
    E. dilates the pupil

15. The fovea centralis of the retina

    A. is the area of most acute vision
    B. contains the highest concentration of rods
    C. has the lowest threshold for excitation in the retina
    D. is specialized for excitation in the dark
    E. is not used during routine vision

16. Complete dark adaptation of the visual system

    A. is accomplished within 10 minutes after entering a dark room
    B. involves a shift from rod vision to cone vision
    C. occurs only in people who are emmetropic
    D. occurs in two stages, an early cone phase and a later rod phase
    E. occurs in rods but not cones

17. Ganglion cells depolarize in response to a spot of light being turned on at the center of their receptive field because of

    A. excitation of receptors
    B. excitation of bipolar cells
    C. disinhibition of bipolar cells
    D. inhibition of bipolar cells
    E. inhibition from bipolar cells

18. Hypercomplex cells in the visual cortex

    A. respond best to stimuli that are stationary
    B. respond best to corners or moving bars of optimum length
    C. respond only to circular stimuli of small radius
    D. respond best to moving parallel lines of any length
    E. fail to respond to stimuli that reflect "yellow light"

AUDITION AND BALANCE

Audition

Sound waves in air are regions of compression and rarefaction. Velocity of sound is the product of frequency in Hertz (cycles/sec) times wave length. The **decibel** is a logarithmic notation for sound intensity, 10 dB for a 10-fold change in amplitude, but **20 dB** for a 10-fold change of the more commonly used unit of **sound pressure** (amplitude squared). The ear is sensitive over a range of a 10^7 change of sound pressure or 140 dB. (A whisper is 30 dB, conversation is 70 dB, and noisy traffic is 100 dB.)

The middle ear matches the acoustic impedance of air with that of endolymph. It achieves this partly by the mechanical linkage system of ossicles but mainly by a large tympanic membrane compared with a small oval window; total amplification is 20-fold. The linkage of bones may be protected from damage due to loud sounds by an **attenuation reflex**, which is mediated through an **efferent inhibitory** system that reflexly contracts the tensor tympani and stapedius muscles. However, the reflex has a 40 msec latency, so there is no protection for fast transient sounds of high intensity (eg, a thunder clap). The Eustachian tube is a pressure relief system to maintain equal pressure on both sides of the ear drum.

Movement of the **oval window** sets up **traveling waves** in the basilar membrane which becomes more stiff towards the apex. High frequencies cause maximum displacement of the membrane near the oval window; low frequencies cause maximum displacement at the apical end, 2 1/2 turns away from the oval and round windows. The elastic **round window** allows the incompressible endolymph to respond to movements of the stapes. Vibration of the basilar membrane causes a shearing action between it and the tectorial membrane that is attached at one edge. The cell bodies of hair cells are fixed in the basilar membrane; the hairs extend into the collagenous tectorial membrane and are sheared to one side or the other as the tectorial membrane moves across the basilar membrane. This is the mechanical stimulus for increased Ca^{2+} permeability, seen as a receptor potential. Sounds are detected almost entirely by the inner hair cells. The outer hair cells are contractile and may modulate sensitivity of the system.

The **endocochlear potential** is +70 mV in the scala media relative to the scala tympani. This is established by an active electrogenic Na^+, K^+-ATPase pump in the stria vascularis. Since hair cells have intracellular potentials of -70 mV, their transmembrane potential relative to the scala media fluid around them is about -140 mV. Consequently, the driving force across the membrane is doubled, so increased Ca^{2+} conductance due to bending of hairs will result in <u>double</u> the current flow of less polarized sensory receptors. The **cochlear microphonic** is thc first detectable electrical change in response to a sound and probably results from summated hair cell receptor potentials.

Auditory **sensitivity** tested by audiometry is maximum at about 3,000 Hz, with a high frequency limit of 20,000 Hz for young persons but decreasing with age (**presbycusis**). In **conduction deafness** the vibration of a tuning fork placed on the mastoid process can be heard, while sensation via air and the middle ear is attenuated, suggesting that the tympanic membrane or ossicles are malfunctioning (Rinne test). In **sensorineural deafness** even such bone conduction cannot be heard; the loss is often in the hair cells. The Weber test applies a tuning fork to the middle of the forehead. If sound is louder in the poorer ear, it is a conductive loss. If sound is louder in the better ear, it is a sensorineural loss. Presbycusis is caused by years of noise pollution damaging the hair cells closest to the oval window. This explains the progressive loss of sensitivity to the highest frequencies. The central connections for audition involve least one more neuron than in the visual and somatosensory systems, and there is bilateral input at all levels after the cochlear nuclei. Testing for the locus of **central deafness** is done by far-field recording of the **brain stem auditory evoked response** (BAER). This requires computer averaging of the sound-evoked electrical signals of ascending action potentials recorded with surface electrodes (Fig. 2-7, p. 62).

A particular auditory neuron responds best to a limited range of frequencies with an optimal center frequency and a wider frequency response as intensity increases (a plot of frequency response versus stimulus intensity forms a V shape). Central pitch discrimination begins when a sound within that range of frequencies displaces the particular region of the basilar membrane innervated by that neuron (the **Place Theory**). Neurons also fire action potentials in phase with low frequencies of sound. Pools of neurons can follow frequencies up to 1,000 Hz by **frequency following** with sequential action potentials across the pool, but this system is not useful at higher frequencies. Auditory neurons are spontaneously active and show surround inhibition. Similarly to the visual system, most auditory neurons have **binaural** input, often inhibitory from one ear and excitatory from the other. Detection of phase and intensity differences of sound arriving at the two ears is used to localize sounds in space. At higher relay centers neurons are more sensitive to **changes** of pitch and intensity and respond only to increasingly complex sound patterns. **Masking** is the suppression of the response to a sound, due to convergence of the same frequency from both ears or adjacent frequencies in the same ear. The **olivocochlear** system is an efferent control system. Cochlear hair cells receive inhibitory synaptic input from olivary neurons. This may help in detection of less intense simultaneous tones. The hair cell receptors of the vestibular system also have efferent inhibitory synapses.

The **auditory cortex** receives information from both ears, although input is predominantly from the contralateral ear. As is typical of sensory cortex, primary auditory cortex shows columnar organization with topographical representation of auditory sensory space. Frequency is coded along the anterior-posterior axis of the top of the temporal lobe (low frequency anterior), while intensity is coded along the medial-lateral axis (low intensity medial). In general, the auditory cortex mediates discrimination of pitch, recognition of patterns, and localization of sound sources in space, while the inferior colliculus mediates reflexes designed to orient the body to the direction of the sound. Overall, the auditory system can **discriminate** 1 dB sound pressure differences at low intensities; at higher intensities the difference threshold is even less. **Adaptation** to a constant sound intensity and frequency results in an increase in auditory threshold at that frequency.

Balance

The **vestibular system** employs one set of receptors, the utricle and saccule, for detecting head orientation and linear acceleration and a second set, the semicircular canals, for detecting angular acceleration. The macula of the utricle, fixed horizontally in the skull, and the macula of the saccule, fixed vertically, detect **linear acceleration** and change in the direction of **gravity** relative to the head (head orientation). High density otoliths (granules of $CaCO_3$) are imbedded within a gelatinous substance. Hair cell bodies are fixed in place with hairs (cilia) extending into the gelatinous substance.

The whole otolith substance slides over the hair cells in response to linear forces. The hair cells are spontaneously active; movement is excitatory in one direction and inhibitory in the other.

Three pair of **semicircular canals** detect **angular acceleration** or rotation (around any of three axes). Relative inertial movement of endolymph bends the **cupula**, thereby bending hairs imbedded in the **cristae ampularis**. As in the maculae, hair cells in the cristae are spontaneously active, so endolymph movement in one direction is excitatory and in the opposite is inhibitory. Afferent fibers from utricle, saccule and semicircular canals synapse in vestibular nuclei in the brainstem, which in turn distribute inputs to several CNS areas including the cerebellum, reticular formation, neocortex and spinal cord. In particular, the vestibular nuclei have major connections to 1) the oculomotor system, for stabilizing the visual image; 2) neck muscles, for stabilizing the head; and 3) postural muscles, for body balance. Human orientation in space is primarily by visual input (detection of horizontal), secondarily by contact with the ground and only tertiary by vestibular sensations.

Nystagmus is an alternating slow movement of the eyes in a particular direction (usually laterally), with a quick recovery movement in the opposite direction. It is an alternation of a smooth pursuit movement and a saccade. Nystagmus is designated clinically by the direction of the quick movement (eg, to the left if the person's eyes move toward <u>their</u> left in the quick movement). **Optokinetic** nystagmus is a following reflex of the eyes to movement of the visual field. If the visual field moves to the right, the eyes follow the field to the right (slow movement), and then recover to the left (fast movement), so optokinetic nystagmus is to the left. **Rotatory** nystagmus (with eyes closed) is in the same direction as the initial angular acceleration, but in the opposite direction to that of movement after rotation stops (and the cupula bends in the opposite direction than it did initially). The **caloric** nystagmus test is performed by placing warm or cool fluid in the external ear. The resultant thermal convection currents produced by cool water stimulate the semicircular canals, which results in nystagmus to the opposite side as the ear being tested. Warm water produces nystagmus to the same side. The mnemonic is COWS; cold, opposite; warm, same.

Review Questions

19. The slow movement of the eyes during nystagmus is

    A. used to define its direction clinically
    B. in the direction of endolymph movement
    C. in the direction of the direction of the saccade
    D. opposite to the direction of a preceding rotation of the patient
    E. in the direction of head acceleration

20. Optimal hearing requires

    A. equal static air pressure on both sides of the ear drum
    B. a negative potential within the scala media
    C. fixation of the malleus, incus, and stapes
    D. continuity between scala vestibuli and scala media
    E. rigidity of the tympanic membrane

21. Deafness limited to high tones is associated with a lesion in the

    A. tympanic membrane
    B. ossicular chain
    C. apical portion of the basilar membrane
    D. basal portion of the basilar membrane
    E. medial geniculate body

22. "Sensorineural" deafness could result from

    A. blocked Eustachian tubes
    B. punctured ear drums
    C. destruction of cochlear hair cells by antibiotics
    D. immobilized stapes
    E. caloric nystagmus

GUSTATION AND OLFACTION

Taste and smell are both chemical senses, and the systems are similarly organized. **Flavor** depends on both smell and taste. Substances exciting gustatory and olfactory receptors must first **dissolve** in saliva or in olfactory mucus (water soluble), which surrounds the cilia of receptor cells. To be smelled substances must also be in a gaseous state to mix with air inspired over the receptors. The olfactory receptor is a specialized neuron. In contrast, gustatory receptors are specialized, non-neural cells within the taste buds; they "age" over a 10-day period and are replaced by new receptors of the same taste "modality." Gustatory nerve fibers innervate several receptors within a taste bud and may innervate receptors in several different buds. They have a lower threshold for one taste modality, but they will respond to higher concentrations of other modalities. Receptor responses of both chemical senses are proportional to stimulus concentration. Adaptation occurs within seconds to a few minutes for both senses. Because of the overlapping responsivity to the submodalities, stimuli may be confused at threshold concentrations (eg, 2x threshold [NaCl] may taste sweet).

Discrimination of stimulus intensity for the chemical senses is relatively poor with a 30% change in intensity required for a subjective "just noticeable difference." The overall **intensity range** that can be signaled is less than a 100-fold increase. (This contrasts with audition and vision where the range is more than a million-fold). However, their absolute sensitivities are remarkable; denatonium saccharide (sweet) can be tasted at a 10^8 dilution and quinine sulfate (bitter) at 10^6. Methyl mercaptan added to natural gas can be smelled at a 10^{11} dilution.

Discrimination of stimulus quality in both systems depends upon the chemical and spatial structure of molecules. Smell perceptions have been variously divided into seven to nine classes (eg, flowery, etheric, musky, camphorous, sweaty, rotten, peppery), but there is no corresponding specialization of olfactory receptors. In contrast, there are four primary taste sensations organized as shown in Table 2-2.

Table 2-2. Characteristics of taste sensations.

Taste Quality	Characteristic chemical inducers	Localization of taste buds on the tongue	Absolute sensitivity
Sweet	Sugars, organic substances (saccharine)	Front	1 part/200 (1 part/10^7)
Salt	Ionized salts (NaCl)	Front	1 part/600
Sour	Acids, especially inorganic (acetic acid)	Sides	1 part/10^5
Bitter	Alkaloids, long chain organic substances (quinine sulfate)	Back	1 part/10^7

Anterior taste buds (sour, salty, sweet) are innervated by the facial nerve and posterior (bitter) by the glossopharyngeal nerve. Damage to the facial nerve results in the detection of only bitter substances, while damage to the glossopharyngeal reduces only bitter detection. Important taste and smell-induced reflexes include 1) serous gland secretion, 2) saliva secretion and composition, 3) secretion of gastric juices, and 4) vomiting to certain taste sensations.

Review Questions

23. The taste of sweet

    A. is associated with some organic chemicals
    B. is localized on the sides of the tongue
    C. is associated with alkaloids and some plant-derived poisons
    D. may stimulate a sneeze
    E. is a combination of two basic tastes, minty and salty

24. The olfactory system

    A. adapts slowly
    B. senses only substances that are lipid soluble
    C. has four primary smells
    D. requires a 1% change in odorant concentration to detect intensity differences
    E. requires detected substances to be gases initially

SPINAL MOTOR CONTROL

The motor system controls movement. The neural systems that control movements are arranged in a hierarchy. The highest level determines the general intention of an action. The middle level determines the actual postures and movements necessary to carry out the intended action. In some cases stored **motor programs** are utilized, and sensory information arising during the ongoing movement fine tunes the actual movement to match the intended movement. The lowest level determines which motor neurons, and thus which muscles, will be activated.

Muscles

The motor system ultimately controls which muscles will be stimulated to contract. Muscles that work together to move a limb in the same direction are called **agonists** or **synergistic** muscles. **Antagonistic** muscles pull a limb in opposite directions. **Flexors** are muscles that cause flexion, and **extensors** are muscles that cause extension of a limb. Muscles that oppose gravity are **physiological extensors** or **anti-gravity** muscles. The biceps muscle is a physiological extensor, because flexion of the arm opposes gravity. Movement of limbs involves **reciprocal innervation.** For movement to occur at a joint, the nervous system causes contraction of one muscle and inhibition of its antagonists. Reciprocal innervation also operates across the spinal cord to control muscles on opposite sides of the body, and between upper and lower limbs.

The **motor unit** is a single motoneuron and the skeletal muscle fibers it innervates. The **innervation ratio** is the number of muscle fibers innervated by a single motoneuron. The innervation ratio ranges from about three (in ocular muscles) to several hundred (in some postural muscles). Strength of contraction is graded primarily by the **number of motor units** activated and secondarily by the frequency of firing of motor units. Smaller motoneurons are more excitable and so are **recruited** first; larger motoneurons (with larger innervation ratios) are recruited later and generate larger increments of tension. This concept is known as the **size principle**. Small motoneurons innervate "slow," fatigue-resistant muscles (eg, those involved in posture), and large motoneurons innervate "fast" muscles (eg, muscles involved in rapid movements).

Spinal Organization

Neurons in the CNS make synaptic contact in a variety of patterns. They show **divergence**, branching to many neurons, and **convergence**, input from many neurons onto a single neuron. **Facilitation** is an enhanced response by spatial and temporal summation of activity from more than one input. **Occlusion** is an apparently reduced response due to different inputs converging upon the same pool of neurons.

Skeletal muscle fibers are innervated by **α-motoneurons**, which are located in the ventral horn of the spinal cord and secrete ACh to excite muscle fibers. Excitation of an α-motoneuron causes its motor unit to contract. Because of this, α-motoneurons are called the **final common pathway**; that is, motor pathways ultimately converge onto α-motoneurons. Smaller motoneurons, termed **γ-motoneurons**, are also located in the ventral horn; these do not innervate skeletal muscle fibers. Instead they innervate small intrafusal fibers located within muscle spindles, as discussed later.

When motoneurons fire, they also excite **Renshaw cells** in the ventral horn via recurrent collaterals. The Renshaw cell is an inhibitory interneuron that synapses both on the motoneurons that excite it and on adjacent motoneurons. Renshaw cells are excited by ACh and release glycine, an inhibitory transmitter. When one Renshaw cell synapses on another Renshaw cell, the result is **disinhibition**, increased activity by **release** from inhibition. Renshaw cells provide stability of motoneuron output to synergistic muscles and have a focusing effect on motor output due to lateral inhibition.

Muscle Receptors

Muscle spindles signal muscle stretch and are oriented in parallel with skeletal muscle fibers. Spindles contain **intrafusal fibers**, which do not generate force, whereas skeletal muscle fibers are **extrafusal fibers**, which do generate force. Two types of intrafusal fibers make up the spindle. **Nuclear bag** fibers have a bulge in the middle and signal mainly phasic stretch. **Nuclear chain** fibers signal mainly tonic stretch. The center region of both types of intrafusal fibers is innervated by **Ia primary endings** (also called **annulospiral** endings) (see Table 2-1, p. 31). These are excited as the center regions of the fibers are stretched. The nuclear chain fibers also are innervated by group **II endings** (also called **flower spray** endings); these are located away from the center. The ends of the intrafusal fibers are innervated by **γ-motoneurons** or **fusimotor** neurons. Excitation of γ-motoneurons stretches the center part of the spindle; they control the amount of stretch on the spindle and hence can control its sensitivity. **Golgi tendon organs** are located in tendons, are oriented in series with skeletal extrafusal muscle fibers, and signal muscle tension. They are innervated by group **Ib fibers** (see Table 2-1). Passive stretch of muscles excites both spindles and tendon organs. Active contraction of muscles inhibits spindles but excites tendon organs.

Spinal Reflexes

The **stretch reflex** (myotatic or tendon jerk reflex) is the fastest reflex in the body, utilizing fast conducting sensory (Ia fibers) and motor (α-motoneuron) pathways and only one spinal synapse. Stretch of annulospiral endings initiates muscle contraction that opposes the lengthening stimulus, a negative feedback response. This is the reflex elicited by the tendon tap, which stretches muscle spindles, not Golgi tendon organs. The stretch reflex provides a smoothing or damping effect on muscle movement and maintains appropriate muscle length. **Muscle tone** is produced by a continuous mild stretch reflex in muscle. In extensor muscles this makes muscles ready for movement and helps to maintain an erect posture.

The **inverse myotatic reflex** (tendon reflex or lengthening reaction or clasp-knife reflex) inhibits muscle contraction by activating Golgi tendon organs. It is disynaptic with one inhibitory interneuron. It regulates muscle tension, dampens the force of stretch reflexes, and may protect tendons from tearing in the face of strong muscle contractions.

The **flexor reflex** (withdrawal reflex) is multisynaptic; activation of nociceptors causes withdrawal of the limb from the stimulus. This reflex has a protective function. Flexion of a limb is accompanied by the **crossed extensor reflex** which produces extension of the contralateral limb in order to support the body. Activation of touch endings may produce a multisynaptic **extensor thrust reflex** (positive supporting reaction) that helps to bear weight as a limb meets the ground. Note that the flexor reflex and extensor thrust reflex produce opposite effects (flexing and extending the limb, respectively) in response to contact with a surface. In one case the stimulus is noxious (activated by nociceptors), and in the other it is innocuous (activated by touch receptors).

The Gamma Motor System

Excitation of γ-motoneurons causes contraction of the ends of muscle spindles, thereby stretching the center and activating Ia fibers. In effect, excitation of the gamma system **elicits a stretch reflex**. The function of the γ-efferent system is shown in Figure 2-4 along with activity of stretch (annulospiral) and tension (Golgi tendon organ) receptors in muscles. A muscle "at rest" (Fig. 2-4A) exhibits slight muscle tone, so there is some firing of annulospiral endings and some mild reflex activation of extrafusal muscle fibers. When the muscle receives an "external stretch" (Fig. 2-4B), both annulospiral and Golgi tendon organs increase their activity. If a muscle were to contract without operation of the γ-efferent system (Fig. 2-4C)), then the annulospiral ending would be unloaded and stop firing. The muscle spindle unloads, because it is **in parallel** with extrafusal muscle fibers. Golgi tendon organs would fire maximally, because they are **in series** with the contracting extrafusal muscle fibers. In fact, γ-motoneurons normally fire along with α-motoneurons (Fig. 2-4D). Consequently, intrafusal muscle fibers also contract and take up the slack, so annulospiral endings continue firing during muscle contractions.

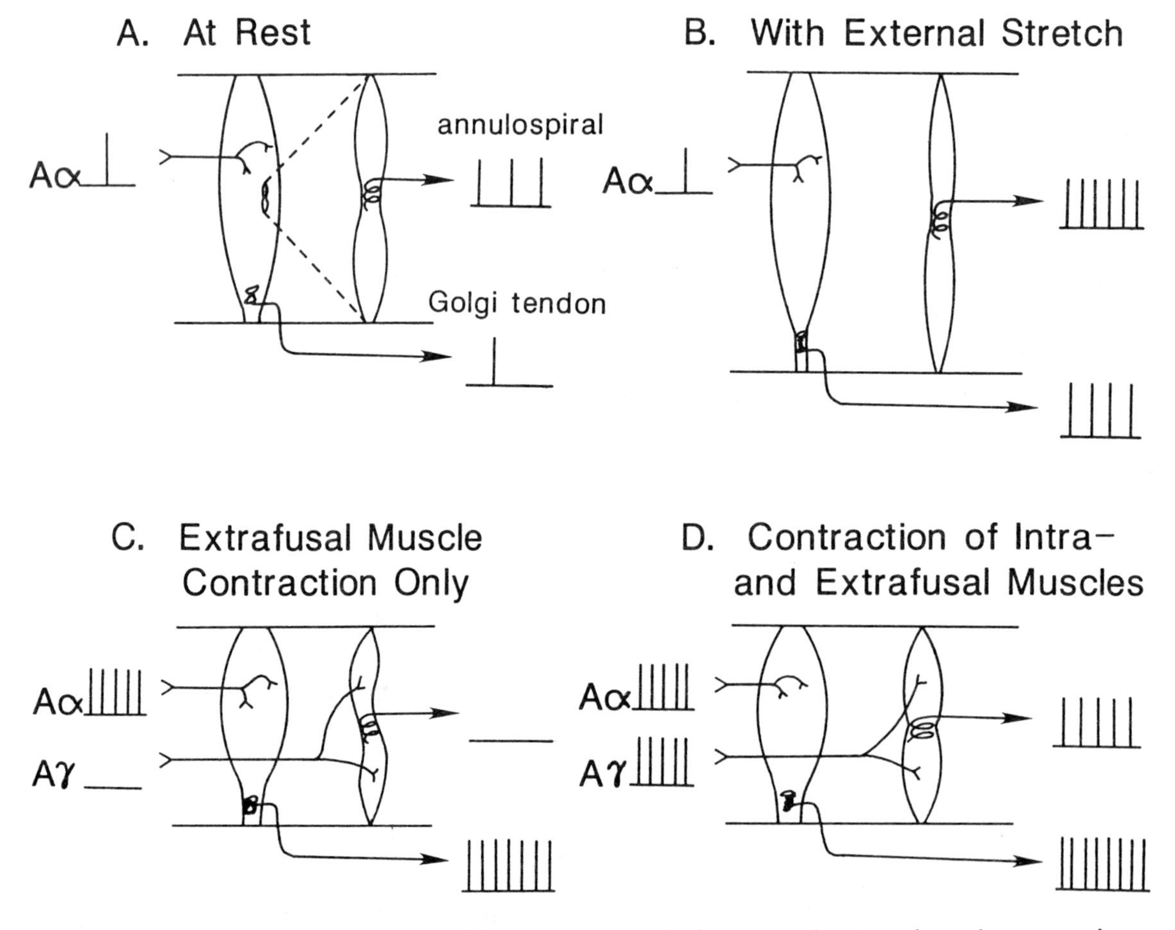

Figure 2-4. Role of the gamma efferent system during muscle stretch and contraction.

The gamma motor system has three functions in motor control. 1) It maintains the **sensitivity** of muscle spindles during muscle contractions, such as tendon jerk reflexes. Continuing activity of gamma motoneurons **controls** the sensitivity of muscle spindles to stretch from moment to moment. 2) Firing of gamma motoneurons can initiate a muscle contraction by the **spindle loop** pathway. This consists of the following sequence: γ-firing, contraction of intrafusal muscle fibers, firing of annulospiral endings, excitatory synaptic activation of α-motoneurons, and contraction of extrafusal muscle fibers. 3) Normally, α-motoneurons are not activated alone; voluntary movements involve the activation of both α- and γ-motoneurons; this is called **alpha-gamma coactivation.** Probably the most important function of the gamma motor system is its function as a **servo-assist system.** In effect, the gamma motor system compares the muscle stretch required for a particular movement with the muscle stretch actually achieved by the activation of α-motoneurons. If the stretch is inappropriate, then the γ-motoneurons could (for example) increase muscle force of contraction by the spindle loop. This response is more likely to occur for postural movements and in learning new movements.

Review Questions

25. A hemisection of the right upper thoracic (T_1-T_5) spinal cord would cause all of the following **EXCEPT**

    A. reduction of sympathetic neural outflow to the heart
    B. impairment of motor control of the toes on the right side
    C. impairment of pain sensation of the toes on the left side
    D. impairment of motor control of the fingers on the right hand
    E. eventual atrophy of upper thoracic chest muscles

26. Muscle relaxation evoked by the inverse myotatic reflex

    A. is initiated monosynaptically
    B. is initiated by annulospiral endings in the relaxing muscle
    C. requires inhibition of motoneurons supplying the relaxing muscle
    D. supports the weight of the body
    E. occurs simultaneously with a crossed flexor reflex in the opposite limb

27. The action potential frequency in primary (Ia) afferent fibers from the muscle spindle signals

    A. muscle length only
    B. the rate of change of muscle length only
    C. both muscle length and rate of change of muscle length
    D. muscle tension only
    E. rate of change of muscle tension only

28. Rate of firing in Ia fibers from annulospiral endings decreases in response to

    A. firing of alpha motoneurons without coactivation of gamma motoneurons
    B. a sharp tap applied to the muscle tendon
    C. lengthening of a synergistic muscle
    D. stimulation of gamma efferent fibers to the muscle spindle
    E. isometric contraction of a muscle

29. A patient presents with the following signs/symptoms: 1) loss of the sensations of pain and temperature from the right leg; 2) loss of the sensation of touch from the left leg; 3) loss of voluntary control of the left leg; 4) normal sensations in the arms. The lesion causing these signs/symptoms likely involves a hemisection of the

    A. left sacral spinal cord
    B. right sacral spinal cord
    C. left lumbar spinal cord
    D. right lumbar spinal cord
    E. left lower thoracic spinal cord
    F. right lower thoracic spinal cord
    G. left cervical spinal cord
    H. right cervical spinal cord

CENTRAL MOTOR CONTROL

Brainstem Control of Movement

There are four major descending motor pathways from the brainstem (Fig. 2-5). The **rubrospinal tract** (from the **red nucleus**) is a crossed pathway that travels in the dorsolateral spinal cord. The **vestibulospinal tract** originates in the **lateral vestibular** (or **Deiters**) **nucleus**, and it is an uncrossed pathway that travels through the ventromedial spinal cord. The origin of the **medial reticulospinal tract** is the **pontine reticular formation**. It is mainly uncrossed and travels in the ventromedial spinal cord. Cell bodies of the **lateral reticulospinal tract** lie in the **medullary reticular formation**, and their axons are both crossed and uncrossed and travel in the ventrolateral spinal cord. The **corticospinal** (or **pyramidal**) **tract** is a fifth pathway that originates in the **cerebral cortex**, is crossed, and travels in the dorsolateral spinal cord. Functionally these pathways can be divided into a **medial** system and a **lateral** system based on spinal projections. The medial system mainly influences posture, and the lateral system controls fine movements. The medial system (vestibulospinal and medial reticulospinal tracts) mainly excites extensors and inhibits flexors; it influences axial and girdle muscles. The lateral system (rubrospinal and corticospinal tracts) mainly excites flexors and inhibits extensors; it controls distal muscles and extremities. The lateral reticulospinal tract is an exception to this general rule. It travels in the lateral portion of the spinal cord, but it is considered part of the medial system. It mainly excites flexor and inhibits extensor axial and girdle muscles, thereby providing a balance to the vestibulospinal and medial reticulospinal influences.

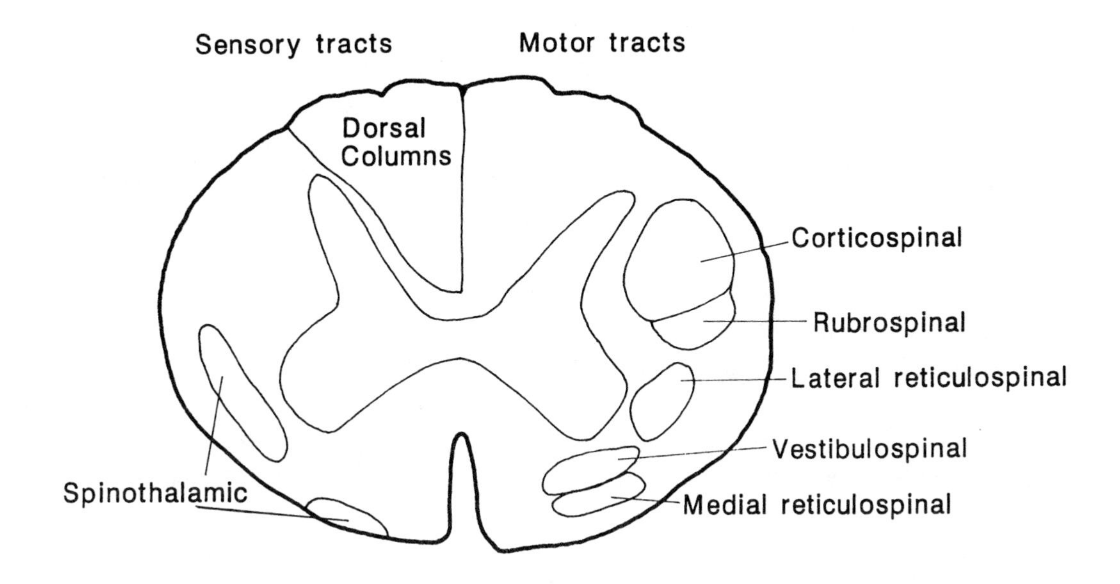

Figure 2-5. Diagram of locations of major ascending (sensory) and descending (motor) tracts in the human cervical spinal cord.

Decerebrate rigidity results from transection between the inferior and superior colliculi in the midbrain. It is characterized by increased tone in extensor (anti-gravity) muscles, largely due to enhanced activity of the vestibulospinal tract. Normally the rubrospinal and corticospinal tracts facilitate lateral reticulospinal activity which inhibits extensors. Damage to the midbrain removes this influence to the medulla; so, extensor tone increases with no compensation for vestibulospinal excitement of extensors. Decerebrate rigidity is also termed **γ-rigidity**, because it is due to relative overexcitation of γ-motoneurons. This rigidity is removed if the sensory roots are cut, indicating the importance of the spindle loop in maintaining the rigidity. If the cerebellum is removed after the midbrain is cut, then the rigidity becomes worse, because the cerebellum normally has an inhibitory influence on vestibulospinal neurons. However, in this case the rigidity is largely reduced by blocking α-motoneurons but not by cutting the dorsal roots, indicating **α-rigidity**. Destruction of the vestibular nuclei prevents decerebrate rigidity, indicating that the vestibulospinal tract mediates the rigidity. Brainstem motor mechanisms regulate postural responses which occur largely in anti-gravity muscles. Decerebrate rigidity has properties similar to clinical **spasticity**. Spasticity is **unidirectional**, since resistance to passive movement is greater in anti-gravity muscles. It is **velocity dependent**, and there is a **hyperreflexive tendon jerk**. On the other hand, clinical **rigidity** is **bidirectional**, not velocity dependent, and there is no hyperreflexive tendon jerk.

There are several types of **postural reflexes**. **Tonic reflexes** help distribute appropriate tone to trunk muscles in order to support the head. **Righting reflexes** permit trunk muscles to support the head starting from a supine position. **Statokinetic reflexes** are those related to head- and eye-turning reactions that permit tracking a target; they also maintain balance while moving.

Cerebellar Control of Movement

The cerebellum manages information for the motor cortex. It receives rapid input of length and tension information from all muscles of the body by **ipsilateral** projections. It projects to the cerebral cortex via the thalamus and to spinal neurons via reticular relays. Damage to the cerebellum does not produce sensory deficits or muscle weakness. Instead, cerebellar lesions produce disturbances in coordination of movement, muscle tone, and/or posture. The cerebellum is involved with planning and coordination of movements; it is a repository for learned movement "programs."

The cerebellum is divided into three major lobes. The **anterior lobe** and the **posterior lobe** both have somatotopic organization in their more medial regions. The **flocculonodular lobe** is involved with postural movements and has connections with the lateral vestibular nucleus. Mediolaterally, the cerebellum is divided into three regions. Each of these regions projects to specific cerebellar **sub-cortical nuclei**. 1) The **vermis** is located centrally and projects to the **fastigial nucleus**. 2) The **intermediate zone** projects to the **interpositus nucleus**, consisting of the globus and emboliform nuclei. 3) The **lateral zone** projects to the **dentate nucleus**. The axial regions of the body are represented in the vermis, and the limbs and face in the intermediate zone. The lateral zone does not receive direct information from muscles nor send projections to them. Its connections are with the cerebral motor cortex.

Spinal inputs to the cerebellum from the legs include the **dorsal spinocerebellar tract** (cell bodies in **Clarke's column**) and the **ventral spinocerebellar tract**. Input from the arms includes the **cuneocerebellar tract** and the **rostral spinocerebellar tract**. Information from the cerebral cortex and red nucleus aids the cerebellum in comparing desired movements with actual movements. The cerebellum receives cerebral cortical input indirectly via the **corticopontocerebellar pathway**. The **inferior olive** and **lateral reticular nucleus** in the medulla also relay information to the cerebellum.

All output from the cerebellar cortex is via inhibitory **Purkinje cells** which project to the sub-cortical, or deep cerebellar, nuclei. Consequently, the cerebellar cortex produces a pattern of inhibition onto neurons in three sub-cortical nuclei whose axons, in turn, are the output of the cerebellum. 1) The fastigial nucleus projects to the lateral vestibular nucleus. 2) The interpositus nucleus projects mainly to the red nucleus, but some fibers project to the motor regions of the thalamus (ventroanterior and ventrolateral nuclei). 3) The dentate nucleus projects to the motor thalamus, and the thalamus relays this information to the motor cortex.

All input to the cerebellar cortex is via **climbing** and **mossy fibers**. Climbing fibers originate exclusively from neurons in the inferior olive; all other inputs use mossy fibers. Each climbing fiber excites only about 10 Purkinje cells. Mossy fibers excite **granule cells**, the only excitatory interneurons of the cerebellar cortex. Climbing and mossy fibers also excite the sub-cortical nuclei. The parallel fibers of granule cells excite all the **inhibitory neurons** of the cerebellar cortex: Purkinje cells, basket and stellate cells, and Golgi cells. Input from mossy fibers excites a group of granule cells, which then excite their local population of Purkinje neurons. Due to the structure of parallel fibers, a **beam** of Purkinje cells is excited, and this is said to be **on-beam**. At the same time stellate and basket cells are also excited, and these inhibit **off-beam** Purkinje cells, which are lateral to on-beam neurons. This form of lateral inhibition by inhibitory interneurons is termed **feedforward inhibition**. Golgi cells cause feedforward inhibition of granule cells when excited by mossy fibers, but cause **feedback inhibition** when excited by parallel fibers of the granule cells themselves. This is analagous to Renshaw cell inhibition of α-motoneurons in the spinal cord. Basket and stellate cells **disinhibit** neurons of sub-cortical nuclei by inhibiting inhibitory Purkinje cells. Considering the whole circuit, mossy fibers and climbing fibers first excite the sub-cortical nuclei. One or two synapses later, Purkinje cells are excited which inhibit the sub-cortical nuclei. After another few synapses the Purkinje neurons are turned off. This rapid switching on and off of Purkinje neurons helps to control movements. In particular, it provides for damping of movements, which prevents limbs from overshooting their targets.

The **functions** of the cerebellum include producing synergy in movements, coordinating muscle contractions, modifying cerebral cortical movement commands as a function of the current position of the body and limbs, giving smoothness to movements, and controlling the "braking" or stopping of movements. More specifically, the flocculonodular lobe and vermis influence posture and muscle tone. The intermediate zones control ongoing movements and help insure that the intended movement is in fact the actual movement. The lateral zones are mainly involved in planning and initiation of movement. The cerebellum and the frontal cerebral cortex store complex motor programs, such as those involving hand-eye coordination. Since the cerebellum mainly controls coordination of movement, damage to the cerebellum produces various types of uncoordinated movements, depending on the site of the cerebellar lesion. Cerebellar damage affects muscles on the ipsilateral side; symptoms include hypotonia, asynergia, dysdiadochokinesia, dysmetria, pendular reflexes, intention tremor, nystagmus, vertigo, dysarthria, and ataxic (drunken) gait.

Basal Ganglia in Movement

The **basal ganglia (caudate nucleus, putamen, globus pallidus)** set the postural background for limb movements, so they are a bridge between the medial and lateral descending systems. The caudate nucleus and putamen comprise the **neostriatum**, and the putamen and globus pallidus are the **lenticular nucleus**. The **subthalamic nuclei**, **substantia nigra**, and **ventroanterior** and **ventrolateral nuclei of the thalamus** function with the basal ganglia. The primary input and output of the basal ganglia is to and from the cerebral cortex via long negative feedback loops. All regions of the cerebral cortex (motor, sensory, association) project to the neostriatum. The neostriatum projects to the globus pallidus and substantia nigra, and the substantia nigra projects to thalamus and back to neostriatum. The globus pallidus has reciprocal connections with the subthalamic nucleus. In addition, the globus pallidus provides the output of the basal ganglia, and projects to the thalamus, midbrain, and reticular formation. The thalamic nuclei relay back to the cerebral cortex. Due to these multiple relays, the basal ganglia can influence movement only indirectly. The basal ganglia participate in the initiation of movements, with patterns that are elicited by the motor cortex. They also control gross movements and adjust body position appropriately for a given task (such as getting the limb into a position for the fingers to perform a delicate task). They help to determine the direction, force, and speed of movements.

Disorders of the basal ganglia produce unusual movement deficits. **Parkinson's disease** is due to degeneration of **dopaminergic** neurons in the substantia nigra. Its symptoms include resting tremor, bradykinesia, plastic (or lead-pipe) rigidity of limbs, and a shuffling gait. Loss of nigrostriatal neurons reduces **dopamine** in their inhibitory terminals in the neostriatum. This can be treated with massive doses of a dopamine precursor, L-desoxyphenylalanine (L-DOPA). Surgical destruction of the ventroanterior and ventrolateral thalamic nuclei also reduces signs of Parkinsonism. **Huntington's disease** is due to loss of both cholinergic neurons in the neostriatum and inhibitory GABAergic striatonigral neurons. Its symptoms include chorea (flicking movements), dementia, and hypotonia. Since it is due to an autosomal dominant gene, half the children of an afflicted parent inherit it, and it has an adult onset. Parkinsonian patients receiving too much L-DOPA develop signs of Huntington's disease. Excessive firing of inhibitory dopaminergic neurons from the substantia nigra to the neostriaum is similar to loss of neostriatal neurons projecting to the substantia nigra. This demonstrates the reciprocal inhibition between neostriatum and substantia nigra. Damage to basal ganglia may cause other involuntary movements, such as ballistic or athetoid movements, seen in patients with cerebral palsy.

Cerebral Cortical Control of Movement

The motor cortex has a **somatotopic** representation of the contralateral body surface. The **primary motor area** lies directly anterior to the central sulcus, mostly in precentral gyrus. It has a large representation for fingers and thumb as well as the muscles used in speech. The primary motor area controls contralateral movements, mainly of the distal limbs and digits. Bilateral movements involve the **supplemental motor area**. This region is located on the medial surface of the precentral gyrus, is somatotopically organized (but controls larger muscle groups than the primary motor area), has a bilateral representation of the body, and influences mainly axial and girdle muscles. The **premotor area** lies anterior to the primary motor area and controls complex movements as well as integrating sensory and motor information; it is considered to be the motor association area. Its axons project to the primary motor area, not to lower levels. The motor cortex modulates patterns of movement generated by the basal ganglia and the cerebellum, where movement-related electrical activity can be detected before that in the motor cortex when initiating movement. The output of the motor cortex is the **corticospinal** (or **pyramidal**) tract. It gives off collateral branches to virtually all motor regions of the brain as the axons course to the spinal cord. All regions of the motor system are thereby updated about what the motor cortex is doing; this is termed **efferent copy** or **corollary discharge**. Efferent axons of the large Betz cells, which synapse monosynaptically on spinal motoneurons, make up only three percent of the pyramidal tract. Much of the pyramidal tract consists of unmyelinated axons, and many (40%) come from cortical areas other than primary motor cortex.

The activity of corticospinal neurons correlates with the **force** required for movements, not with position or direction of movement. Small lesions of the cerebral cortex may produce an initial hypotonia or paralysis (like spinal shock), loss of fractionation of movement, and loss of strength and rate of muscle contraction. The most common corticospinal lesion is damage to the internal capsule from a cerebrovascular accident (CVA or **stroke**); this damages the basal ganglia as well as the pyramidal tract. The result is hemiplegia, an extensor plantar response (Babinski sign) and spasticity. Spasticity is often accompanied by **clonus**, an alternation of stretch reflexes in agonist and antagonistic muscles at a joint, due to lack of relaxation of a muscle when its antagonist contracts.

Traditionally the motor system has been divided into a **pyramidal system** and an **extrapyramidal system**. The former includes only the corticospinal tract, and the latter consists of all other portions of the motor system. Lesions of the motor system have often been described as producing pyramidal signs or symptoms or extrapyramidal signs or symptoms. This terminology is outdated, because designation of a symptom as "extrapyramidal" provides no information about the locus of the lesion. Furthermore, few if any lesions only affect the pyramidal tract; thus, a pure lesion of the pyramidal tract virtually does not occur.

Motor lesions are often divided into damage to **upper** or **lower** motoneurons. Lower motoneuron refers to α-motoneurons or nerves; thus, they are located in the spinal ventral horn and cranial motor nuclei. Damage to lower motoneurons is characterized by decreased muscle tone, muscle atrophy, fasciculations of single muscles, absence of tendon reflexes, and no Babinski sign. Upper motoneuron refers to corticospinal neurons or other motor system neurons that do not exit from the brain or spinal cord. Lesions of upper motoneurons are most often characterized by increased muscle tone, although tone can also decrease (depending on region damaged). Other signs are derangements of groups of muscles, enhanced tendon reflexes, presence of a Babinski sign, and muscle atrophy (rarely).

Review Questions

30. Lower motor neuron disease is characterized by

    A. exaggerated reflexes
    B. enhanced recurrent activation of Renshaw cells
    C. being a later stage from the development of upper motor neuron disease
    D. atrophy of the affected muscles
    E. spasticity

31. Purkinje cells of the cerebellar cortex are excited by

    A. stellate cells
    B. basket cells
    C. granule cells
    D. Golgi cells
    E. input from deep cerebellar nuclei

32. Severing dorsal (sensory) spinal roots results in

    A. increased muscle tone
    B. exaggerated myotatic reflexes
    C. increased awareness of the position of limbs in space
    D. a tendency for movements to undershoot the mark (hypometria)
    E. increased jerkiness of movements

33. Hyperkinetic syndromes, such as chorea and athetosis, are usually associated with damage to

    A. primary motor areas of the cerebral cortex
    B. pathways for recurrent inhibition in the spinal cord
    C. those portions of the reticular formation that modulate motoneuron excitability
    D. anterior hypothalamus
    E. basal ganglia

34. A person has the following signs/symptoms: 1) loss of pain and temperature sense on the left side; 2) loss of pain, temperature, and touch sensations on the face; 3) paralysis of chewing muscles on the right side, and the jaw deviates toward the right; and 4) intention tremor, ataxia, hyporeflexia, hypotonia, and dysdiadochokinesia on the right. The most likely site of this lesion is the

A. cerebellum on the right side
B. cerebellum on the left side
C. thalamus on the right side
D. thalamus on the left side
E. lateral pons on the right side
F. lateral pons on the left side
G. lateral medulla on the right side
H. lateral medulla on the left side

35. A cerebrovascular accident involving the middle cerebral artery

A. produces permanent hypotonia
B. produces abnormalities similar to cutting the corticospinal tract at the medullary pyramids
C. produces spastic paralysis
D. affects the ipsilateral side of the body
E. primarily damages lower motor neurons

36. A patient presents with ataxia of the trunk, instability of posture, ataxia of gait, a tilt of the head, nystagmus, and ocular dysmetria. A lesion in the ______ could cause all these effects.

A. basal ganglia
B. ventral (anterior) gray portion of the spinal cord
C. red nucleus
D. midline (vermis) region of the cerebellum
E. corticospinal tract

CENTRAL AUTONOMIC CONTROL

Regulation of the autonomic nervous system and visceral functions is performed by certain brain stem nuclei, the hypothalamus, limbic system, and cerebral cortex. The medulla contains areas that regulate circulation and respiration; their functions are described in Chapters 3 and 4. The **limbic system** integrates autonomic and somatic motor activity to express of emotional behavior; it also modulates activity of emotion and behavior-related areas of the hypothalamus. The **cerebral cortex** coordinates specific autonomic and somatic adjustments (eg, the central control of specific muscle blood flow during movement). Frontal areas of the cortex participate in the coordination of cognitive activity with autonomic responses.

The **hypothalamus** selectively activates components of the endocrine, autonomic and somatic nervous systems to maintain homeostasis. It initiates appropriate behavioral responses to homeostasis-threatening stimuli such as stress, reproduction, exercise, and thermal regulation. It also monitors body water and plasma glucose levels and provides motivation for energy expenditure in locomotion and postural adjustments required for food-seeking behavior. The hypothalamus exerts important control over posterior pituitary (oxytocin, vasopressin) and anterior pituitary (growth hormone, ACTH, TSH, FSH, LH and prolactin) secretions. See Chapter 7 for neuroendocrine functions of the hypothalamus.

Hunger is controlled by a reciprocal interaction between the lateral feeding center (stimulate: hyperphagia; lesion: anorexia) and the ventromedial satiety center (stimulate: anorexia; lesion: hyperphagia). The **satiety center** transiently inhibits the chronically active **feeding center** after ingestion of sufficient food. The satiety center is likely noradrenergic; amphetamines stimulate this center. Prolonged inactivation of the satiety center may produce **hypothalamic obesity syndrome** with an initial rapid weight gain followed by a period of slower weight gain. Hypothalamic centers regulate the set point for body weight rather than food intake. The ventromedial nucleus contains glucose receptors which are sensitive to changes in blood glucose; these "**glucostats**" probably initiate feeding behavior.

The anterior hypothalamus controls water balance by creating the sensation of **thirst** and controlling the excretion of water by the kidney. Both hypothalamic **osmoreceptors**, excited by increased osmotic pressure in extracellular fluid, and pressure-sensitive receptors in the great veins control salivary output and the release of ADH. Angiotensin activates the subfornical organ located in the diencephalon but outside the blood-brain barrier. In turn, this organ excites hypothalamic areas involved with thirst.

Temperature regulation is controlled by the hypothalamus. It maintains body temperature at a set point of 37°C to optimize the rate of chemical reactions. Core temperature is regulated by 1) heat production derived from chemical reactions and muscle activity (shivering), 2) heat conservation through vasoconstriction of the blood vessels, and 3) heat loss through vasodilation (conduction, radiation and convection) and evaporation of sweat. The **preoptic** region and adjacent anterior nuclei of the hypothalamus contain a thermostat for establishing the "**set point**"; that core temperature which the system attempts to maintain.

Figure 2-6 on the next page shows how hypothalamic nuclei interact to control **heat loss** (A) and **heat conservation and production** (B). The **posterior** hypothalamus is **spontaneously active**, but it contains no neurons to directly sense temperature. The preoptic area contains temperature sensitive cells that increase their **discharge rate** when the blood is warmer than "set point". This leads to increased activity of the anterior hypothalamus and inhibition of cells in the posterior hypothalamus (Fig. 2-6A). When blood is cooler than the "set point", preoptic neurons decrease their firing, leading to decreased activity of the anterior hypothalamus. Cells in the posterior hypothalamus become active because the inhibition is removed (Fig. 2-6B). The "set point" of temperature-sensitive cells of the preoptic area is lowered by afferent input from warm receptors and raised by input from cold receptors.

A. Blood Temperature <u>Above</u> "Set Point"

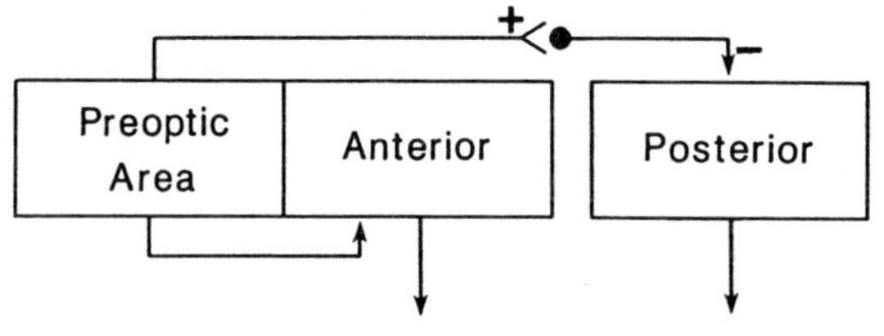

B. Blood Temperature <u>Below</u> "Set Point"

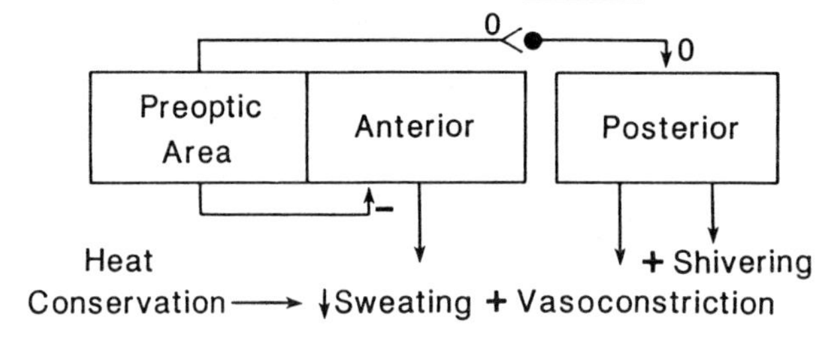

Figure 2-6. Responses of the hypothalamus for heat loss (A) and heat conservation and production (B). Key: + = excitation, - = inhibition, 0 = no activity.

Fever is produced by endogenous **pyrogens** (eg, interleukin 1) which act to reset the "set point" to a higher temperature. Since body temperature is then cooler than the "set point", body temperature will increase (heat production and conservation of heat loss) until it stabilizes at the new, elevated "set point" temperature. After the fever **breaks** and the "set point" returns to 37°C, the patient will vasodilate and sweat to lose heat until body temperature returns to normal. Endogenous pyrogens increase prostaglandin synthesis which in turn stimulates the thermoregulatory center. Aspirin is effective therapy for suppressing fever, because it interrupts **prostaglandin** synthesis.

Review Questions

37. Chronic stimulation of the hypothalamic satiety center would

    A. produce the hypothalamic obesity syndrome
    B. produce anorexia
    C. cause hyperphagia
    D. activate the feeding center

38. When body heat increases during exercise, core temperature is held constant by

    A. increasing the set point of the hypothalamus
    B. increasing the activity of sympathetic adrenergic fibers
    C. decreasing the activity of glucose receptors
    D. vasodilation
    E. reducing the activity of the subfornical organ

39. Which of the following would result from stimulation of the anterior hypothalamus?

A. Decreased gut motility
B. Increased blood pressure
C. Increased secretion of vasopressin
D. Sympathetic activation
E. Vasodilation

40. Hypothalamic centers controlling body temperature

A. respond to changes in osmoreceptor activity
B. have efferent outflow only through parasympathetic and sympathetic systems
C. respond to information from peripheral temperature receptors
D. respond only by altering heat loss rather than heat production

CEREBRAL CORTEX

Electrophysiology

Since the brain is essentially a saline gel, electrical potentials generated in the neocortex are **volume conducted** to the surface where they may be recorded from dura or scalp. **Spontaneous** electrical activity appears as wavelike fluctuations in electrical potential at frequencies of 1 to 40 Hertz and amplitudes of 2 to 300 μV. Electrical **evoked responses** elicited by synchronous artificial stimulation of afferent pathways to the neocortex are recorded as large amplitude, short latency waves followed by long latency, increasingly damped waves of varying polarity. Computerized summation or "averaging" of repeated evoked responses is required to extract the low amplitude "time-locked" evoked responses from "random" background spontaneous activity. Both spontaneous and evoked responses are generated mainly by synaptic activity rather than action potentials.

Primary evoked responses have the following characteristics: 1) they are elicited by brief, intense stimulation of sensory receptors or primary afferent pathways to the neocortex; 2) generated by many pyramidal cells forming vertically-oriented dipoles; 3) recorded as **positive-negative** potential waves of several msec duration at a latency of 15-25 msec if cutaneous or external stimuli (eg, light flashes of high intensity, short duration tones) are used. Primary afferent fibers synapse massively in Layer IV, generating a deep negative "sink" of depolarization. Current from superficial and very deep layers flows into Layer IV, causing the cortical surface to become relatively positive, thus generating the early positive deflection of the primary evoked response. The later negative wave may be generated by 1) deep recurrent inhibition (deep positive source), 2) longer latency thalamic non-specific afferent excitation of pyramidal cell apical dendrites, and 3) more superficial excitatory synaptic depolarization. The largest amplitude **primary evoked response** occurs over the primary sensory cortex of the stimulated modality and is highly resistant to neuronal depressants, such as anesthetics. Most primary responses are followed by **diffuse secondary responses** which last for tens of msec. These are generated in part by non-specific relays in the thalamus, are recordable over large areas of the cortex, and are easily decreased in amplitude by neuronal or synaptic depressants.

Sensory evoked responses are used to test the integrity of sensory projection pathways, such as visual, auditory, and somatosensory. Most evoked responses are generated by neocortical electrical events, but brain stem integrity can be tested by surface recording of evoked responses of < 1 μV amplitude. The **brain stem auditory evoked response** (BAER), recorded between the skull vertex and mastoid prominence, is produced by averaging 1000-2000 responses to clicks (Fig. 2-7). Waves I through V are probably generated by successive activation in the auditory nerve, axonal propagation through the levels of the cochlear nucleus, olivary complex, lateral lemniscus, and inferior colliculus. Waves VI and VII may result from activity in the medial geniculate. These brain stem signals are highly resistant to neuronal or synaptic depressants and changes in arousal level, which suggests that they are of axonal rather than synaptic origin. The BAER and comparable short latency evoked responses from somatic or visual sensory nerves are used to 1) test nerve and CNS integrity after trauma, 2) diagnose demyelinating diseases, or 3) monitor possible spinal cord or brain stem ischemia during surgical procedures.

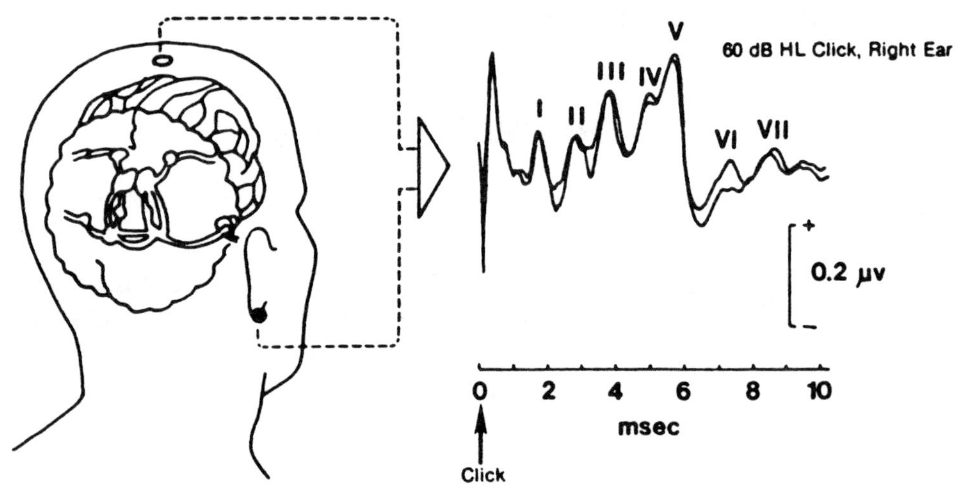

Figure 2-7. The brain stem auditory evoked response (BAER) produced by clicks to the right ear at 60 dB above normal hearing threshold.

Electroencephalography (EEG)

Spontaneous electrical activity is routinely recorded from the scalp by the techniques of electroencephalography. EEGs are described by **amplitude** and **frequency**. EEG waves result from temporal summation of volume-conducted cortical postsynaptic potentials (not action potentials). Thus **EEG amplitude** varies as a function of 1) the number of active synapses, 2) synchronicity of postsynaptic potentials, and 3) the distance of these potentials from recording electrodes. The predominant EEG frequency is correlated with the state of CNS **arousal**, being low in deep sleep, cortical depression, hypoxia, and anesthesia, and high during mental alertness and excitation. During arousal a pattern of **EEG desynchrony** of high frequency, low voltage activity is observed. The standard nomenclature identifies various EEG frequency bands; from lowest frequency to highest these are Delta (2-4 Hz), Theta (5-7 Hz), Alpha (8-13 Hz), and Beta (>13 Hz).

The **alpha rhythm** is generated by **thalamocortical systems** (primarily non-specific) at their natural resonant frequency and is observed optimally over the occipital areas of the relaxed, supine patient with eyes closed. **Alpha blocking** occurs when a relaxed patient is aroused by sensory stimulation or any cognitive activity. **Delta** range waves are seen in cortical depression or deep sleep when the cortex is almost functionally deafferented or isolated from the thalamus.

The clinical EEG is used for diagnosis of localized or generalized brain lesions induced by trauma, infection, and other disease processes. It is a primary tool in detection and location of epileptic seizure foci. General CNS or cortical depression from metabolic disorders, anesthetics, encephalitis, or hypoxia produces slowing and synchrony. Space-occupying lesions (eg, tumors, abcesses, sub- or extradural hematomas) produce slowing and amplitude alterations which may be focal or generalized depending upon the site of the lesion. Cortical trauma may produce increased neuronal excitability, associated with cortical scarring, leading to later seizures. Seizure processes may be provoked during EEG recording by hyperventilation (causing hypocalcemia and subsequent membrane hyperexcitability) or photic stimulation.

The Reticular Activating System

The reticular formation receives information from **all afferent systems** through multisynaptic pathways (eg, spinoreticular) and from more rostral areas. Lesions of the **pontine-mesencephalic** portion of the reticular formation are associated with **reduced** levels of consciousness or **coma.** This region facilitates sensory transmission to the neocortex by increasing background depolarization in sensory relay nuclei, so it is called the ascending reticular activating system. This is part of global CNS arousal during "orienting" responses or the human analogue of "fight-or-flight" conditions. Stimulation of the reticular activating system increases the amplitude of sensory evoked potentials, lowers thresholds for sensory input, enhances the ability to discriminate between sensory stimuli closely spaced in time (eg, two-flash discrimination), and facilitates descending motor activity.

More caudal areas of the reticular formation, predominantly located in the medulla, have apparently opposite functions. Lesions of the **medullary reticular formation** are associated with insomnia in experimental animals. The pontine-mesencephalic and medullary portions of the reticular formation can inhibit each other.

The reticular formation is sensitive to both cholinergic and adrenergic agents; atropine may block excitation directed at the neocortex, causing synchrony of the EEG. Amphetamines directly excite the reticular formation; LSD and other hallucinogens may act as false transmitters to excite it. Tranquilizers, such as chlorpromazine, are thought to have inhibitory effects on the reticular formation. General anesthetics, such as barbiturates, are synaptic depressants and have their greatest effect on slow-conducting, multisynaptic systems such as the reticular formation, while there is still normal transmission in primary sensory pathways.

Review Questions

41. The primary evoked cortical response elicited by cutaneous shock

    A. has an area of maximal response on the cortex
    B. results from electrical changes in the cortex due to non-specific thalamic excitation
    C. has a normal latency of 75 msec for the first positive wave
    D. demonstrates a normal resting rhythm of 10 Hz
    E. has its greatest amplitude over the ipsilateral somesthetic sensory cortex

42. Which of the following does NOT affect the amplitude of the electroencephalogram (EEG) recorded from the scalp?

    A. Number of active synapses
    B. Synchronicity of postsynaptic potentials
    C. Distance of active synapses from the recording electrodes
    D. Amplitude of propagating action potentials
    E. Temporal summation of synaptic potentials

43. The alpha rhythm of the EEG

    A. results from cortical desynchronization during mental excitation
    B. is in the frequency range of 4 to 7 Hz
    C. is indicative of cortical hypoxia
    D. is the intrinsic resonant frequency of the cortex in deep sleep
    E. is prominent during supine relaxation with eyes closed

44. The reticular formation of the brain stem has

    A. a tropic influence on the reticuloendothelial system
    B. a purely excitatory influence on the cerebral cortex
    C. an influence on the levels of excitability of neurons in the cerebral cortex
    D. a purely inhibitory influence on the cerebral cortex
    E. control of the autonomic nervous system

45. Activation of the reticular activating system causes

    A. a decrease in systemic blood pressure
    B. a decrease in EEG frequency
    C. facilitation of spinal reflexes
    D. increased thresholds for the detection of sensory stimuli
    E. decreased muscle tone

SLEEP AND WAKEFULNESS

Sleep is a periodic and reversible state of decreased ability to interact with the external environment; it is accompanied by a significant reorganization of the neural, endocrine, and somatic systems. The physiological causes of sleep onset, sleep maintenance, and waking are unknown; the need for sleep is unexplained. Sleep is conventionally divided into four **slow wave** (non-REM, synchronized) sleep stages plus the **REM** (**R**apid **E**ye **M**ovement, paradoxical) state. These five sleep stages are identified by polygraphic (EEG, EMG, and electro-oculographic or EOG) and behavioral criteria.

STATE W (wakefulness): The EEG alpha activity associated with quiet wakefulness or low voltage desynchronized activity of active wakefulness. The EMG has a high level of activity, and there are frequent voluntary, conjugate eye movements (EOG).

Sleep Stages

STAGE 1: The EEG shows <50% alpha waves. Occasional slow, rolling eye movements occur, and there is a slight decrease in EMG amplitude.

STAGE 2: The EEG shows "K"-complexes and bursts of a 12-15 Hz rhythm appearing as "sleep spindles" and contains < 20% delta activity. No conjugate eye movements in this or Stages 3 and 4.

STAGE 3: The EEG shows delta activity 20-50% of the time.

STAGE 4: The EEG shows delta activity over 50% of the time. In Stages 3 and 4 heart rate and blood pressure are reduced, pupils are miotic, and monosynaptic reflexes are slightly to moderately depressed. Arousal threshold increases with depth of sleep stages.

STAGE REM: This stage is characterized by 1) desynchrony of the EEG, 2) atonia of all skeletal muscles except those supporting eye and respiration-related movements, 3) difficulty in being awakened, and 4) penile or clitoral erection and lubrication. Against this tonic background several phasic events occur simultaneously. These include 1) rapid conjugate eye movement, 2) abrupt release of atonia, allowing myoclonic twitches especially in distal limb musculature, and 3) large variations in blood pressure, heart, and respiration rates, called "autonomic storms". Stage REM is also called **dream sleep** because of the many dream reports during this stage. Myocardial ischemia (heart attacks) often occurs in the early morning hours during REM sleep.

In normal sleep a person proceeds down through Stages 1 to 4 in order, ascends back to Stage 2 and enters Stage REM about 90 min after sleep onset; Stage REM is also exited via Stage 2. Repeated cycles are made through the stages, and Stage REM is entered 4-6 times/night at an average inter-REM **period of 90 min.** Total sleep time per day decreases with age (16 hrs at birth to 6 hrs in old age) mainly at the expense of Stage REM, which is reduced from 50% to about 25% of total sleep time in adults. The elderly usually have little or no Stage 4 and awaken frequently. The fetus approaches 100% Stage REM in sleep time in the pre-30 weeks of gestation, 70-80% during the 30-35 week period, and 50% at birth. Consequently, Stage REM is probably important for early maturation of the brain.

Sleep onset is aided by **passive** factors, such as isolation, relaxed posture, and voluntary reduction of sensory input, and by **active** factors involving activation or inhibition of certain brain stem nuclei. **Slow wave** and REM sleep may be partially controlled by the neurotransmitter serotonin (5-HT) secreted by the midline anterior raphe nuclei of the pons. Chemical blockade of serotonin or lesions of raphe nuclei cause decreased slow wave and REM sleep; loading with the 5-HT precursor, tryptophan, reverses this chemical blockade.

Moderate to severe **sleep deprivation** (60 hrs) is characterized neurologically by hand tremor, weakness in neck flexion, awkwardness, ptosis, dysarthria, poverty of facial movement, short attention span, reading difficulty, and apathy. Extreme deprivation of all sleep (100-200 hrs) is marked by loss of ocular convergence, slurred speech, inability to concentrate, episodic disorientation in time, immediate memory loss, and occasionally frank paranoid delusions. Recovery sleep shows a marked increase in slow wave sleep time. Selective **REM deprivation** is **not** accompanied by detectable behavioral or neurological changes in normal subjects; REM time may transiently "rebound" to greater than normal levels after several nights of REM deprivation.

Primary sleep disorders include REM narcolepsy and hypersomnia. **Narcolepsy** may result from the inability of the wakefulness system to suppress Stage REM sleep. Stage REM may be the first stage of nocturnal sleep. The etiology of narcolepsy is unknown; the only active treatment is pharmacological stimulants. **Hypersomnia** (pseudonarcolepsy) is characterized by episodic or continual extraordinary sleepiness which may induce slow wave sleep. However, sleep does not relieve this sleepiness.

True physiological **insomnia** is rare; most "self-reported" insomnias have normal sleep onset latencies and are usually resolved by psychotherapy. Some sleep disorders receive clinical attention. Arousal disorders occur in Stage 4; **somnambulism (sleep walking)** and night terrors are common in children but are often diagnostic of psychological disturbances in adults. **Enuresis**, when it occurs, is often correlated with the transition from Stage 4 to 2 and Stage REM; micturition occurs after the transition. **Sleep apnea** may occur in adult "snorers," many of whom are overweight; it may also occur in neonates (up to 2 years) with risk of death. Adult sleep apnea victims may develop a severely decreased arterial P_{O_2}. Voluntary **sleep deprivation** is a major public health problem that is especially common in students.

Review Questions

46. Paradoxical or rapid eye movement (REM) sleep is characterized

    A. by an EEG pattern of low amplitude desynchronized activity
    B. by a cortical EEG pattern of delta range (2-4 Hz) activity
    C. by a significant increase in muscle tone
    D. as paradoxical, because it occurs only once during the night while other sleep stages occur several times
    E. by an arousal threshold comparable to that determined during waking

47. A patient with a brain stem infarction involving primarily the midline raphe nuclei of the pons would probably show which of the following sleep-related signs?

A. Continuous somnolence
B. Normal latency for sleep onset
C. Marked increase in stages 3 and 4 (slow-wave sleep)
D. Transient or chronic insomnia
E. Increased secretion of serotonin

48. During which stage of sleep is there the greatest variability of autonomic excitability?

A. Stage 1
B. Stage 2
C. Stage 3
D. Stage 4
E. Stage REM

49. Stage 4 sleep

A. time is greater in the elderly than in young adults
B. is often accompanied by growth hormone secretion in young children
C. is the state where enuresis occurs most often
D. is marked by minimal or no EMG and great variability in heart and respiratory rates
E. shows a desynchronized EEG

CEREBRAL CORTICAL FUNCTIONS

Neocortical areas not committed to primary or secondary sensory and motor functions are called **association** areas and include portions of the frontal, temporal and parietal lobes.

The **frontal** lobes receive information from sensory, limbic, and autonomic systems and engage in complex cognitive functions. Damage to the frontal lobes may result in deterioration of higher intellectual abilities, initiative, planning, and problem solving abilities. The basis for **prefrontal lobotomy** is the apparent loss of anxiety resulting from disconnection of perceptions from normal emotional responses.

The **temporal** lobes function in memory, speech and language, and emotional behavior. Temporal lobe lesions may cause one or more signs of the **Klüver-Bucy Syndrome.** Classically, this results from bilateral anterior **temporal** lobectomy and includes 1) placidity, loss of fear, reduced aggressiveness, 2) visual agnosia, 3) excessive tendency to examine objects (hypermetamorphosis), 4) hypersexuality and 5) orality (including copraphagia). The temporal lobes have a low threshold for seizures, and these may be associated with hallucinations, amnesia, and acute emotional disturbances. **Psychomotor seizures** (seizures of thought processes) may result from temporal lobe trauma. Direct electrical stimulation of the temporal lobe may produce reports of complex memories, like eliciting "memory traces".

The **parietal** areas support complex association functions between the visual, somesthetic, and auditory systems, such as recognition of words or symbols; they are primarily concerned with organization in space. Unilateral parietal damage results in defects in spatial orientation and loss of body image (neglect) on the contralateral side.

The dominant hemisphere is the left side in 90-95% of right-handed persons as defined by speech and language capabilities. In those 10% of the population who are left-handed, the left side is still dominant in about 70%. The non-dominant **right hemisphere** has executive control for **spatial abilities** and is more adept at tasks requiring accurate spatial orientation. Unilateral damage to right parietal or posterior cortex results in lack of coordination in complex fitting tasks and defects in non-verbal tasks (eg, music, map reading, place learning). Cortical damage prior to adolescence may allow reorganization of the cortical areas, so that spatial or language defects may be compensated for in adulthood. The **corpus callosum** provides communication between hemispheres, allowing information received on one side to be shared by both sides.

Review Questions

50. Defective social behavior, urinary frequency, and early personality changes suggest a lesion in

    A. temporal lobes bilaterally
    B. the dominant parietal lobe
    C. frontal lobes bilaterally
    D. one or the other occipital lobe
    E. the right insula

51. A right-handed patient with a tumor in the right parietal lobe is most likely to exhibit

    A. speech defects
    B. right homonymous hemianopsia
    C. psychomotor seizures
    D. neglect of the left extremities
    E. recent memory loss

52. Which of the following remains after removal of the human cerebral cortex?

    A. Postural control
    B. Conditioning to patterned tonal stimuli
    C. The capacity to regulate body temperature
    D. Patterned vision
    E. Hypertonia in the extensor muscles

Aging of the Nervous System

The brain slowly **loses weight** after age 20. The sulci broaden and gyri flatten as the cortical area is reduced during the aging process. Cells are lost faster than fibers, so the ratio of gray to white matter decreases. Inclusions and changes in microtubule structure occur in neurons. The most common inclusion is the "aging pigment", lipofuscin. After 60 years of age the number of neuronal microtubules decrease linearly. This is often accompanied by **neurofibrillary tangles**, which are microtubules densely arranged in double helices. These tangles are a marked feature of **senile dementia** of the Alzheimer's type. Distortion of dendrites and loss of dendritic spines occurs in aging neurons, perhaps just before cell death. The rate of perfusion of the brain decreases 20% from 30 to 70 years, which is still sufficient to maintain neuronal functions. But in the "older old" (over 85 years old) perfusion is barely adequate.

Reduced velocity of conduction of nerve impulses is accompanied by increased synaptic delays. Axonal transport of materials from soma to axonal endings falls by a factor of ten. This loss appears to be fundamental to diseases of peripheral degeneration and loss of nerve cells in the cerebellum. Impaired axonal transport may be responsible for loss of synapses and neurons with aging. The concentration of norepinephrine falls 40% to 50% between adulthood and 70 years. **Dopamine**, the neurotransmitter that provides communication between the cells of the substantia nigra and the corpus striatum, declines with age. Symptoms of Parkinson's disease occur when 80% is lost.

Among sensations, pressure, touch and tactile discrimination are impaired, but vibratory sense is severely impaired. Pain and thermal sense are both diminished despite the fact that there is no loss of free nerve endings of C fibers. Perception from joint receptors is lessened, so older persons require greater angular movements of joints for perception to be achieved, especially of the lower limb.

ANSWERS TO NEUROPHYSIOLOGY QUESTIONS

1. E. This is its definition.

2. B. Generator potentials are decrementing, electrotonic potentials that are not propagated but may trigger APs (A, B, C). They are depolarizing and due to increased gNa (D, E).

3. E. Spontaneous activity occurs normally in some sensory fibers, especially for special senses.

4. B. Area I receives little ipsilateral information (A), but it does receive contralateral joint information (C), which Area II does not. The anterolateral system also projects to this area (D). The parietal cortex mediates visual and somatic stimuli (C).

5. C. Velocity detectors are not activated by noxious heat (nociceptors) (B) or by static displacement (D), since they are rapidly adapting (A). Acceleration detectors respond when rate of displacement is changed (E).

6. C. Position sense requires slowly adapting receptors (A, B). The dynamic phase discharge may code the direction and speed of joint rotation (C, D). Nociceptors do not respond to normal joint positions (E).

7. D. Neither receptor threshold (A) nor afferent axon conduction velocity (B) is relevant. Maximum overlap of fields makes discrimination difficult (C). Low density of receptors implies larger areas for each receptive field (E).

8. B. Stereognosis is mediated by the dorsal column system (A). Enkephalins are released by excitation of spinal interneurons via descending, not ascending pathways (C). The ascending system is arranged somatopically in the ventral position lateral thalamus (D). The system transmits somatic visceral sensations (E).

9. D. Parietal pain is triggered by nociceptors in the body wall, not in the viscera (A). These are not Pacinian corpuscles (B). Answer C describes splinting pain; answer E is the basis of some referred pain.

10. D. Pricking pain is fast in onset compared to slow burning pain (B) and is conveyed centrally in the anterolateral system (E) to at least the thalamus (A) (and possibly to sensory Area I cortex) by Group III fibers (D). "Gating" of pain is mediated by Groups I and II stimulation, not group IV (C).

11. A. Pain is carried in Group III fibers also (D). Pain sensation is derived from receptors specialized for pain transduction, ie, the nociceptors (B) whose adequate stimulus is tissue injury (A). Pain can be blocked by stimulation of posterior columns (C).

12. B. Pain is sensed in the contralateral side of the body because the spinothalamic tract axons cross the midline of the spinal cord near their origin. The dorsal column (A) and corticospinal tract (D, E) are not involved directly in pain sensation.

13. D. In hyperopia the eyeball is too short, or the lens is too flat (D). In either case the closest focusable visual point is beyond the normal near point for a person's age. In myopia the near point is closer than normal (E), and a negative, concave lens would be needed to move the near point farther from the eye (A). Cylindrical lenses are used to correct for astigmatism (B). Presbyopes also have their near point farther from the eye than normal, but none of the choices considers presbyopia.

14. D. When the ciliary muscles contract, they **reduce**, not increase (B) tension on the suspensory ligaments, allowing accommodation. The sphincter and dilator muscles control the pupil (C, E). Only trauma or changes in interocular pressure should change the shape of the eyeball (A).

15. A. The fovea has the highest concentration of cones (B). Consequently, eye movements (saccades) normally tend to place targets on the fovea for the greatest visual acuity (E), away from the low threshold rods (C) used under low levels of illumination (D).

16. D. Complete dark adaptation involves both rods and cones (B). Cone dark adaptation occurs within 10 min, but rod dark adaptation requires 30 min (A, E). Optical defects do not affect the normal pattern of dark adaptation (C).

17. C. Bipolar cells are disinhibited (C, D) after hyperpolarization of receptors in response to light (A). Bipolar cells then excite "on-center, off-surround" ganglion cells (E).

18. B. Hypercomplex cells fail to see stationary targets and respond poorly to small circular images even if moving (C). Color is immaterial (E). They respond best to "optimum length" (B) rather than to any length (D).

19. B. The fast eye movement in nystagmus defines the direction of clinical nystagmus (A), and the fast movement is a saccade (C). The slow eye movement is opposite to head acceleration (E) and in the direction of previous rotation after a person stops.

20. A. Unequal air pressure across the tympanic membrane occurs with consequent poorer hearing before ears "pop" when changing altitude (A). The scala media has a **positive** potential relative to extracellular space (B); communication between it and the scala vestibuli would eliminate the endocochlear potential (D). Either rigidity of the eardrum or fixation of the middle ear bones would decrease hearing sensitivity (C, E).

21. D. Perforation of the tympanic membrane causes more loss of low frequencies (A). Damage to the middle ear bones would cause general deficits (B). The apical portion transduces low tones (C). Lesions of the medial geniculate body would produce deficits dependent on the exact site of the lesion (E).

22. C. Blocked Eustachian tubes, punctured tympanic membranes, or immobilized stapes would produce conduction deafness (A, B, D). Caloric nystagmus is a clinical test for vestibular pathology (E).

23. A. Sweet is a primary taste quality for receptors located at the front of the tongue (B). Alkaloids taste bitter (C); no tastes normally cause sneezes (D). Minty is not a basic taste (E).

24. E. The olfactory system adapts quickly (A) and requires water soluble odorants (B). Odorants may attach to external receptor sites on the receptor membrane and not require lipid solubility (B). Taste has four primary sensations (C). A 30% intensity change is required (D).

25. D. A hemisection of the upper thoracic spinal cord will damage the dorsal horn, intermediolateral nucleus, and ventral horn. Due to damage to the intermediolateral nucleus, sympathetic neural outflow to the heart will be reduced (A). Motor control of the toes will be reduced on the side ipsilateral to the lesion (B), because all the motor pathways have already crossed by the time they reach the thoracic spinal cord. Pain sensation in the contralateral extremities below the lesion will be impaired (C), because the spinothalamic tract crosses near the segment of origin. Since the lower motor neurons supplying the chest musculature are damaged, the muscles will eventually atrophy (E). There will be no impairment of motor control of fingers (D), because the lesion is below the level of exit of the nerves controlling the hand.

26. C. The reflex is initiated by Golgi tendon organs (B) and is at least disynaptic through an inhibitory interneuron to motoneurons (A). The body would collapse if supported by the affected limb (D). A crossed extensor, not flexor, reflex could support the body with the opposite limb (E).

27. C. Ia afferents that arise from annulospiral endings primarily signal change in muscle length, but also have a static response for muscle length. Golgi tendon organs signal muscle tension responses (D, E).

28. A. Firing of α-motoneurons alone unloads the spindles and decreases annulospiral excitation (A). Sharp tap, lengthening of a synergistic muscle, and excitation of γ-motoneurons increase the rate of firing (B, C, D). Since the muscle length does not change during isometric contraction, activity in Ia fibers would not change (E).

29. E. Hemisections of the spinal cord produce deficits in sensations of pain and temperature on the contralateral side below the lesion, but loss of the sensation of touch on the ipsilateral side below the lesion. Descending pathways for voluntary control cross in the brainstem, spinal lesions produce voluntary movement deficits on the ipsilateral side. Together, the signs/symptoms indicate a lesion on the left side. Since the function of the arms is normal, and their innervation arises from the cervical spinal cord, the lesion must be below the cervical spinal cord. Neural deficits occur in the legs, which are innervated by lumbar segments, so the lesion must be in the thoracic spinal cord.

30. D. Upper motor neuron diseases release spinal lower motor neurons from descending inhibition, resulting in spasticity and exaggerated reflexes (A, E). Pathology of lower motoneurons results in muscle atrophy (D) and less activation of recurrent pathways (B). Pathology of upper and lower motoneurons usually have different causes (C).

31. C. Purkinje cells are inhibited by all other cortical interneurons (A, B, D). Purkinje neurons project to deep nuclei; they do not receive input from them (E).

32. E. Loss of sensory information from receptors would decrease muscle tone (A), abolish myotatic reflexes (B), and decrease awareness of limb position (C). Movements would tend to overshoot (D) in a jerky fashion.

33. E. Increased, uncontrolled, but stereotyped movement is seen with descending medial system and basal ganglia pathology (E). Cerebral motor pathology would cause hypotonia (A). Removal of recurrent inhibition would cause muscle hyperactivity, as in "lockjaw" from tetanus poisoning (B). The anterior hypothalamus is not relevant (D).

34. E. These deficits are due to a lesion of the dorsolateral portion of the rostral pons on the right side. They could be caused by infarctions due to thromboses of branches of the basilar artery or the superior cerebellar artery. The loss of sensation from the body is caused by damage to the spinothalamic tract which courses in this region. The loss of sensation from the face and the motor deficits involving the jaw are caused by damage to the trigeminal system (sensory and motor). The cerebellar-like deficits are due to damage to the superior cerebellar peduncle which

joins the rostral pons. A lesion localized to the cerebellum itself would be unlikely to cause all of the cerebellar deficits listed here.

35. C. Spastic paralysis is characteristic of a stroke (C), on the contralateral side of the body from the lesion (D). Brief hypotonia gives way to hypertonia (A). Cutting the pyramidal tract does produce permanent hypotonia (B). Upper motor neurons are affected by this lesion (E).

36. D. This is the only lesion that will produce all these effects. Lesions in the spinal cord would produce motor deficits only in the muscles innervated by that segment (B). Lesions of the corticospinal and rubrospinal tracts produce contralateral motor deficits, not bilateral deficits, and usually affect the limbs much more than the trunk (C and E). Basal ganglia lesions normally affect the limbs more than the trunk and are usually not associated with nystagmus (A).

37. B. Stimulation of the satiety center depresses food intake and inhibits the feeding center (B, D). Stimulation of the **feeding center** would override phasic satiety center inhibition and produce hyperphagia (C) with consequent obesity if stimulation were chronic (A).

38. D. "Set point" normally only changes with fever, circadian rhythm and activation of peripheral receptors (A). Sweating through eccrine glands is activated by sympathetic cholinergic fibers which are centrally controlled from the anterior (and generally parasympathetic) hypothalamus (B). Activation of glucose receptors may promote feelings of hunger during exercise, but they have no direct influence on core temperature (C). Vasodilation results from decreased sympathetic tone (D). The subfornical organ is involved with thirst mechanisms of the hypothalamus (E).

39. E. Posterior hypothalamic stimulation activates the sympathetic ANS (D), which would cause the other consequences listed (A, B, C).

40. C. Efferent pathways for temperature control include outflow to spinal motoneurons to mediate shivering from the dorsal medial posterior area of the hypothalamus (B). The hypothalamus can initiate mechanisms for heat production as well as for heat loss (D). Changes in osmoreceptor activity involve thirst (A).

41. A. A primary cortical evoked response in the somesthetic system has its greatest amplitude over the contralateral somesthetic cortex (E) by excitation of specific afferents (B). Latency of the first positive response is less than 25 msec (conduction time from site of stimulation plus synaptic delays) (C). There is no intrinsic frequency in the primary response (D).

42. D. Propagating action potentials do not affect EEG amplitude because of their fast conduction velocity and small electrical fields.

43. The alpha rhythm is a normal response of the neocortex (C) in the frequency range of 8-13 Hz (B), seen when relaxed (A); predominant frequency is lower in deep sleep (D).

44. C. Both excitatory and inhibitory afferents from the reticular formation project to all levels of the neuraxis (B and D). The reticular formation affects excitability of **other** brain areas that control the autonomic nervous system more directly (E). The "reticuloendothelial system" is not in the brain (A).

45. C. Blood pressure increases due to increased sympathetic tone (A). Decreased EEG frequency (B) and increased sensory detection thresholds (D) would be more likely with activation of the medullary reticular formation. Muscle tone increases by direct excitation of motoneurons (E).

46. A. Delta activity is characteristic of Stages 3 and 4 (B). There is a significant decrease in muscle tone (C); the EMG may be flat. REM is cyclical with other sleep stages (D) and arousal threshold is high (E).

47. D. Somnolence or coma would result from lesions in the pontine-mesencephalic reticular formation lesions (A). Sleep onset would be protracted (B), and one would expect less time in Stages 3 and 4 (C). The midline raphe would secrete less serotonin if damaged (E).

48. E. Autonomic excitability decreases with the deeper non-REM stages (C, D) but becomes very high during Stage REM (E).

49. B. The elderly spend little or no time in Stage 4 (A). Enuresis occurs when going from deeper sleep to lighter sleep stages (C); somnambulism occurs in Stage 4. Minimal EMG, variable heart and respiratory rates, and desynchronized EEG characterize Stage REM (D, E).

50. C. These symptoms result from frontal lobe damage; signs of temporal damage include A, C and E. Body image is a parietal function (D).

51. D. Body image is maintained by the contralateral parietal cortex in parallel with contralateral somesthetic sensory projection; the information about preferred hand is not relevant. A, C, and E are temporal problems, B is occipital.

52. C. Decorticate animals with an intact hypothalamus adequately regulate core temperature (C). Although posture is at least partially controlled through the basal ganglia, postural control will be impaired by destruction of cortex (A). Behaviors conditioned to patterned tones require associational cortex (B), as does mediation of patterned vision (D). Decerebration would produce hypertonia (E).

CARDIOVASCULAR PHYSIOLOGY

CARDIAC MUSCLE

Structure. The heart is composed of three types of cardiac muscle; atrial muscle fibers, ventricular muscle fibers, and other fibers specialized for initiation and conduction of action potentials rather than for generating force. The morphology of cardiac muscle cells is similar to skeletal muscle cells, since they both contain the contractile proteins, **actin** and **myosin**, and the regulator proteins, **tropomyosin** and **troponin**. Large numbers of elongated mitochondria are located close to myofibrils. **Transverse tubules** aid in conducting electrical activation toward the interior regions of cells. There is an extensive **sarcoplasmic reticulum**, which serves as an intracellular store of calcium. The **sarcolemmal membranes** of adjacent cells are joined by **intercalated disks** to form cardiac muscle fibers. Intercalated disks provide a strong mechanical connection between cells, allowing axial mechanical forces to be transmitted from one cell to another. There is no protoplasmic continuity between cells. However, **gap junctions** provide low resistance pathways for spread of electrical excitation to adjacent cells. Thus, cardiac muscle behaves as a **functional syncytium**. Action potentials initiated in one cell will spread to all other cells through the latticework of cellular interconnections. By the **all-or-nothing principle** an effective stimulus anywhere in the heart always activates the entire heart.

Cardiac Action Potentials. The resting membrane potential of atrial and ventricular muscle cells (-80 mV) is slightly less than that for skeletal muscle cells. However, cardiac action potentials last 20 to 50 times longer than those in skeletal muscle. This prolongation of the action potential is due to the presence of a plateau phase. Cardiac muscle and cardiac pacemaker action potentials are described in Chapter 1 (Fig. 1-5).

Excitation-Contraction Coupling. The process of translating membrane electrical activity into mechanical contraction in cardiac muscle is similar to that in skeletal muscle. Calcium ions are released from the **sarcoplasmic reticulum** to initiate the contractile process. However, in cardiac muscle calcium ions enter the cell during the plateau phase of the action potential and trigger the release of **internal calcium** stores. Extracellular Ca^{2+} accounts for about 5% of the total amount of Ca^{2+} necessary to initiate contraction.

Mechanical Properties. The mechanical properties of cardiac muscle are like those of skeletal muscle with two important differences. First, the cardiac action potential persists for approximately 70% of the duration of mechanical contraction, lasting past the time when peak contractile force is generated. During this time muscle cells are refractory to further stimulation, so cardiac muscle cannot undergo tetanic contraction. The duration of contraction of cardiac muscle is primarily a function of the duration of the action potential. Second, because there are more elastic components in cardiac muscle than in skeletal muscle, cardiac muscle exhibits a significant amount of **passive tension** at normal resting length. In skeletal muscle the force of contraction is determined by the number of fibers that contract. On the other hand, in cardiac muscle contraction strength is graded by direct adjustment of the intracellular contractile mechanism of myocardial cells. This adjustment is controlled by mechanical, humoral, and neural factors. Force of contraction also varies with the length of cardiac fibers at the moment of initiation of contraction. Passive stretching of muscle fibers causes an increase in contractile strength which is seen to a more limited extent in skeletal muscle (length-tension curve, Fig. 1-3). This relationship is known as the **Frank-Starling Mechanism** and should not be confused with the concept of contractility. **Contractility** refers to a change in the ability to generate force at a **given** initial length. Thus, there are two ways to increase the force of cardiac contraction; 1) increase the initial muscle **length** (Frank-Starling mechanism), or 2) increase the **strength** of contraction at a given muscle length. **Digitalis glycosides** increase contractility presumably by inhibiting Na^+-K^+ ATPase of cell membranes. **Catecholamines** increase contractility via β-adrenergic receptors and cyclic AMP.

Metabolism. Cardiac muscle requires an abundant supply of oxygen-rich blood, since it depends almost exclusively on **aerobic** metabolism to supply the ATP for its contractions. During hypoxic states about 10% of its energy can be provided by **anaerobic** metabolism. But if anoxia occurs, sufficient energy for contraction cannot be supplied. At rest the heart uses oxidation of fatty acids for its ATP; only small quantities of glucose are utilized. ATP is rapidly replenished from a larger pool of **creatine phosphate** in equilibrium with ATP. When cardiac workload is increased, cardiac muscle removes lactic acid from coronary blood and oxidizes it directly.

When the heart converts chemical energy into mechanical work, it performs both internal (muscle shortening) and external (moving blood) **work**. The units of work and energy are the same. External stroke work performed can be estimated as the product of mean aortic pressure and the stroke volume, with an additional small (5%) kinetic energy component. Total external work = pressure energy + kinetic energy. Cardiac efficiency represents the total energy of contraction that appears as mechanical work; normally it is 15-20%.

$$\text{Efficiency} = \frac{\text{Cardiac work/min}}{\text{Myocardial } O_2 \text{ consumption/min}} = \frac{\text{Work performed}}{\text{Chemical energy utilized}}$$

Myocardial oxygen consumption is approximated by the product of peak systolic aortic pressure and heart rate (Rate-Pressure Product). External work/min is approximated by the product of cardiac output and mean aortic (arterial) pressure. If aortic pressure increases at a constant cardiac output, then oxygen consumption is elevated as minute work increases, but efficiency may not change. If the **inotropic state** (contractility) and heart rate (sympathetic response) do not increase during mild exercise, then cardiac efficiency may increase slightly. β-adrenergic blockade in a patient with angina pectoris may facilitate cardiac output by the Frank-Starling mechanism rather than increasing cardiac efficiency by increased heart rate and contractility.

Review Questions

1. Under normal physiological conditions the strength of contraction of cardiac muscle fibers is controlled by

   A. the number of fibers that contract
   B. direct adjustment of intracellular contractile mechanisms
   C. extracellular chloride concentration
   D. the membrane potential during the plateau phase
   E. the amplitude of the fast response action potential

2. In cardiac muscle

   A. actin is absent from myofibrils
   B. action potentials may last several hundred milliseconds
   C. strength of contraction is unaffected by neural activity
   D. transverse tubules are absent
   E. extracellular calcium is not required to initiate contraction

3. The all-or-nothing principle, as applied to cardiac muscle, states that all cardiac muscle cells

   A. are capable of generating action potentials
   B. contract when directly stimulated by ACh released from nerve terminals
   C. are stimulated to contract if an action potential is generated anywhere in the heart
   D. can be tetanized if the duration of the action potential is sufficiently long
   E. are refractory for the entire period of the action potential

4. Cardiac glycosides, such as digitalis, enhance the contractile performance of cardiac muscle fibers by

   A. inhibiting the Na^+-K^+ ATPase of cell membranes
   B. stimulating intracellular production of cyclic AMP
   C. decreasing the amount of Ca^{2+} available to myofibrils
   D. stimulating cardiac β-adrenergic receptors
   E. blocking cardiac β-adrenergic receptors

EXCITATION OF THE HEART

The Conduction System

Functional Anatomy. The cardiac cycle has two distinct time periods, **diastole** when the chambers of the heart relax and fill with blood, and **systole** when the chambers contract. Atrial contraction completes filling of ventricles, and ventricular contraction ejects blood to systemic and pulmonary circulations. The electrical signals that initiate cardiac contraction are generated within the heart by **pacemaker cells**, which are modified muscle cells, located in the **excitation and conduction system**. Since cardiac tissue shows automaticity and rhythmicity, no external signals are necessary for cardiac contraction, but external signals from the nervous system **modulate** the electrical activity of the heart. The cardiac fibers that act as pacemakers exhibit a characteristic slow, spontaneous depolarization during diastole. The pacemaker and conduction system consists of four major portions: 1) the **sino-atrial (SA) node** which generates the excitatory pulse, 2) the **atrio-ventricular (AV) node** which conducts but delays the impulse before it passes to the ventricles, 3) the **bundle of His** and 4) the **left and right bundle branches** which conduct the impulse to all parts of the ventricles through terminal Purkinje fibers. Inherent rhythmicity is noticible in the SA node; it is less pronounced in the AV node, bundle of His, and Purkinje fibers.

Activation and Conduction

Pacemaker Activity. The heart has its own pacemaker region, the SA node, located in the right atrium. These specialized cardiac muscle cells have the unusual property of automaticity due to phase 4 depolarization (Fig. 1-5B). The SA nodal cell membrane potential difference decreases slowly from its depolarized value of -90 mv due to a net cation influx ("potassium leak").

The rate of change of depolarization ("slope") is modified by cardiac autonomic activity through changes in SA nodal ion conductances (Fig. 3-1). Sympathetic activity increases the slope of depolarization, and parasympathetic activity decreases it. When the SA nodal membrane potential reaches threshold (about -60 mv), an action potential is generated. Increasing the slope of depolarization increases heart rate by causing more depolarizations per min (Fig. 3-1A); decreasing the slope slows the heart rate (Fig. 3-1B).

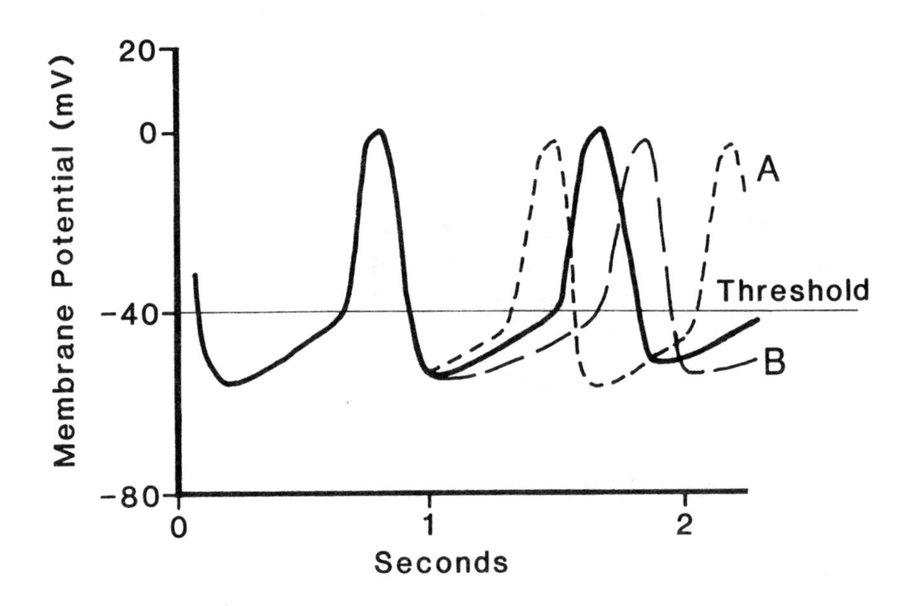

Figure 3-1. Changes in the slope of diastolic depolarization affect the frequency of heart beats.
A - Increased slope of depolarization produces a shorter interval between beats;
B - Decreased slope of depolarization produces a longer interval between beats.

Control of Pacemaker Activity. Parasympathetic nerve fibers are distributed mainly to the SA and AV nodes; **sympathetic nerve fibers** are nearly uniformly distributed to all parts of the heart, including the SA and AV nodes. **Acetylcholine**, the parasympathetic neurotransmitter, slightly hyperpolarizes pacemaker cells and decreases their rate of diastolic depolarization, thereby slowing heart rate. **Norepinephrine**, the sympathetic neurotransmitter, increases heart rate by increasing the slope of diastolic depolarization. The intrinsic SA depolarization rate of approximately 100/min is normally slowed to a typical rate of 65-75 beats/min by the dominant cardiac parasympathetic (vagal) activity. Heart rates higher than 65-75/min are caused by a simultaneous decrease in cardiac vagal activity and an increase in cardiac sympathetic activity; both systems innervate the SA node. Even at rest the time between heartbeats is rarely constant, shortening with inspiration and lengthening with expiration. This is principally due to a respiratory-linked change in cardiac vagal activity.

Atrial Activation and Conduction. The **SA node** is the primary pacemaker of the heart, because cardiac fibers in this region have the fastest inherent rhythm of any region of the heart. **Atrial conduction** requires no specialized conduction tissues; the wave of excitation travels at 1 m/sec over **atrial muscle** fibers.

AV Node Conduction. The **AV node** is critical for cardiac function. The delay of the cardiac impulse in this region gives the contracting atria adequate time to empty their contents into the ventricles before ventricular contraction is initiated. This slowed conduction through the AV node is due to both 1) thin fibers which produce slow conduction and 2) fewer gap junctions for impulse transmission from cell to cell. Conduction disturbances are most common at the AV node, because it is the "weakest" point of the functional syncytium. The AV node has one-way conduction (towards the ventricles); under normal conditions the AV node is the only pathway for electrical activity to pass from atria to ventricles. AV node conduction is also influenced by autonomic activity. Increased vagal activity not only decreases the rate of SA node discharge but also slows conduction of the resulting action potentials through the AV node. Sympathetic activity increases AV node conduction velocity.

Ventricular Activation and Conduction. Cardiac action potentials leaving the AV node enter the bundle of His and subsequently the left and right bundle branches. Conduction velocity is approximately 4 m/sec in these fibers, which results in a rapid spread of the cardiac action potential toward the ventricular musculature. The bundle branches terminate in Purkinge fibers which are composed of large cells with numerous gap junctions. The terminal Purkinje fibers enter the endocardium and extend about one-third of the way into the muscle mass. It takes approximately 100 msec for the wave of depolarization leaving the AV node to spread over the Purkinje fibers and reach all parts of the ventricles. The wave of depolarization spreads first from the septum toward the apex of the heart, and then turns from the apex and moves toward the base of the heart. The action potential then leaves Purkinje fibers and enters **ventricular muscle fibers**, moving generally from endocardium to epicardium. The velocity of conduction in ventricular muscle fibers is about 0.3 m/sec. This sequence of activation and conduction provides symmetry of contraction, especially in large hearts, and an orderly contraction sequence from apex to base for efficient ejection of blood from the ventricles.

Ventricular action potentials look different from SA nodal and other cardiac action potentials. This is most evident during phase 2 where prolonged calcium influx is primarily responsible for maintaining the ventricular electrogram above 0 mV for several hundred msec (Fig. 1-5A). Ventricular tissue is very easily re-excited at the end of phase 3 just before complete return to the resting membrane potential. This easily excited or supernormal period is a narrow time interval where a low intensity stimulus can cause another cardiac electrical and mechanical response. Just prior to the supernormal period, a larger than normal stimulus can elicit an electrical response (relative refractory period). Still earlier, no response can be initiated no matter how large the excitation (absolute refractory period).

The Electrocardiogram. Electrical activity recorded from the surface of the body represents the vector sum of action potentials recorded from many cardiac muscle fibers oriented in many different directions. A deflection away from the baseline observed on the **surface electrocardiogram** (EKG or ECG) indicates an electrical imbalance between the pair of recording electrodes, as a wave of activity moving within the heart produces current flow between the distant areas being monitored by the electrodes. The configuration of the electrocardiogram bears little resemblance to single fiber action potentials. An electrocardiogram furnishes information about heart rate, conduction pathways, cardiac excitability, cardiac refractoriness, and some anatomical irregularities. It does not provide any information about mechanical actions of the heart, such as valve movements or contractile force of the myocardium.

Lead I of the EKG is recorded between the right arm (+) and left arm (-), **Lead II** between the right arm (-) and left leg (+), and **Lead III** between the left leg (+) and left arm (-). The **augmented leads, avR, avL, and avF,** are various combinations of the limb leads. These six leads record activity in the frontal or vertical plane of the thorax. The precordial **Leads V_1-V_6**, placed on the chest and left side, record electrical activity from the transverse or horizontal plane. A wave of depolarization moving toward a (+) lead causes an upward deflection on the EKG.

The standard EKG tracing is composed of a **P wave** (atrial depolarization), a **QRS complex** (ventricular depolarization), and a **T wave** (ventricular repolarization) (top tracing of Fig. 3-4 on page 84). Repolarization of the atria is masked by the QRS complex. The P wave lasts about 100 msec. The electrically silent **(isoelectric)** period between the P wave and the QRS complex lasts about 80 msec. During this time the wave of depolarization is travelling through the AV node into the bundle of His. When excitation breaks out of the A-V node, the initial portion of the QRS complex is recorded. The Q wave is usually small; the S wave is easily seen in leads II and III but is small in lead I. The QRS complex lasts about 100 msec. The isoelectric S-T segment corresponds to the plateau phase of the cardiac action potential when all fibers are simultaneously depolarized. The T wave commonly deflects in the same direction as the R wave, indicating that ventricular repolarization occurs in the opposite direction within the heart from ventricular depolarization. The time between consecutive EKG complexes is expressed as the R-R interval; its value is 800 msec for a person with a heart rate of 75/min.

Abnormal Cardiac Rhythms

The normal cardiac EKG is affected by many diseases. Women and children have slightly faster heart rates than 65-75/min, and men and elderly persons have slightly slower ones. Normal resting heart rate is a reference for either abnormal cardiac slowing (**bradycardia**; less than 50 beats/min) or speeding up (**tachycardia;** greater than 90 beats/min). Highly trained and healthy athletes may have normal resting heart rates of 45/min. Fever in normal individuals can accelerate heart rate to 100/min with no underlying cardiac pathology. In all cases noted above, an atrial systole is always followed by a ventricular systole.

Abnormal cardiac rhythms can be caused by many mechanisms, such as the following: 1) abnormal rhythms of the SA node, 2) occurrence of pacemaker-type activity in other parts of the heart, 3) blockage of action potential transmission anywhere in the conduction system, 4) abnormal pathways of impulse transmission through the heart, and 5) spontaneous generation of sporadic abnormal impulses anywhere in the heart. **Ectopic foci** are observed in the case of excitable, diseased hearts. These foci discharge spontaneously and increase the heart rate (atrial or ventricular **extrasystoles** or premature beats).

Atrio-Ventricular Block. A variation from the normal 1:1 atrial/ventricular depolarization is seen with AV block. This occurs when the AV node slows its conduction of action potentials to the point where transmission of atrial depolarization is irregular or does not occur at all. **First degree AV block** occurs physiologically in athletes with a high degree of cardiac vagal activity and very slow AV nodal conduction, but no inherent cardiac disease. This type of AV block presents simply as an increased P-R interval. A more serious condition is **second degree AV block** where there is intermittent failure of AV conduction into the ventricle (Fig. 3-2A). Note the very irregular R-R intervals when block occurs. Second degree block can present with every other atrial impulse being conducted into the ventricle (2:1 rate) or other ratios such as 3:1 or 4:1. The resultant heart rate can be either close to normal or very low depending on the number of atrial depolarizations that are conducted into the ventricle per minute. Complete or **third degree AV block** is present when no atrial depolarizations are conducted into the ventricle (Fig. 3-2B). In this case there is no association between the timing of the P waves and the QRS complexes. The lower His bundle and Purkinje fibers exhibit a slow phase 4 depolarization rate and can act as pacemaker regions. The resulting R-R interval is approximately 2000-3000 msec (30-20 beats/min). AV block can also occur pathologically after myocardial infarction when the dead tissue includes part of the AV conduction pathway. Artificial pacemakers which "sense" the absence of QRS complexes and initiate ventricular depolarizations are used in patients with second or third degree AV block.

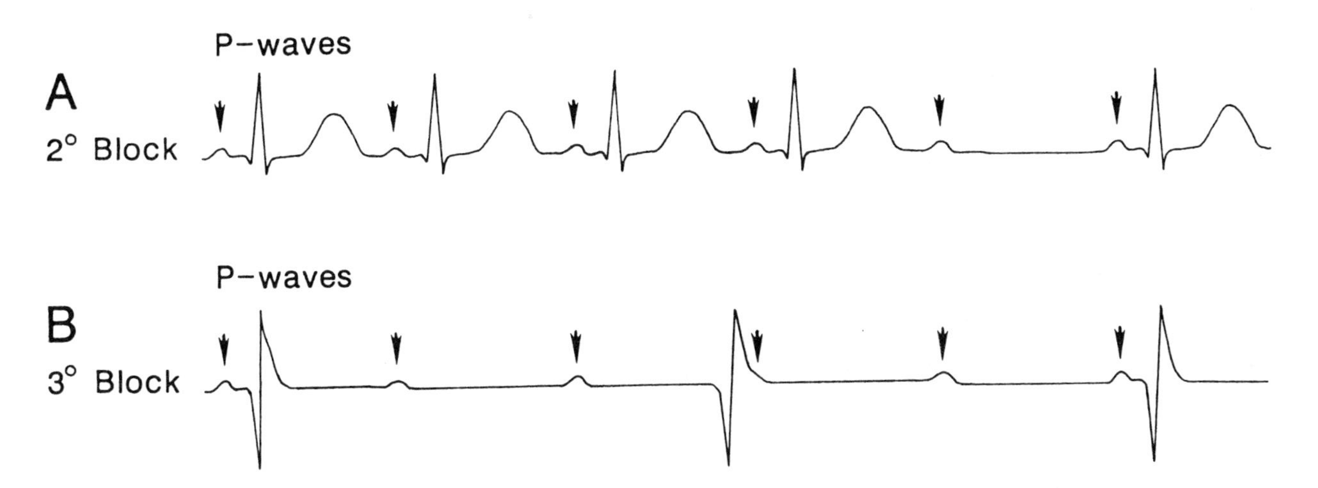

Figure 3-2. Second degree (A) and third degree (B) atrio-ventricular (AV) blocks.

AV conduction abnormalities are present when there are extra or accessory cardiac tissue pathways that electrically connect from atrial to ventricular tissues. This effectively "short circuits" the normal conduction pathway around the AV/His region. Then atrial depolarizations are rapidly conducted into the ventricle, since no AV nodal delay is present. Serious ventricular arrhythmias occur when the normally conducted AV depolarization meets ventricular tissue that is already partially depolarized or is repolarizing and in the brief supranormal period. The classic clinical example of this malfunction is Wolff-Parkinson-White syndrome, named for the physicians who first described it. Effective treatment for this syndrome requires surgical destruction of these accessory AV pathways.

Atrial tachycardia or fibrillation can exist with a normal ventricular rhythm and vice versa. Atrial tachycardia or fibrillation results in ineffective atrial pumping, but ventricular filling occurs normally, so cardiac output is virtually unaffected as one of the many atrial beats is propagated to the ventricles. Yet atrial tachycardia is not benign; ventricular rate will usually increase. Atrial fibrillation can result in ventricular tachycardia.

Ventricular Arrhythmias. If these become cyclical, they are potentially life threatening. One arrhythmia causes another until mechanical dysfunction occurs, and no blood is pumped. Single aberant ventricular beats occur at low rates in healthy people, but the frequency of these arrhythmias increases with cardiac disease. The most common ventricular arrhythmia is **premature ventricular contractions** (PVC). These usually occur when a part of the ventricle improperly conducts action potentials or a small area of the ventricle starts depolarizing by itself (ectopic focus). Because the depolarization does not travel along the normal Purkinje conduction system, there is a large dispersion of ventricular tissue depolarization. Note that a typical PVC (Fig. 3-3A) has no P wave, a long ventricular depolarization interval, and abnormal repolarization. This aberant electrical event will generate a mechanical systole, but the strength of contraction is reduced, since the ventricle does not contract uniformly. When three or more PVCs occur together (ventricular rate of 100-280 beats/min), it is called **ventricular tachycardia** (Fig. 3-3B). Blood is pumped by the ventricles during ventricular tachycardia, but cardiac output is greatly reduced, since time for ventricular filling is minimal at high heart rates. Ventricular tachycardia often leads to **ventricular fibrillation** (Fig. 3-3C), where ventricular rates exceed 280 beats/min, and the EKG becomes completely chaotic in appearance.

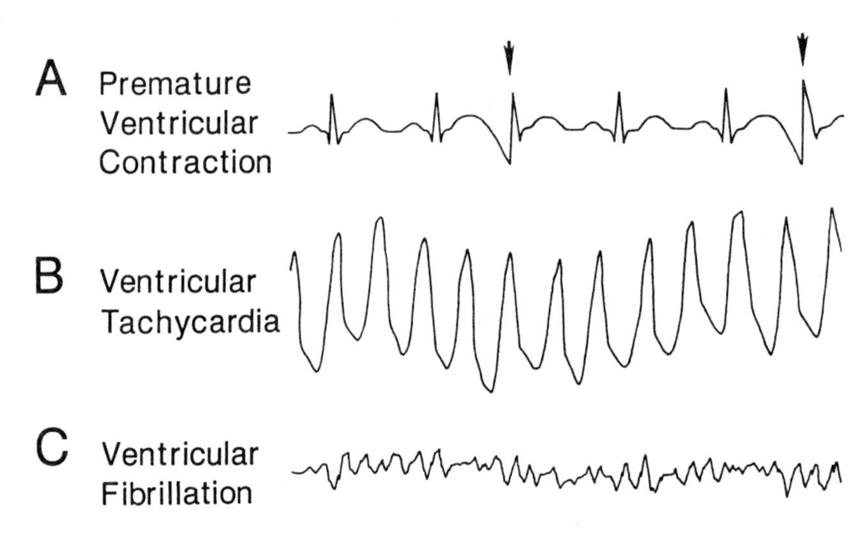

Figure 3-3. Types of ventricular arrhythmias.

Review Questions

5. The delay of spread of electrical activation from atria to ventricles is

   A. an aid in ventricular filling
   B. recorded on the electrocardiogram as the S-T segment
   C. recorded on the electrocardiogram as the Q-T interval
   D. due to the small number of large diameter conducting fibers in the AV node
   E. the function of the slow conducting fibers in the bundle of His

6. The portion of the heart with the fastest conduction velocity for action potentials is the

    A. left atrium
    B. right atrium
    C. left ventricle
    D. right ventricle
    E. bundle of His

7. The T wave of the electrocardiogram is related to

    A. atrial repolarization
    B. atrial depolarization
    C. bundle of His conduction
    D. ventricular repolarization
    E. ventricular depolarization

8. The electrocardiogram provides information about each of the following **EXCEPT** cardiac

    A. conduction pathways
    B. excitability
    C. refractoriness
    D. contractility
    E. repolarization sequence

THE HEART AS A PUMP

Functional Anatomy of the Heart. The heart is a four-chambered pulsatile pump that generates the energy necessary to move blood through the circulatory system. Blood flow between the atria and ventricles is controlled by the **A-V valves**, which establish a unidirectional flow pattern. The **semilunar valves** are located at the outlet of each ventricle and establish a unidirectional flow into the aorta and pulmonary artery.

The Cardiac Cycle

Figure 3-4 shows the time course of many important variables in the cardiac cycle. The cardiac cycle consists of two separate phases, **diastole** (the relaxation phase) and **systole** (the contraction phase). Diastole consists of three periods; rapid filling, slow filling, and atrial contraction. Systole also consists of three periods; isovolumic contraction, ejection, and isovolumic relaxation. The sequence of events in the cardiac cycle is typically described in terms of pressure changes on the left side of the heart. This sequence is similar for the right side of the heart, but right ventricular pressures are about 20% of those generated in the left heart.

During diastole cardiac musculature is relaxed, the atrioventricular valve (mitral valve) is open, and the aortic valve is closed. When the SA node initiates an action potential, the spread of this electrical activity over atrial muscle fibers causes **atrial contraction**. This produces a small (10-20%) increase

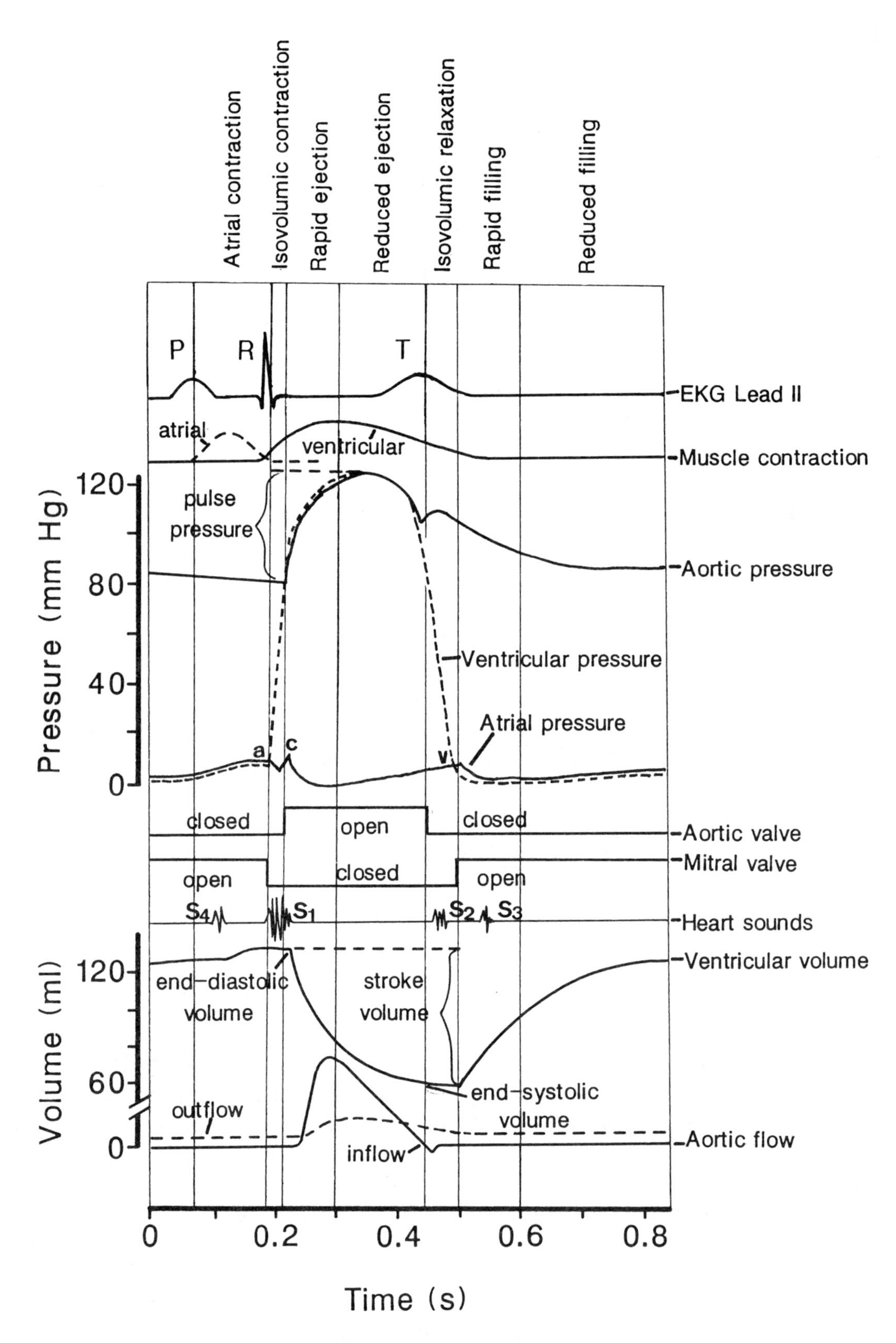

Figure 3-4. Sequence of pressure and volume changes during the left ventricular cardiac cycle.

in the volume of blood in the left ventricle. Although atrial contraction is not essential for normal persons, in some diseases (eg, mitral stenosis) atrial contraction may be vital for effective ventricular filling.

The beginning of ventricular contraction occurs at the peak of the R wave of the EKG. Initially, the increasing ventricular pressure tends to move a small amount of blood back across the mitral valve into the atrium. However, this movement causes leaflets of the mitral valve to close, producing the **first heart sound.** Then both mitral and aortic valves in the left ventricle are closed, but muscular contraction is continuing. This is called the **isovolumic contraction period**, and ventricular pressure increases very rapidly. When ventricular pressure exceeds aortic pressure, the aortic valve opens, ejecting blood from the ventricle into the aorta. During the early part of ejection, blood flows rapidly into the aorta (the **rapid ejection period**), and ventricular volume decreases sharply. This rapid inflow of blood into the aorta exceeds the runoff from the aorta toward peripheral vascular beds, and aortic pressure increases. Later during ejection, blood enters the aorta less rapidly as the ventricle nears the end of its contraction phase. Peripheral runoff then exceeds the inflow of blood into the aorta, and aortic pressure falls. As left ventricular pressure continues to fall, there is a tendency for a small amount of blood to flow back across the aortic valve into the ventricle, but this movement closes the leaflets of the aortic valve, producing the **second heart sound.**

Closure of the aortic valve causes the **incisura**, which is immediately followed by a secondary small vibration, the dicrotic notch. Aortic valve closure allows aortic pressure to be maintained at a level much higher than ventricular pressure during the course of the subsequent diastole. Once the aortic valve closes, the period of **isovolumic relaxation** begins. The aortic valve has closed, and the mitral valve that closed at the beginning of systole has not yet opened. During isovolumic relaxation, ventricular pressure falls rapidly and soon falls lower than atrial pressure, when the mitral valve opens. Blood rushes from the left atrium into the left ventricle, atrial pressure falls, and ventricular volume increases rapidly (the **rapid filling phase**). During the later stage of diastole, ventricular filling continues but at a much slower rate, the **slow filling phase**. At this point a cardiac cycle has been completed, and the heart is awaiting the next electrical depolarization from the conduction system.

Atrial pressure changes continuously during ventricular systole and diastole. The small increase in atrial pressure during atrial systole is designated the **a-wave** (Fig. 3-4). The **c-wave** is observed during ventricular isovolumic contraction and is attributed to closing of A-V valves. Following the c-wave, atrial pressure decreases due to mechanical distortion of the atrial chamber by the twisting motion of ventricular contraction. Atrial pressure increases during the later stages of ejection, because filling of the atria continues from the systemic and pulmonary venous systems. The **v-wave** is the highest pressure at the end of filling of the atria just prior to the opening of the A-V valves.

Pressure-Volume Diagrams. Construction of pressure-volume diagrams (Fig. 3-5) provides another means of graphically representing the sequence of events in the cardiac cycle. They are analogous to length-tension diagrams of skeletal muscle (compare with Fig. 1-4). Each complete cardiac cycle is described by one counter-clockwise rotation around the closed path labelled with arrows. The area encompassed by the loop represents **stroke work.**

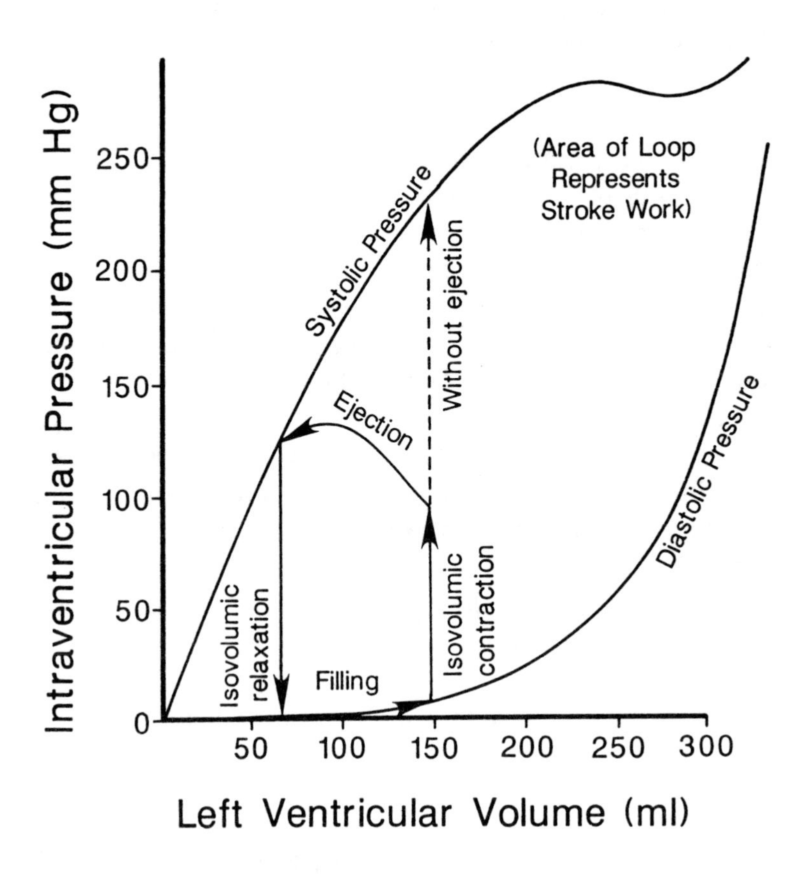

Figure 3-5. Pressure-volume diagram of the cardiac cycle.

The amount of blood ejected from the ventricle with each contraction (about 60-70 ml during limited activity) is called the **stroke volume**. It is calculated as the end-diastolic volume minus the end-systolic volume, when these two parameters are known. **Cardiac output** is the quantity of blood pumped per minute by each ventricle of the heart. It can be calculated as the product of stroke volume and heart rate. The **ejection fraction** is the percentage of ventricular end-diastolic volume that is ejected during a given contraction. It is calculated as stroke volume divided by end-diastolic volume and is typically 50-60% in normal hearts. Work output of the left ventricle and myocardial oxygen consumption can be measured simultaneously. The ratio of these two quantities is **cardiac efficiency**, typically 15-20%. During exercise cardiac efficiency increases slightly.

Review Questions

9. Each of the following statements regarding the cardiac cycle is true **EXCEPT**

   A. c-wave of the atrial pressure curve is mainly caused by closing of the A-V valves
   B. first heart sound is produced by closure of the A-V valves
   C. v-wave in the atrial pressure curve occurs just before the opening of the A-V valves
   D. ventricular systole occurs a few milliseconds after the beginning of the QRS complex
   E. atrial contraction is initiated by pacemaker cells in the AV node

10. For which of the following events does the second event immediately follow the first event during one cardiac cycle?

    A. Rapid ventricular filling, isovolumic relaxation
    B. Rapid ventricular ejection, isovolumic contraction
    C. Rapid ventricular filling, opening of the A-V valves
    D. Reduced ventricular ejection, closing of the semilunar valves
    E. Reduced ventricular ejection, closing of the A-V valves

11. During the isovolumic contraction phase of the cardiac cycle

    A. left atrial pressure increases slowly
    B. aortic pressure increases slowly
    C. aortic pressure increases rapidly
    D. left ventricular pressure increases rapidly
    E. left atrial pressure increases rapidly

12. The second heart sound is associated in time with the each of the following **EXCEPT**

    A. aortic incisura
    B. onset of isovolumic relaxation
    C. end of ventricular ejection
    D. onset of atrial contraction
    E. opening of the A-V valves

CONTROL OF CARDIAC FUNCTION

Autoregulation

The performance of the heart can be autoregulated by two types of factors; 1) those which alter conditions at the time of contraction, such as left ventricular end-diastolic volume and initial length of the cardiac muscle fibers, and 2) those which change the activity of intracellular contractile processes.

Alterations in Initial Conditions. The **Frank-Starling Mechanism** describes the ability of the heart to modify its systolic output on a **beat-to-beat** basis in response to variations in the amount of diastolic filling. Within limits, the more the heart is filled during diastole, the greater the quantity of blood (stroke volume) ejected during the subsequent systole. Filling the ventricles during diastole **stretches** relaxed muscle fibers that make up the walls of these chambers. Muscle fibers of the heart increase their force of contraction as initial fiber length increases. During a particularly long diastole, ventricles fill with extra blood, increasing the initial length of individual fibers, resulting in a more forceful ventricular contraction to expel the additional amount of blood. On the pressure-volume diagram (Fig. 3-5) increased filling results in increased ejection (and proportionally increased stroke work). The **ventricular function curve**, where stroke work or cardiac output is plotted as a function of left ventricular end-diastolic pressure, is used to describe this relationship ("Control" in Fig. 3-6). **Stroke**

work is calculated as the product of stroke volume and mean arterial blood pressure; it measures the amount of external work performed by the heart. **Left ventricular end-diastolic pressure** is related to the end-diastolic volume and is also an index of the length of the cardiac muscle fibers just before contraction begins. It is frequently referred to as **ventricular preload.** This type of autoregulation has two important aspects; 1) it takes place on a beat-to-beat basis, and 2) it is a property of the cardiac muscle fibers themselves and is not dependent on innervation.

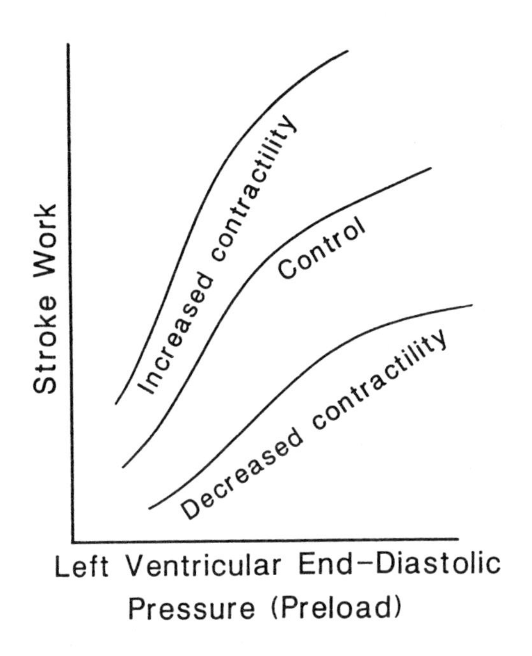

Figure 3-6. Ventricular function curves at different levels of cardiac contractility.

Intracellular Contractile Processes. The performance of the heart can also be altered by affecting the efficiency of the intracellular contractile machinery, a change in **contractility** that is independent of changes of initial length of muscle fibers. A sudden increase in the amount of blood returning to the heart initially causes an increase in end-diastolic volume. However, if the ventricles are subjected to this extra filling for a period of 30 seconds or more, continued stretching of muscle fibers initiates **internal metabolic changes** that produce increases in contractile strength **above** those referable to the Frank-Starling mechanism. Gradually the end-diastolic volume returns to normal, and additional stroke volume is produced by decreasing end-systolic volume. The heart is still producing more work but from the same end-diastolic volume that existed before the increased workload was imposed. These changes in the performance of the heart are seen as shifts of the ventricular function curve (Fig. 3-6). **Increased** cardiac performance is indicated by shifts of the curve upward and to the left; decreased performance results in shifts of the ventricular function curve downward and to the right. **Cardiac failure** (such as congestive heart failure) is often associated with loss of these intracellular autoregulation mechanisms, when the heart depends entirely on the Frank-Starling mechanism to modify its performance.

Neural Control. In addition to autoregulation, the autonomic nervous system also regulates the pumping action of the heart. **Parasympathetic** input affects primarily heart rate, with only minor effects on strength of contraction. In contrast, **sympathetic** input affects both heart rate and strength of contraction.

Cardiac Contractility. The term **contractility** describes the functional status of cardiac muscle fibers independent of external influences, such as increased initial length. The perfect index of cardiac contractility would be independent of changes in left ventricular end-diastolic pressure **(preload)**, aortic systolic pressure **(afterload)**, and heart rate. A frequently used index of contractility is the maximum rate of rise of left ventricular pressure **(left ventricular dP/dt)**. Changes in cardiac contractility are also seen in the ventricular function curves (Fig. 3-6). Increases in contractility **(positive inotropic effects)** shift the curve upward and to the left, while decreases in contractility **(negative inotropic effects)** shift the curve downward and to the right. On the pressure-volume diagram (Fig. 3-5) increased contractility causes increased ejection and movement of the curve to the left. ACh has a negative inotropic effect on atrial muscle and a small negative inotropic effect on ventricular muscle. Norepinephrine and epinephrine, by stimulating β_1 **adrenergic receptors**, have positive inotropic effects on both atrial and ventricular muscle.

Cardiac Output

Cardiac output is the amount of blood pumped by either ventricle each minute. **Venous return** is the amount of blood flowing from the veins into either atrium each minute. Transient inequalities of these two parameters can exist, but they must be equal in the steady-state, because the circulatory system is a closed system. **Normal** cardiac output is about **5 liters/min**. However, cardiac output is a function of body size; the **cardiac index**, cardiac output per square meter of body surface area, is more often used clinically. Its normal value is about 3 liters/min/m^2. Cardiac output is largely regulated by the metabolic needs of peripheral tissues. At rest an output of 5 liters/min is adequate, but when demands of the peripheral tissues are increased, the heart must increase its output proportionally. In a young athlete performing vigorous exercise cardiac output can reach 35 liters/min.

Control of Cardiac Output. The factors that determine cardiac output can be summarized as a cascade of paired factors onto each determinant. **Cardiac output** is the product of heart rate and stroke volume. **Stroke volume** is the difference between end-diastolic and end-systolic volumes of the ventricle, and both heart rate and stroke volume are modulated by the autonomic nervous system (ANS). **End-diastolic volume** is determined primarily by atrial pressure and ventricular compliance. **Ejection fraction** is the ratio of stroke volume to end-diastolic volume. Increases in **atrial pressure** cause increased filling of the ventricle during diastole and lead to an increase in end-diastolic volume. Atrial pressure is frequently referred to as **ventricular filling pressure**. **Ventricular compliance** describes the relative ease of ventricular expansion with filling of blood from the atrium. After myocardial infarctions, damaged myocardial cells are frequently replaced with fibrous tissue that is less compliant than normal cells. This causes a decrease in ventricular compliance, which can severely limit the ability of the damaged heart to fill adequately during diastole. **End-systolic volume** is determined by ventricular contractility and ventricular afterload. **Ventricular contractility** can be changed by ANS activity or by drugs whose actions mainly mimic or inhibit the effects of sympathetic neurotransmitters. **Ventricular afterload** describes the amount of pressure against which the ventricle ejects blood. This is aortic pressure, or pulmonary artery pressure for the right ventricle. Increases in afterload increase end-systolic volume and therefore reduce stroke volume. In summary, increases in cardiac output can be achieved by increasing ventricular contractility, decreasing ventricular afterload, increasing ventricular compliance, or increasing atrial pressure. Various combinations of these changes are observed in the overall regulation of cardiac output.

Measurement of Cardiac Output. Cardiac output is the product of heart rate and stroke volume, but stroke volume is difficult to measure noninvasively. In clinical situations cardiac output is measured, and stroke volume is determined by dividing cardiac output by heart rate. In the **Fick Technique** blood flow (Q) is calculated by $Q = \dot{V}_{O_2} / (A\text{-}V)O_2$, where $\dot{V}_{O_2}$ is the total oxygen consumption of the body per minute, and $(A\text{-}V)O_2$ is the difference between systemic arterial and mixed venous (pulmonary artery) blood oxygen content. Clinically, the Fick Technique is most accurate when the cardiac output is low or normal. Oxygen uptake by the lung is measured by a comparable equation (Equation 2, p. 134). In the **indicator dilution technique** a known amount of dye (indicator) is injected into a vein, and the concentration of dye as a function of time is recorded in a major artery. Cardiac output (Q) can then be computed from the relationship

$$Q = \frac{60A}{Ct}$$

where A is the number of milligrams of dye injected, C is the average concentration of dye in arterial blood, and t is the amount of time to carry out the dye concentration sampling. The indicator dilution technique is accurate at normal or high cardiac outputs. The **thermodilution technique** is similar to the indicator dilution technique. With this technique a known amount of cold saline is injected into the right atrium, and the resulting temperature change of pulmonary artery blood is recorded.

Review Questions

13. Positive inotropic agents

A. decrease maximum left ventricular dP/dt
B. shift the ventricular function curve downward and to the right
C. are mediated by muscarinic receptors
D. are mediated by β_1-adrenergic receptors
E. increase left ventricular end-diastolic pressure

14. A decrease in stroke volume results from each of the following **EXCEPT**

A. decrease in ventricular contractility
B. decrease in ventricular compliance
C. increase in ventricular afterload
D. decrease in atrial pressure
E. decrease in heart rate

15. When using an indicator dilution technique to measure cardiac output,

A. the result may be inaccurate if the patient is in congestive heart failure
B. the indicator is injected into the arterial system
C. oxygen consumption of the total body is measured by spirometry
D. cold saline is injected into a peripheral artery
E. stroke volume is measured directly

For Questions 16-17. After a complete workup a patient is sent to the cardiovascular catheterization laboratory for evaluation of diminishing exercise tolerance. The following data are obtained during the catheterization procedure:

Left ventricular end-diastolic pressure	=	20 mm Hg
Peak left ventricular pressure	=	190 mm Hg
Mean left atrial pressure	=	7 mm Hg
Peak aortic pressure	=	110 mm Hg
Peak right ventricular pressure	=	25 mm Hg
Peak pulmonary artery pressure	=	23 mm Hg
Mean right atrial pressure	=	6 mm Hg
Cardiac output	=	3.2 L/min
Arterial O_2 content	=	19 mmO_2/100 ml blood
Mixed venous O_2 content	=	15 mmO_2/100 ml blood

16. The most likely cause of this patient's low cardiac output is

    A. a decrease in the number of cardiac β-receptors
    B. inadequate pulmonary ventilation
    C. an elevation in left ventricular end-diastolic pressure
    D. stenosis at the aortic valve
    E. poor cardiac muscle function after a myocardial infarction

17. Which of the following interventions would be most likely to improve the overall cardiac status of this patient?

    A. Administration of an α-blocking agent
    B. Administration of a rapidly-acting diuretic
    C. Administration of a β-blocking agent
    D. Breathing 100% oxygen for 1 hr
    E. Replacement of the aortic valve

CIRCULATORY HEMODYNAMICS

Physical Characteristics of Blood. The **hematocrit ratio** is the percent (by volume) of whole blood that is composed of cells (normal = 45%). The remaining fluid is called **plasma. Viscosity** indicates the internal friction of a fluid and affects resistance to flow. The viscosity of blood is 3 to 4 times that of water, mainly due to the presence of formed elements (cells). The viscosity of plasma is 1.5 to 2 times that of water. In non-Newtonian fluids such as blood, the term **apparent viscosity** represents viscosity measured under specific physical conditions. Apparent viscosity decreases as shear rate increases (shear thinning), causing erythrocytes to accumulate in axial laminae at high flow rates. Conversely, at low shear rates erythrocytes form aggregates, increasing viscosity.

Blood Flow Through Vessels

Flow (Q) through a blood vessel is entirely determined by two factors; 1) the **pressure gradient or perfusion pressure** (ΔP) tending to push blood through the vessel and 2) the **resistance** (R) to blood flow through the vessel. These quantities can be related by

$$Q = \Delta P/R$$

Blood flow within a vessel can be laminar or turbulent. **Laminar** or **streamlined flow** (volumetric) occurs when blood flows in concentric layers. Maximum flow velocity (v) occurs in the center of the stream; minimum velocity occurs at the wall of the vessels. When blood passes through a small vessel at moderate velocity, red blood cells tend to concentrate towards the middle of the stream. **Turbulent flow** occurs in the absence of streamline flow when eddy currents are formed. **Reynold's number, N_R**, a dimensionless measure of the tendency for turbulent flow to occur, can be calculated if the vessel diameter, blood viscosity, blood density and flow velocity are known. Turbulence is highly likely to occur when $N_R > 3000$. High flow velocities (eg, through the narrow orifice of aortic stenosis) also reduce the lateral, distending pressure in vessels.

Resistance to blood flow cannot be measured directly but must be calculated from

$$R = \Delta P/Q$$

If pressure is expressed in **mm Hg**, and flow is measured in **ml/sec**, then resistance is defined arbitrarily in **peripheral resistance units** (PRU) of mm Hg/ml/sec. The **total peripheral resistance** of the systemic circulation (TPR) is also called the **systemic vascular resistance** and is approximately 1 PRU. The organ systems represent parallel resistances of major vessels, and the resistance of any one is always greater than the total peripheral resistance. Additionally, for a given pressure difference and for cylindrical vessels of given dimensions, blood flow will vary inversely with the viscosity; abnormally viscous blood impedes flow. The principal determinant of resistance to blood flow is the diameter of the vessel. Small changes in radius cause large changes in R. Therefore, minute by minute sympathetic vasoconstrictor mechanisms in the small arteries and arterioles are the primary mechanism for regulating blood flow.

Pascal's Law. The pressure (P) at the bottom of a column of fluid is equal to the height (h) of the column times the density of the fluid (ρ) times the gravitational constant (g). In the CGS (Centimeters, Grams, Seconds) system this term of dynes/cm^2 is too cumbersome, so the height of a column of mercury is used as a relative standard.

$$P = h\rho g$$

Poiseuille's Law. The volume flow (Q) through a cylindrical tube is related to the driving pressure and the resistance to flow as expressed by Poiseuille's equation (for ideal fluids)

$$Q = \Delta P \times \pi r^4/8\eta L$$

where ΔP is the pressure difference between the two ends of the tube, r is the radius of the tube, η is the viscosity of the fluid, and L is the length of the tube. Resistance can be expressed as

$$R = 8\ \Delta L/\ \pi r^4$$

Since the cardiovascular system is a closed system, most of these variables can be considered constant. Thus resistance to flow is inversely proportional to the **fourth** power of the radius of the tube.

Bernoulli's Principle states that the total energy in fluids with streamlined flow is constant and is equal to the sum of its potential energy (pressure) and kinetic energy (flow velocity). Normally, the kinetic energy of flowing blood is only a small portion of the total energy. At sites of vascular constriction where flow velocity is increased, the corresponding lateral, distending pressure is decreased due to this conservation of energy principle.

Laplace's Law states that the tension (T) in the wall of a blood vessel is proportional to the product of the transmural pressure (P) and the radius (r), so that

$$T = P \times r$$

The transmural pressure is the difference between the pressures inside and outside the blood vessel. For a given transmural pressure the force required in the wall of a **large** blood vessel to keep the blood vessel from distending is higher than in a **smaller** vessel. A pathologically large diameter blood vessel (eg, an aortic aneurysm) tends to rupture spontaneously because of high tension in the wall even at normal arterial pressure. In contrast, capillaries need very little wall tension and remain intact with a wall thickness of less than 1 µm. This law also shows that a large, dilated heart must do more work than a normal-sized heart to generate a given intraventricular pressure.

Critical Closing Pressure. Blood vessels may collapse and be unable to maintain blood flow if the intravascular distending pressure falls below some critical value (20 mm Hg in many vascular beds). The pressure where this collapse occurs is called **critical closing pressure**. The vessel will collapse when elastic and muscular forces in the vascular wall, coupled with the extravascular tissue pressure, exceed intravascular distending pressure. This mechanism is important in severely hypotensive patients where arteriolar collapse can lead to tissue ischemia.

Vascular Compliance. The viscoelastic properties of blood vessels are described by pressure-volume characteristics which result from collagen, elastin and connective tissue components. **Compliance** (C, or capacitance)) is defined as

$$C = \Delta V/\Delta P$$

where ΔP is the change in transmural pressure that produces a change in volume, ΔV of a vessel. Note that compliance describes the distensibility of blood vessels and is the slope of the **vascular pressure-volume curve**. Veins are about five times more compliant than arteries. The compliance of a given vessel is not constant; it varies with pressure or volume. As a vessel becomes more distended, it becomes less compliant. So, vascular pressure-volume curves are typically non-linear. Compliance is altered by age, disease processes, autonomic nervous system activity, and drugs. With aging arterial walls become infiltrated with fibrous tissue which is less compliant; this is usually accompanied by increased resistance and results in increased arterial pressure. The rapid ejection phase of ventricular systole is significantly prolonged with decreased aortic capacitance. Sympathetic control is important in changing vascular compliance; sympathetic stimulation decreases arterial but especially venous compliance. It provides a major mechanism for shifting blood volume from one portion of the circulatory system to another.

Review Questions

18. Volume flow of a liquid through a rigid tube

    A. doubles when the pressure difference between the two ends of the tube is doubled
    B. increases eight times when the radius of the tube is doubled
    C. increases with an increase in viscosity
    D. is unchanged by doubling the length of the tube if the pressure difference is unchanged
    E. is an excellent model for volume flow in distensible vessels

19. When blood passes through a small blood vessel at moderate velocity,

    A. flow is more rapid along the wall than in the middle
    B. erythrocytes spin along the wall
    C. the speed of flow is determined by the Law of Laplace
    D. red blood cells are concentrated towards the middle of the stream
    E. turbulence occurs

20. According to Bernoulli's Principle

    A. total energy of streamline flow is constant
    B. total energy is the sum of potential and 1/2 kinetic energy
    C. lateral pressure distending a vessel is increased by constriction
    D. valve leaflets tend to repel each other during rapid ventricular filling and emptying
    E. there is greater lateral pressure with greater velocity of flowing blood

21. The critical closing pressure of a blood vessel

    A. decreases during adrenergic α_1-receptor activation
    B. does not change if the blood vessel becomes less compliant
    C. decreases if extravascular pressure decreases
    D. increases during adrenergic β_2-receptor stimulation
    E. does not change unless a patient is critically ill

For Questions 22-24: Match a drug or physical mechanism listed below with each physiological event in the questions.

A. Nitrate vasodilator drug
B. Beta-adrenergic blocking drug
C. Acetylcholine blocking drug
D. Decreased aortic compliance
E. Decreased aortic/arterial afterload

22. Causes increased rate of rise of thoracic aortic blood pressure.

23. Causes indirect lowering of arterial pressure by reducing cardiac output.

24. Causes increased heart rate by decreasing the effects of vagal nerve activity.

THE SYSTEMIC CIRCULATION

Distribution

Functional Anatomy. The systemic circulation extends from the aorta to the junction of the superior and inferior vena cava at the level of the right atrium. This portion of the circulatory system carries well-oxygenated blood from the heart to the peripheral organs and tissue beds and returns deoxygenated or venous blood back to the heart. The elastic **arteries**, both large and small, serve as high pressure conduits for oxygenated blood from the heart to the periphery. The aorta especially acts as a Windkessel vessel (hydraulic filter) to contain the stroke volume, promote flow during diastole, reduce the pulsatile blood pressure and produce more continuous flow. Their order of vascular compliance is veins > aorta > arteries > arterioles. Compliance decreases with age and is a major cause of increased pulse pressure in older adults. The arteries bifurcate as they continue peripherally until the **arterioles** (resistance vessels) are reached. The arterioles act to control blood flow into the capillary beds by markedly changing their diameter. They are the major site of controllable resistance in the systemic circulation (Fig. 3-7A). The **capillaries** (exchange vessels) facilitate the exchange of fluid and nutrients between the blood and the interstitial space. The **venules** are small veins that collect blood from the capillaries. These venules combine to form **veins** (capacitance vessels), expandable reservoirs, that return blood back to the heart.

Blood Distribution. The percent distribution of total blood volume at a given instant in the systemic circulation is shown in Fig. 3-7B. The systemic circulation contains about 80% of total blood volume. The pulmonary circulation contains 9-12%; the heart contains 8-11%. The volume of blood flow is the product of blood flow velocity and vascular cross-sectional area. Cross-sectional area and blood flow velocity are inversely related. The total cross-sectional area of the systemic circulation increases from 2.5 cm^2 at the central aorta to 2500 cm^2 in the capillaries (Fig. 3-7C). The collecting venules and veins reduce cross-sectional area to 8 cm^2 at the level of the right atrium. Blood **flow velocity** is maximal in the central aorta, reaches its lowest value in the capillary beds, and returns to a moderate velocity at the input to the right atrium (Fig. 3-7D).

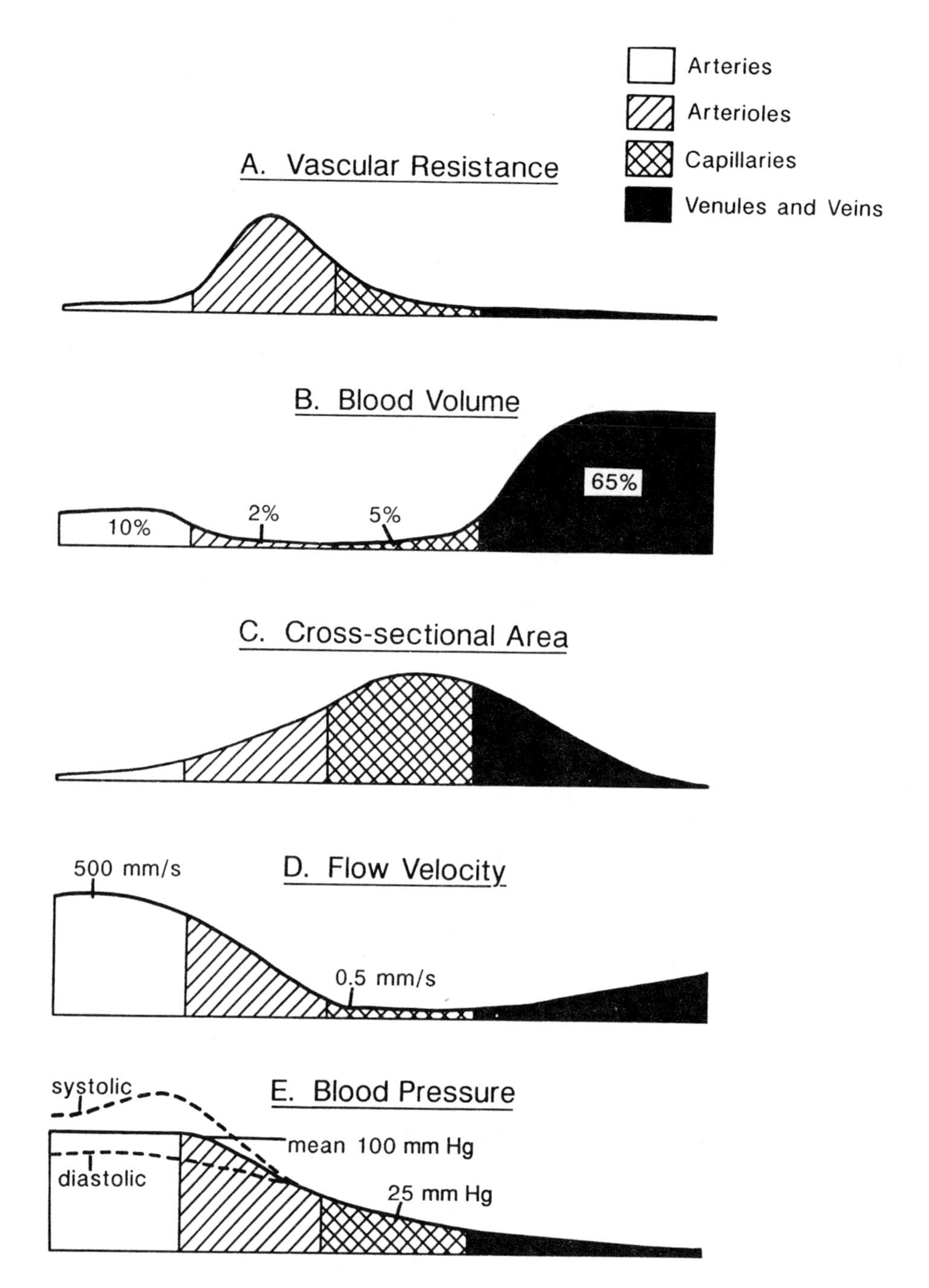

Figure 3-7. The distribution of blood in the systemic circulation and the physical variables related to blood flow, such as resistance, cross-sectional area, velocity and pressure.

Pressure and Resistance

The arterial blood pressure is determined directly by two major physical factors: 1) the arterial blood volume and 2) the arterial compliance. These physical factors are affected in turn by certain physiological factors, primarily the heart rate, stroke volume, and peripheral resistance. Therefore, arterial pressure is determined by two physical factors, each of which is affected by multiple physiological factors.

Systolic arterial pressure is the highest aortic (arterial) pressure observed during cardiac systole (See Fig. 3-4). **Diastolic pressure** is the lowest pressure observed in the aorta during diastole. A blood pressure of 120/80 mm Hg describes a systolic pressure of 120 mm Hg and a diastolic pressure of 80 mm Hg (Fig. 3-7E). Arterial pressure is measured indirectly in humans by an auscultatory method (sphygmomanometer and stethoscope), or directly by a catheter-tip pressure transducer or indwelling cannula filled with saline attached to an external pressure transducer.

Pulse pressure is the difference between the systolic and diastolic pressures. Two major factors affect arterial pulse pressure; 1) stroke volume of the ventricle and 2) compliance of the arterial system. If stroke volume increases, the compliant aorta has to accommodate more blood with each beat. Systolic pressure rises, thus increasing pulse pressure. Stroke volume will be affected by changes in 1) heart rate, 2) total peripheral resistance, 3) blood volume within the circulatory system and 4) the strength of ventricular ejection. If compliance of the aorta decreases, then pulse pressure will increase for any given stroke volume. Arterial compliance will be decreased by 1) increases in mean arterial pressure, 2) pathologic alterations of the vessel wall or 3) the aging process. The time available for diastolic run off into the arteries also affects pulse pressure. Slower heart rates increase diastolic run off and lower the diastolic pressure. The **arterial pulse** is a **pressure wave**, produced by the ejection of blood into the aorta, which travels down the arterial system. This pressure wave can be felt in peripheral arteries because of their low compliance. The velocity of the pressure pulse wave (5 to 8 m/sec) is much faster than the velocity of blood flow (0.5 to 1 m/sec). The actual blood ejected by the left ventricle travels only a few centimeters in the same time that it takes the pressure pulse caused by that ejection to reach the radial artery (about 0.1 sec). The pulse is strong when stroke volume is increased, as during exercise. In shock the pulse is weak and sometimes cannot be palpated at the radial artery. Aging tends to increase systolic and decrease diastolic pressures. **Transmural pressure** is the pressure difference between the inside and outside of the vessel wall ($P_t = P_i - P_o$). This is the pressure used in the application of **LaPlace's Law** for cylindrical vessels.

Mean arterial pressure is the average pressure measured throughout the cardiac cycle and is about 95 mm Hg. At rest, diastole occupies about 2/3 of the cardiac cycle and systole occupies about 1/3, so mean arterial pressure is less than their arithmetic mean. True mean pressure is determined by integrating the area under the arterial pressure curve. In the central aorta mean aortic (arterial) pressure (MAP or $\bar{P}_a$) can be estimated from

$$\textbf{MAP = Diastolic Pressure + 1/3 Pulse Pressure}$$

Mean arterial pressure is directly related to cardiac output and total peripheral resistance. At the arterial side of capillaries, pressure is about 35 mm Hg and is non-pulsatile. Most of the pressure drop in MAP occurs across the arterioles (Fig. 3-7E). At the venous end of the capillaries pressure is about 15 mm Hg, and pressure drops to almost 0 mm Hg at the level of the right atrium.

Veins not only function as conduits for flow of blood to the heart but also passively relax or actively constrict under sympathetic control, thus varying the volume of "stored blood". Mobilization of blood from venous reservoirs occurs when O_2 consumption is greater, such as in exercise. Veins are also important in regulation of cardiac output, because increased venous return increases cardiac output by the **Frank-Starling Mechanism**.

Since all veins ultimately enter the right atrium, right atrial pressure is frequently called the **central venous pressure**. The zero reference point for all clinical pressure measurements is at the level of the right atrium. Right atrial pressure is regulated by a balance between the heart pumping blood out of the right atrium and blood flowing into the right atrium from peripheral vessels. Thus, central venous pressure affects the pressure gradient for venous return and, ultimately, ventricular filling and stroke volume. Under normal circumstances this balance is quite precise. However, in the case of right heart failure, more blood is returned to the heart than the right ventricle can pump into the lungs, so peripheral veins become engorged with blood.

Gravity and Venous Pooling. Hydrostatic pressures are superimposed on the pressures generated by the heart, because the three-dimensional vascular network is subjected to gravity. When standing some blood vessels are above the level of the heart and some are below. The gravitational effect is 0.77 mm Hg per cm of height. Thus, with a mean aortic pressure of 100 mm Hg, the mean pressure in a large artery in the head 50 cm above the heart is 62 mm Hg. Similarly, the mean pressure in a large artery in the foot, 120 cm below the heart, is 192 mm Hg. Blood flow in veins is unidirectional due to the presence of valves. Muscular contraction, particularly in the legs, compresses veins and propels blood toward the heart. This valvular venous mechanism (**muscular milking**) is important for returning blood to the heart, and also for keeping venous pressure less than 25 mm Hg in the feet of a standing person. Venoconstriction and negative intrathoracic pressures during respiration also assist venous return.

Review Questions

25. The pressure pulse wave in systemic arteries

    A. propagates at about 0.7 m/sec
    B. is initiated by the ejection of blood into the aorta
    C. is palpable at the wrist 1 second after ventricular ejection
    D. has the same velocity as arterial blood flow
    E. is unaffected by arterial vasoconstriction

26. The pulse pressure

    A. is 1/3 of the difference between systolic and diastolic pressures
    B. increases if the compliance of the blood vessel decreases
    C. declines in the arteries (compared to the aorta)
    D. is always present in capillaries
    E. is independent of age

27. When a person is standing, mean arterial pressure

   A. at the level of the heart is approximately 60 mm Hg
   B. in a large artery in the brain is less than 93 mm Hg
   C. in a large artery in the foot is nearly 60 mm Hg
   D. in a large artery in the foot does not differ from the supine position
   E. is unaffected by changes in emotional state

28. All of the following statements about blood flow velocity are true **EXCEPT** velocity is

   A. not a major determinant of pulse pressure
   B. greatest in the arterial capillaries
   C. increased if vessel compliance is decreased
   D. inversely related to the cross-sectional area of a vessel
   E. inversely related to the hematocrit (viscosity) of blood

For Questions 29-31. Match the correct part of the systemic circulation in column B with each descriptor in column A.

Column A	Column B
29. Major site of vascular resistance.	A. Aorta
	B Arterioles
30. Most compliant vessels of the circulation.	C. Capillaries
	D. Veins
31. Contain about 5% of the total blood volume (at rest).	

MICROCIRCULATION AND LYMPHATIC SYSTEMS

Functional Anatomy. Blood enters capillary beds from **arterioles**, resistance vessels about 20 μm in diameter. It next passes into **metarterioles** that allow blood either to enter capillaries or to bypass them as they merge with venules. **True capillaries**, where the exchange of gases, nutrients and waste products takes place, arise from metarterioles. At the entrance of capillaries smooth muscle fibers are arranged to form **precapillary sphincters**. Capillaries are about 4-8 μm in diameter and about 0.5 mm in length. The total surface area of muscle capillaries is 6,000 m^2. The capillary wall is a unicellular layer of endothelial cells surrounded by a thin basement membrane on the outside; it is about 0.5 μm thick. Small slit-like spaces, minute channels from the interior of the capillary to the interstitial space, are called **pores or clefts**. These pores, about 8 nm in width, are spaces between adjacent endothelial cells. Water and many dissolved substances pass through capillary walls in either direction via these pores. Smooth muscle cells of arterioles and metarterioles are often innervated by sympathetic nerves, but precapillary sphincters and capillaries are not innervated.

Upon leaving the capillary bed, blood enters venules that are typically larger than arterioles. Small venules have no smooth muscle; larger ones do. **Arteriovenous anastomoses** are direct, non-nutritive channels communicating between arterioles and venules. They have the capability of producing **shunt flow** and are found in many tissues but especially in skin. The total volume of flow through metarterioles and capillaries is controlled by the arterioles, whereas the fraction of this blood that flows through true capillaries is controlled by precapillary sphincters. Arterioles are heavily innervated and do not respond to locally released metabolites. In contrast, metarterioles and precapillary sphincters are largely controlled by local metabolites.

Control of Blood Flow. In resting skeletal muscle only one capillary in 50 is open at any one time. The ratio of active to inactive capillaries depends upon the specific tissue and its state of metabolic activity. Variations of blood flow in true capillaries are regulated by contraction and relaxation of precapillary sphincters and metarterioles. These contractions, called **vasomotion,** occur at 30 sec to several minute intervals. They are responses of vascular smooth muscle (especially pre-capillary sphincters) to local **metabolic** products (vasomotion is also observed in denervated regions) or to autonomic neural influences (**sympathetic vasoconstrictor fibers**). Increased formation of metabolites depresses activity of smooth muscle and reduces vasomotion, leading to increased capillary flow. Local hypoxia develops even at rest when precapillary sphincters are closed, and various metabolic products accumulate in tissues. This depresses the activity of smooth muscles, thus reducing their degree of vasoconstriction (causing vasodilation) and restoring capillary flow. Vasomotion is also observed in arterioles, but these are mainly affected by sympathetic activity and not by local metabolites. Endothelial cells interact with blood components to partially regulate their own flow. For example, stimulation of endothelium by ACh produces a vascular smooth muscle relaxation from **endothelial-derived relaxing factors (EDRFs).** The first identified EDRF was nitric oxide.

Capillary Exchange of Various Substances

Diffusion is the major mechanism for exchange of fluid and solutes between blood and interstitial space. The flux (F), or amount of substance moved per unit time, can be described by the relationship

$$F = -D\,A\,\frac{\Delta C}{\Delta X}$$

where D is a diffusion constant that depends on the size of the molecule and the temperature, A is the area available for diffusion, ΔC is the concentration difference, and ΔX is the distance (thickness) for diffusion. Diffusion is an effective mechanism for moving substances over short distances in capillary beds.

Diffusion of **lipid-soluble substances** occurs directly through cell membranes of the capillary wall and does not depend on the presence of pores. The rate of movement of these materials is several hundred times faster than movement of lipid-insoluble material. **Oxygen** and **carbon dioxide** are lipid soluble, as are various anesthetic gases and alcohol.

Water soluble, lipid-insoluble substances move through the pores of capillary walls by diffusion. The **permeability** of the capillary wall measures the ease of diffusion of these various materials relative to water. Small molecules and ions (Na^+, Cl^-, glucose, urea) move rapidly through capillary pores, so the mean concentration gradient of these substances across the capillary endothelium is small. As molecular size increases, diffusion through the capillary pores becomes more restricted; substances with molecular weights greater than 60,000 show little diffusion.

Exchange by Filtration (Ultrafiltration). Most **fluid movement** at the capillary level takes place through the pores of the capillary as a result of the balance between hydrostatic and osmotic forces (pressures). Less than 2% of the plasma volume flow filters across capillary walls into tissues. The net movement of fluid across a capillary wall is regulated by the balance between the total filtration pressure and total reabsorption pressure. Fluid movement is predicted by the **Starling Hypothesis** expressed by the following equation:

$$\textbf{Fluid movement} = k\ [(P_c + \pi_i) - (P_i + \pi_c)]$$

where P_c is capillary hydrostatic pressure, P_i is interstitial fluid hydrostatic pressure, π_c is plasma protein (colloid) osmotic pressure, π_i is interstitial fluid osmotic pressure, and k is a filtration constant for the capillary membrane. If the net result of the above equation is positive, then filtration into the interstitial spaces occurs; if negative, then reabsorption occurs.

Capillary hydrostatic pressure (P_c) is the most important factor in transcapillary filtration of fluids. It is the only force that varies significantly between the proximal and distal ends of the capillary. The mean capillary blood pressure is about 25 mm Hg; it is about 35 mm Hg at the arterial end, and about 15 mm Hg at the venous end. Capillary hydrostatic pressure varies with 1) changes in arterial blood pressure, 2) capillary flow, and 3) the ratio of the resistance in arterioles to that in venules. For instance, when inflow resistance is smaller than outflow resistance, hydrostatic pressure in the capillaries is increased, and filtration enhanced. Inflow resistance to the capillary is regulated by the arteriolar and precapillary sphincter tone. Outflow resistance is controlled by venous resistance and venous pressure.

Interstitial fluid hydrostatic pressure (P_i) is determined by the volume of interstitial fluid and by distensibility of the interstitial space; it is only 1 to 2 mm Hg. Therefore, the contribution of tissue hydrostatic pressure to transcapillary exchange is small, except with lymphatic blockage or increased capillary permeability. P_i is probably negative in the pulmonary circulation.

Plasma osmotic pressure (π_c) results from the presence of plasma proteins in the bloodstream. Common electrolytes are plentiful in plasma and can exert great osmotic pressure, but they are not physiologically important because of their rapid interchange across capillary walls. Plasma proteins (6 gm/100 ml) are osmotically active because of their relative impermeability. Although the smallest proteins (albumins) permeate capillary walls more easily than the larger globulin and fibrinogen, they are physiologically more important than other plasma proteins, because they are much more abundant. The effective osmotic pressure of plasma is about 25 mm Hg. This pressure is often called **plasma oncotic pressure** or **colloidal osmotic pressure.** Nutritional or metabolic deficiencies can decrease this pressure by decreasing plasma protein concentrations.

Interstitial fluid osmotic pressure (π_i) is proportional to the concentration of plasma proteins that are in the interstitial fluid. An increase in this osmotic pressure enhances the filtration force throughout capillaries. The concentration of proteins in interstitial fluid is usually low, so the tissue osmotic pressure is also low (1 to 2 mm Hg). However, if proteins leak out of the capillaries, then this pressure increases. Physiologically, this happens in the liver and intestines and affects transcapillary exchange in those organs. Pathologically, lymphatic blockage or an increased capillary permeability increase interstitial fluid osmotic pressure.

Since interstitial fluid hydrostatic and osmotic pressures are low, transcapillary exchange of fluids is regulated primarily by the magnitudes of **capillary hydrostatic pressure** and **plasma osmotic pressure**. Net filtration usually exceeds net reabsorption by a slight margin, leading to the formation of **lymph**. Transcapillary exchange is influenced by 1) variations in capillary hydrostatic pressure, 2) level of tissue hydrostatic pressure, 3) plasma or tissue protein concentration, 4) lymphatic drainage, 5) capillary permeability, and 6) total capillary surface area available for diffusion. Increases in **venous** pressure increase filtration into interstitial fluid much more than a comparable increase in arterial blood pressure.

Endocytosis. Some transfer of substances across the capillary can occur by pinching off portions of the surface membrane and forming vesicles containing substances. These **pinocytotic vesicles** cross the membrane and deposit their contents (especially large lipid-insoluble molecules) on the other side.

Lymphatic System

This vascular system provides a supplementary route for fluid to flow from interstitial space back to the circulatory system. All tissues except portions of skin and the bone are penetrated by lymph capillaries. These capillary vessels ultimately join to form small and finally large lymphatic vessels and trunks. About 10% of the fluid that filters out of arterial capillaries returns to the circulation through lymphatic rather than venous channels, which amounts to 21 L in 24 hours. The lymphatic system is most important for the return of high molecular weight materials to the circulation. It is also a major route for absorption of nutrients from the gastrointestinal system, being responsible for absorption of fats. Lymphatic channels have valves similar to veins. These establish unidirectional flow, with a **lymphatic pump** similar to the venous pump.

Review Questions

32. Each of the following statements is true **EXCEPT** that blood flow

    A. is continuous in the arterioles that supply the microcirculation
    B. is continuous in the metarterioles of the microcirculation
    C. through true capillaries is regulated by precapillary sphincters
    D. is discontinuous in the capillaries
    E. is continuous in the capillary beds of skeletal muscle

33. Capillary hydrostatic pressure is 20 mm Hg; interstitial hydrostatic pressure adjacent to the capillary is 2 mm Hg; colloid osmotic pressure of plasma is 25 mm Hg; colloid osmotic pressure of interstitial fluid is 5 mm Hg. The net pressure tending to move fluid in or out of the capillary is

A. 4 mm Hg
B. 2 mm Hg
C. 0 mm Hg
D. -2 mm Hg
E. -4 mm Hg

34. Interstitial fluid osmotic pressure is

A. proportional to the concentration of plasma proteins that enter the interstitial fluid
B. increased if capillary permeability decreases
C. increased by lymphatic drainage
D. determined mainly by the amount of glucose leaving the capillary
E. primarily dependent upon the concentrations of K^+ and Na^+ in interstitial fluid

35. In the microcirculation

A. about 1 in 5 capillaries is open during periods of rest
B. local tissue-released metabolites dilate capillaries
C. all capillary beds are mainly affected by neural mechanisms of blood flow control
D. increased sympathetic activity tends to vasoconstrict the precapillary sphincters
E. vasomotion of precapillary sphincters occurs at intervals from 30 sec to several minutes

36. Movement of fluid across capillary walls

A. is regulated by the balance between total filtration pressure and total reabsorption pressure
B. into the interstitial space is greatest in the distal portion of the capillary
C. increases when large venules dilate
D. decreases when plasma osmotic pressure is below normal
E. is unaffected by lymphatic vessel (lymph) flow rates

For Questions 37-40. Each time a 64 year-old retiree walked to the bus stop he experienced a dull aching in the calf muscle of his right leg. After walking 2-3 blocks the pain was so great he had to stop and rest. After standing quietly for a while, he could continue to go the same distance before the pain reappeared. Physical examination by infrared thermography revealed that the right foot was colder than the left and also was missing the posterior tibial and dorsalis pedis pulses. Skin of the right foot had less color and hair than that of the left. An arteriogram was performed and little of the radiopaque contrast material passed from the femoral to right popliteal artery. During rest and a treadmill exercise tolerance test the calf muscle blood flow was measured; resting flow in right leg was 2 ml/100 gm/min; exercise right calf flow was 8 ml/100 gm/min. The respective flows for the left calf muscle were 3 and 40 ml/100 gm/min.

37. What is the probable cause of this patient's problem?

38. Explain the symptoms and signs using physiological mechanisms.

39. Why was the resting blood flow 2/3 of normal but the exercising flow 1/5 of normal?

40. List some of the factors that influence the transport of oxygen from the blood to the tissues.

CONTROL OF PERIPHERAL CIRCULATION

The regulation of blood flow to various tissues and organ systems is important for total body homeostasis. Blood flow is usually controlled by nutritional needs, primarily for oxygen, and flows as a result of a pressure difference (perfusion pressure or ΔP). Other substances, such as glucose, amino acids and fatty acids, also affect regulation of blood flow. Immediate regulation of blood flow to meet metabolic demands occurs through three major control mechanisms; 1) local (tissue environment) metabolic, 2) neural, and 3) humoral mechanisms. Increased blood flow to meet increased metabolic demand is **active hyperemia.** In addition, several long term mechanisms are used as tissue beds adapt to chronically abnormal situations.

The **arterioles** are vessels mostly involved in regulating the rate of perfusion pressures from blood flow through the body to the microcirculation. With the greatest resistance to flow (Fig. 3-5A) they are important in the maintenance of arterial blood pressure. **Vascular (vasomotor) tone** indicates the general contractile state of an artery or arterial region. Vascular tone is the sum of intrinsic smooth muscle tone plus tonic sympathetic vasoconstrictor influences from nerves and circulating catecholamines. The various organ systems have varying degrees of resting vasomotor tone. Maximal vasodilation depends upon individual regulatory mechanisms of different organs, and it is most profound in salivary glands, skin, myocardium, and the gastrointestinal tract.

Local and Metabolic Control Mechanisms

In many tissues blood flow is relatively constant over a range of perfusion pressures from 75 to 175 mm Hg. Such flow that is independent of arterial perfusion pressure is termed **autoregulation.** If the artery supplying a particular vascular bed is completely blocked for a brief period of time and then released, the flow increases briefly to several times its normal value **(reactive hyperemia).** This is due to accumulation of metabolic products. There are two major mechanisms to explain autoregulatory responses; the metabolic theory and the myogenic theory. In any vascular bed one or both of these mechanisms may operate simultaneously. Additional support of autoregulation may occur from release of EDRFs.

Metabolic processes occurring at the cellular level produce end-products that must be removed from cells. The **metabolic theory** proposes that some of these metabolic products, or their breakdown products, have direct vasodilating effects on blood vessels. As the rate of metabolism of a tissue bed increases, the rate of formation of metabolic end products with vasodilator properties will increase proportionally. Such **vasodilator substances** are released from cells and diffuse to precapillary sphincters, causing vasodilation. This permits more blood to flow into the tissue bed to supply additional amounts of nutrients needed by the tissue bed at its increased metabolic level. Vasoconstriction occurs as metabolites are carried away; thus flow is autoregulated according to metabolic needs. Some substances that have been proposed as triggers for this action are CO_2, lactic acid, adenosine, H^+ and K^+. The direct dilator action of CO_2 is more pronounced in skin and brain than in other tissues. In addition to metabolic influences, other local chemical influences can also affect vascular smooth muscle. **Prostaglandins**, products of the cyclooxygenase pathway of arachidonic acid metabolism, are potent vasoconstrictor or vasodilator substances. **Histamine**, from granules in tissue mast cells, is released by injury or antigen-antibody reactions and is a potent vasodilator. **Bradykinin** is a polypeptide with 10 times the vasodilator potency of histamine. **Serotonin** is a potent vasodilator or vasoconstrictor that is present in high concentrations in platelets and enterochromaffin cells of the gastrointestinal tract.

The **myogenic hypothesis** is based on the observation that vascular smooth muscle responds to passive stretch by contracting. In blood vessels passive stretch is provided by the blood pressure. When this transmural pressure in a local region decreases, the stimulus for vascular smooth muscle contraction is partially removed. The muscle then relaxes, permitting more blood to flow at lower blood pressure. Similarly, increases in pressure elicit more contraction and tend to limit blood flow into the controlled region. Rate of stretch is probably more important for this mechanism than the magnitude of stretch.

Development of **collateral circulation** is a type of long-term local blood flow regulation. When normal blood flow becomes partially or completely blocked, small collateral vessels enlarge and assume the major role in supplying blood to that region. At first these vessels can supply only a small fraction of the normal blood supply, but with sufficient time they can return flow to near normal levels. Regular aerobic exercise may promote collateral vessel development in coronary arteries. Another mechanism for long-term regulation involves changes in tissue vascularity **(angiogenesis)**. In general, a decrease in perfusion pressure results in an increase in the number and size of the vessels supplying local vascular beds.

Neural Control Mechanisms

Sympathetic Vasoconstrictor System. Vasoconstrictor nerves are distributed to almost all parts of the circulation and influence the volume of blood that reaches an organ bed. Innervation is relatively sparse in skeletal muscle, cardiac muscle, and the brain. Extensive innervation is found in vessels of the gut, kidneys, spleen, and skin. The pressor and depressor areas of the **vasomotor center** in the medulla oblongata (Fig. 3-8 on the next page) regulate the discharge rate of sympathetic vasoconstrictor nerves and to a lesser extent cardiac muscle (contractility).

Tonic vasomotor activity comes from this **center** and projects to sympathetic preganglionic neurons in the spinal cord. The vasomotor center is subject to inhibition and reflex modulation by baroreceptor and cardiopulmonary reflexes from higher centers. At rest vasomotor center activity sets the **vasomotor (vasoconstrictor) tone**. The **hypothalamus** exerts powerful excitatory and inhibitory effects on the vasomotor center. Another portion of the medulla inhibits the continuous activity of the vasomotor center. This decreases the amount of vasoconstrictor tone on blood vessels and allows them to passively dilate.

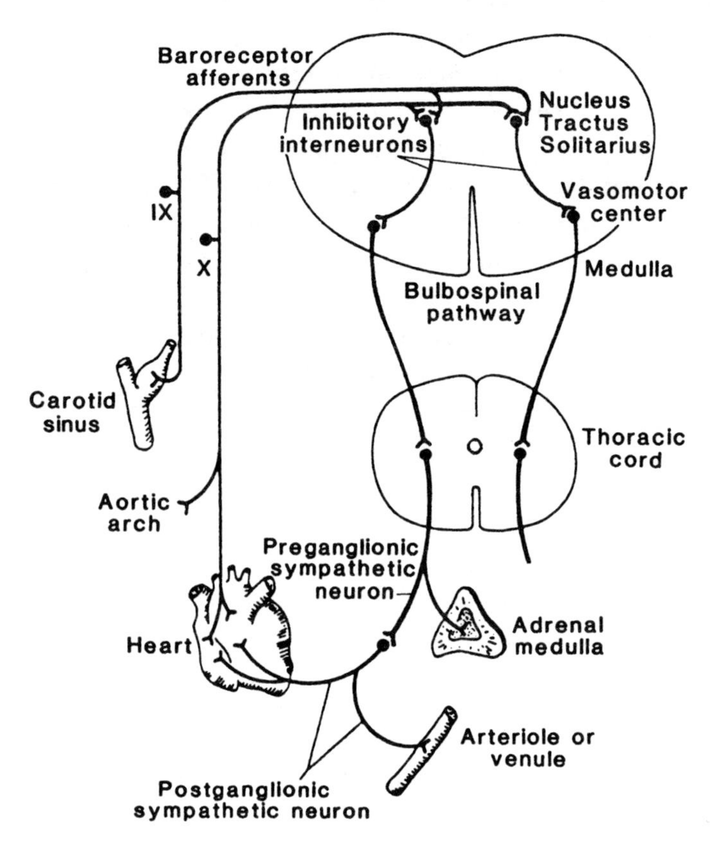

Figure 3-8. Neural pathways for control of blood pressure.

Catecholamine Receptors. Differential responses of vascular smooth muscle to circulating catecholamines (norepinephrine, epinephrine, dopamine) can be explained by α_1 and β_2 adrenergic receptors. Activation of α_1 receptors by norepinephrine elicits vasoconstriction. Excitation of β_2 receptors causes vasodilation. Epinephrine acts on both α_1 and β_2 receptors, so a predominance of β_2 receptors in a tissue will produce vasodilation in the presence of epinephrine. Nevertheless, increase or decrease of α_1-adrenergic vasoconstriction is the predominant mechanism for regulating **vascular smooth muscle tone.** β_1 receptors account for contractility changes in the myocardium, while β_2 receptors account for vascular smooth muscle activity.

Parasympathetic Influence. Only a small portion of the resistance vessels receive parasympathetic nerve fibers (genitalia, colon, bladder and head), so the relative effect of these cholinergic fibers on circulation is small.

Regulation of Arterial Pressure. The local blood flow control mechanisms discussed previously are normally operative under constant, regulated perfusion pressure. Several cardiovascular reflexes control arterial blood pressure within narrow limits. Some of these mechanisms act very rapidly (neural and hormonal mechanisms), and some act very slowly (mechanisms related to blood volume regulation and renal function). Mean arterial pressure is controlled by cardiac output and total peripheral resistance. Any situation that increases either of these parameters will increase arterial blood pressure.

Arterial Baroreceptors. The best known mechanism for arterial pressure control is the **baroreceptor reflex.** Receptors sensitive to **mechanical stretch** are located in the **carotid sinuses** and in the walls of the **aortic arch** (Fig. 3-8). When arterial pressure increases, the aorta and carotid sinuses expand and stretch the baroreceptors, which increase their firing rate. Impulses from baroreceptors **inhibit** the vasomotor center and **excite** vagal motor neurons, causing vasodilation, decreased heart rate, and slightly decreased cardiac contractility. All of these responses **decrease** arterial pressure. Unfortunately, baroreceptors adapt or "reset" in hours to days to any abnormal pressure level. During chronic high blood pressure the rate of firing of baroreceptors gradually decreases. This decreased rate of firing is interpreted as a normal arterial pressure, even though pressure is still elevated. Consequently, the baroreceptor reflex is only a short-term controlling mechanism and is not effective during prolonged periods of abnormal pressure. When arteries lose compliance during aging, baroreceptors are less easily stretched, affecting baroreceptor sensitivity.

Cardiopulmonary Baroreceptors. Numerous stretch (mechano-) receptors and chemoreceptors, located in the atria, ventricles, coronary blood vessels, and lungs, also cause reflex cardiovascular responses. For example, one function of the cardiopulmonary reflexes is to "sense" **atrial pressure (volume)**. This is seen in the **Bainbridge Reflex**, an increase in heart rate in response to stretch of the atria with volume loading. Increased atrial volume also causes decreased sympathetic activity and elicits changes in the **renin-angiotensin-aldosterone system** that reduce blood volume over the course of several days. In general, cardiopulmonary baroreceptors have a tonic sympathoinhibitory (overload protective) influence.

Several other neural reflexes affect the cardiovascular system, including the **diving reflex**, various pain reflex responses, temperature regulation reflexes, and responses to emotional stress and exercise. Mechanical or chemical stimulation of afferent fibers from skeletal muscle causes reflex tachycardia and increased arterial pressure (**exercise pressor reflex**). These muscle-initiated responses may contribute to adaptations in normal exercise. If arterial pressure falls below 50 mm Hg, the brain becomes ischemic, the vasomotor center becomes extremely active, and arterial pressure rises. This **CNS Ischemic Response** (analogous to the **Cushing reflex**) is one of the most powerful activators of the sympathetic vasoconstrictor system.

Chemoreceptor Reflexes. Small chemosensitive structures, known as **carotid and aortic bodies,** contain sensory receptors sensitive to low $[O_2]$. When arterial blood P_{O_2} sensed by these **chemorecepto-rs** falls too low, the vasomotor center is excited, reflexly raising arterial pressure (see Chapter 4, p. 141).

Hormonal Control Mechanisms

Several hormonal mechanisms provide moderately rapid control of arterial blood pressure. Sympathetic stimulation to the adrenal medulla causes release of the vasoconstrictors **epinephrine** and **norepinephrine.** These circulating catecholamines add to the effect produced by direct sympathetic stimulation of vascular smooth muscle. The **renin-angiotensin-aldosterone mechanism** involves the following sequence of steps: 1) a decrease in arterial pressure causes juxtaglomerular cells of the kidney to secrete the enzyme, renin, into the blood, 2) renin catalyzes the conversion of renin substrate (angiotensinogen) into the peptide angiotensin I, 3) angiotensin I is converted into angiotensin II by the action **of angiotensin converting enzyme (ACE)**, present mainly in lung tissue. **Angiotensin II** is the most potent vasoconstrictor known. Angiotensin II also directly causes increased renal retention of salt and water, which aids in expanding blood volume. **Vasopressin (antidiuretic hormone or ADH)** is released by the hypothalamus when arterial pressure falls too low. Vasopressin has a direct vasoconstrictor effect on peripheral blood vessels. It also plays a role in long-term regulation of arterial pressure through its action to decrease renal excretion of water and thus increase blood volume. **Atrial natriuretic peptide** (ANP) is released by increased atrial stretch and increases renal excretion of Na^+ (See Fig. 7-9, page 249). **Bradykinin** is a vasodilator peptide formed in the plasma, and its action resembles that of **histamine.** Both relax vascular smooth muscle, increase capillary permeability and activate cutaneous pain fibers.

Review Questions

41. The metabolic theory for autoregulation proposes that

A. metabolites within the capillary blood act to relax precapillary sphincters
B. vasoconstrictor substances are formed by metabolism
C. adenosine may be important in local regulation of blood flow
D. passive stretch of vascular smooth muscle induces vasodilation
E. a combination of local metabolites and local autonomic control of blood flow jointly regulate blood flow

For Questions 42-45: Match the physiological mechanism, response, substance or site of action in the column below with a statement about the vasomotor center in each question.

A. Acetylcholine
B. Bainbridge Reflex
C. Baroreceptor feedback
D. Cushing Reflex
E. Endothelial relaxing factor
F. Frank-Starling Mechanism
G. Hypothalamus
H. Vasomotor tone

42. Vasomotor center is inhibited by

43. Tonic vasomotor center activity results in

44. Emergency maximal activation of the vasomotor center occurs during

45. Higher CNS influences on the vasomotor center occur from

46. Activation of atrial stretch receptors

A. releases renin from the juxtaglomerular cells
B. increases renal Na^+ excretion
C. triggers the CNS ischemic response
D. causes reflex vasoconstricton of peripheral arteries
E. decreases urine production

47. The baroreceptor reflex

A. has sensors located in the carotid bodies
B. has sensors located in the aortic bodies
C. results in bradycardia after standing up from a supine position
D. involves sympathetic vasodilator (cholinergic) fibers
E. operates in the normal range of arterial blood pressure

For Questions 48-52. A 45 year-old woman was quite anxious when she called her family physician with the relatively sudden feeling of nervousness, sweating and heart "pounding" in her chest (heart rate 125-135/min; blood pressure 172/116 mm Hg). She was seen immediately, and the physician administered an alpha-adrenergic blocking drug because of her dilated pupils and cold hands and feet. Blood pressure subsequently dropped, but heart rate remained high. A urine sample showed above normal levels of epinephrine, norepinephrine, and 3-methoxy-4-hydroxymandelic acid. A blood sample showed blood glucose levels of 180 mg %.

48. What was the most likely diagnosis?

49. Why did she have cold hands and dilated pupils?

50. Why did alpha adrenergic blockade reduce blood pressure but not heart rate?

51. Why did she experience elevated heart rate, mean arterial pressure and pulse pressure?

52. Explain her elevated blood glucose concentration.

PROPERTIES OF SPECIFIC VASCULAR BEDS

Coronary Circulation

Coronary Vascular Anatomy. Blood enters the coronary circulation through the first two branches of the aorta, the left and right coronary arteries. The **left coronary artery** supplies mainly the left ventricle; the **right coronary artery** supplies the right ventricle and also a major portion of the posterior wall of the left ventricle. Large **epicardial coronary vessels** travel over the surface of the heart, then branch to send penetrating vessels to various depths of the myocardial wall. The ventricular walls are supplied by two vascular beds. An **epicardial bed** or **plexus** is formed by the rapid branching of the large epicardial arteries. An **endocardial plexus** is also formed from relatively large vessels that penetrate directly through most of the ventricular wall and then divide rapidly. Seventy-five percent of the coronary venous blood returns to the right atrium via the **coronary sinus**. The small **anterior cardiac veins** return most of the blood from the right ventricle to the right atrium. Some small **thebesian veins** empty directly into all chambers of the heart.

Resting coronary blood flow averages 300 ml/min, about 5% of the cardiac output, and is capable of increasing 5-fold when necessary. Myocardial oxygen consumption ($m\dot{V}_{O_2}$) is about 12% of the total body $\dot{V}_{O_2}$. Most coronary flow to the left ventricle occurs during diastole when the muscular wall is relaxed. The right ventricle receives about equal coronary flow in systole and diastole. The endocardial regions of the left ventricle receive about 20% more blood flow per gram than the epicardial regions, because endocardial muscle fibers generate more force than epicardial fibers and thus need more blood flow. **Fatty acids** are the primary source of energy for the heart; lactate is a secondary source. Without oxygen the heart becomes hypoxic, produces lactic acid, develops an **oxygen debt**, and begins to use glucose for energy.

Coronary blood flow is directly proportional to **oxygen consumption** of the heart, so flow is largely metabolically controlled. However, the heart consumes more oxygen when working against a high pressure with normal cardiac output than when pumping larger quantities of blood against a relatively normal pressure. The oxygen extraction of the heart is greater than any other organ (15 ml O_2/100 ml blood vs. 4 ml O_2/100 ml blood in other organs), leaving little oxygen reserve in venous blood to be used during stress or exercise. During exercise the heart depends upon increased myocardial efficiency (determined by the type of work done by the heart, generally independent of oxygen needs) and increased coronary blood flow. Myocardial oxygen consumption is the main regulator of coronary blood flow. The four major determinants of myocardial oxygen consumption are heart rate, myocardial wall tension, cardiac contractility and systolic pressure. Myocardial oxygen consumption is also increased by other factors, such as digitalis, calcium, and thyroxine. **Local regulatory mechanisms** at the tissue level are most important in controlling coronary blood flow. Possible mediators for blood flow regulation include oxygen demand, CO_2 production, pH, K^+ ions and adenosine, the prime mediator.

Neural Control of Coronary Blood Flow. The direct effects of ANS stimulation on coronary blood vessels are difficult to separate from changes elicited by indirect effects of such stimulation on the myocardial tissue itself. Direct effects result from both sympathetic and parasympathetic activity. **Sympathetic α_1-receptor vasoconstriction** occurs in the coronary circulation and tends to limit coronary blood flow during periods of stress or exercise. However, this vasoconstriction is overriden by increased local metabolic demands of cardiac tissue from β_1-receptor stimulation. **Sympathetic β_2-receptor vasodilation** is also present, but its effects are small. **Parasympathetic vasodilation** only affects small vessels distal to the epicardial arteries.

Cerebral Circulation

Cerebral Vascular Anatomy. The brain is supplied with blood from **internal carotid** and **vertebral arteries.** The **basilar artery** is formed by the convergence of the two vertebral arteries. The two internal carotid arteries and the basilar artery enter the **Circle of Willis,** which then delivers blood to the brain by six large vessels. The vertebral arteries provide very little flow in humans, and there is little cross-perfusion between the two carotid arteries. The **internal jugular** veins provide the majority of the venous drainage via deep veins and dural sinuses. Capillaries of the **choroid plexus** have gaps between the endothelial cells; filtration of fluid through these gaps accounts for about 50% of cerebrospinal fluid production. Capillaries of brain tissue itself have numerous tight junctions between their endothelial cells, and movement of materials out of the brain tissue capillaries is severely restricted. This restriction is unique to brain capillaries and produces the **blood-brain barrier.** Water, oxygen, and carbon dioxide cross the barrier readily, but glucose and ions like Na^+, K^+, Cl^- and others take up to 30 times longer to cross this barrier than in other capillary beds. Very little urea, bile salts, proteins, and catecholamines can enter brain tissue. Understanding the blood-brain barrier is important for drug treatment. For example, antibiotics like penicillin cross the blood-brain barrier with difficulty, but erythromycin penetrates easily.

Cerebral blood flow is controlled within narrow limits and averages about 750 ml/min. The brain receives about 13% of cardiac output and accounts for about 20% of total body oxygen consumption. The cerebral circulation is uniquely contained within a closed rigid structure, the **cranium.** Because of the incompressibility of cranial tissues and fluids, increases in arterial inflow must be accompanied by increases in venous outflow. Otherwise excess pressure develops within the brain, causing edema, decreasing vascular transmural pressure, reducing flow and causing tissue damage.

Regulation of Cerebral Blood Flow. The brain is metabolically active and mainly utilizes glucose. Brain is the tissue least able to tolerate reduction of blood flow. As a whole, brain metabolism is nearly constant. Autoregulation is effective; cerebral blood flow is unchanged at arterial blood pressures as low as 60 mm Hg. Three factors affecting cerebral flow are 1) the arterial and venous pressures at the level of the brain, 2) intracranial pressure, and 3) the relative state of constriction or dilation of cerebral arterioles. Direct **mechanical effects** of cerebral flow occur by varying intracranial pressure. If blood flow is excessive, additional fluid leaves cerebral capillaries, thus elevating intracranial tissue pressure (because of the rigidity of the cranium).

Cerebral blood vessels are innervated by both sympathetic and parasympathetic fibers. α_1-adrenergic receptor stimulation produces minimal vasoconstriction, and there are no sympathetic vasodilator fibers. Stimulation of parasympathetic fibers, especially from the facial nerve, elicits mild vasodilation.

Local metabolic regulation is the most important mechanism for maintaining cerebral blood flow nearly constant. Resistance vessels of the cerebral circulation are sensitive to **local changes in P_{CO_2}**. Increases in arterial P_{CO_2} produce marked vasodilation; decreases in arterial P_{CO_2} induce vasoconstriction. The vessels are less sensitive to P_{O_2} changes. **Hydrogen ions** are also vasodilators, and a decrease in local **pH** of brain tissue causes increased flow. The vasodilating effects of CO_2 may be mediated by local pH changes. Vasoactive metabolites, such as adenosine, may play a secondary role in regulation of cerebral blood flow. Although overall cerebral blood flow remains relatively constant, the distribution of flow to different areas of the brain varies according to their specific second-by-second metabolic needs. For example, the visual cortex receives more blood flow during waking hours than during sleep and a further increase when the visual system is especially active.

Pulmonary Circulation

Functional Anatomy. Pulmonary vessels are highly distensible, and resistance to flow is low. The **transmural pressure** determines the caliber of these vessels. Arterioles cannot constrict as well as systemic arterioles (less smooth muscle). Capillaries are unique, because blood flows in thin sheets between alveoli, and the thickness of the sheets depends upon intravascular and intraalveolar pressures. During pulmonary vascular congestion the width of the sheet is increased several fold; during high alveolar pressures adjacent capillaries may collapse.

Normal pulmonary arterial pressures are about 27/10 mm Hg, pulmonary resistance being 1/10th of the systemic circulation. Since left atrial pressure is about 5 mm Hg, the arteriovenous pressure gradient is about 11 mm Hg. Three types of blood flow patterns exist in the lung of an upright person. These are upper lung zone where alveolar pressure exceeds intravascular pressures; middle zone with alveolar pressure between arterial and venous pressures; and lower zone where intravascular pressures exceed alveolar pressure.

There are parasympathetic and sympathetic nerves to the pulmonary vessels. Alpha receptors predominate and cause vasoconstriction. **Hypoxia** is the most important influence on vasomotor tone, increasing vascular resistance. This regional vasoconstriction helps optimize ventilation and perfusion, reducing blood flow to poorly ventilated alveoli. The mechanism for such hypoxic vasoconstriction is unknown.

Cardiorespiratory interaction occurs in several ways. Negative intrathoracic pressure facilitates venous return of systemic blood to the heart **(thoracoabdominal pump)**, and contributes to cardiac **preload**. Baroreceptor stimulation also dilates pulmonary vessels.

Splanchnic Circulation

A number of abdominal organs including the gastrointestinal tract, spleen, pancreas and liver are collectively supplied by splanchnic blood flow. These organs actively receive about 25% of cardiac output at rest. Since they are involved in digestion and absorption, a large meal can elicit a 30-100% increase in blood flow. Sympathetic nerves provide the dominant control, being able to reduce blood flow to as little as 20% of its resting value. The **hepatoportal system** has a unique anatomy with two inputs to the liver from the intestines (portal vein) and the hepatic artery.

Cutaneous Circulation

Functional Anatomy. Cutaneous blood flow is not controlled by local metabolic factors. The oxygen and nutrient requirements of skin tissue are small. The primary function of the cutaneous circulation is to maintain **body temperature**, so blood flow to the skin responds to changes in ambient and internal body temperatures. In addition to the usual nutritive blood vessels, skin has two structures that are important for its role in temperature regulation; 1) a system of large **subcutaneous venous plexuses** holds large quantities of blood that can heat the skin surface, and 2) **arteriovenous (AV) anastomoses** shunt blood from arterioles to venules, bypassing capillary beds and venous plexuses. These AV anastomoses are especially prominent in the fingertips, palms of the hands, soles of the feet, toes, ears, nose, and lips. They are controlled by sympathetic activity and are easily constricted by epinephrine and norepinephrine. Sympathetic input to these AV anastomoses is controlled from hypothalamic temperature regulation centers. Cutaneous arterioles are primarily controlled by neural activity, with local metabolic control secondary.

At rest, the cutaneous blood flow is approximately 450 ml/min in the average person, about 9% of cardiac output. It accounts for 5% of the total body oxygen consumption. In extreme cold skin blood flow can be reduced to about 50 ml/min, while heat stress causes maximum vasodilation and skin blood flows of 3 liters/min. If blood flow to part of the skin is blocked for some time, a **reactive hyperemia** occurs when flow is re-established.

Skin blood flow at rest is regulated by **sympathetic adrenergic vasoconstriction.** Blood flow to subcutaneous plexuses is limited by constricted AV anastomoses, so minimal amounts of heat are lost to the environment. Heat stress reduces sympathetic tone to skin, permitting large quantities of warm blood to flow into subcutaneous plexuses, thereby enhancing heat transfer out of the body. **Sympathetic cholinergic vasodilation** may also regulate skin blood flow during heat stress. Sympathetic cholinergic nerve terminals activate eccrine sweat glands, promoting the formation of **bradykinin**, a potent vasodilator. Skin blood flow is also important as a **blood reservoir**. During circulatory stress (exercise, anxiety, hemorrhage) further constriction of AV anastomoses redirects large quantities of blood (5-10% of total blood volume) from the skin into the remainder of the circulatory system to supplement circulating blood volume. However, during later stages of exercise this vasoconstriction is overridden by the need for more heat loss with the heat production of muscular work.

Skeletal Muscle Circulation. (See section on Exercise in Chapter 8).

Review Questions

53. Which organ extracts the greatest amount of oxygen from the blood that it receives?

    A. Kidney
    B. Brain
    C. Heart
    D. Skin
    E. Lung

54. Coronary artery blood flow

    A. does not depend on arterial blood pressure
    B. is greatest during systole in the left circumflex coronary artery
    C. is regulated mainly by myocardial oxygen consumption
    D. accounts for about 10% of the cardiac output
    E. is greatest during diastole in the right coronary artery

55. Blood flow to the brain

    A. may account for up to 25% of cardiac output
    B. is unaffected by the magnitude of intracranial pressure
    C. is extremely sensitive to local pH levels
    D. is largely under sympathetic control
    E. is increased following traumatic brain injury (concussion)

56. Each of the following statements about the pulmonary circulation is true **EXCEPT** which one?

    A. Arterial pressure determines the caliber of that vessel.
    B. Flow in the capillaries is in thin sheets.
    C. Arterial pressures are about the same as in the systemic circulation.
    D. There are three basic zones of ventilation/perfusion in the lung.
    E. During hypoxia the vessels vasoconstrict, not vasodilate.

INTERACTION BETWEEN PERIPHERAL AND CENTRAL CONTROL

There are two concepts that are important for understanding how various parts of the cardiovascular system function together. These are variations of the equation: pressure = resistance x flow.

1) **Tissue perfusion** depends on **arterial pressure** and **local vascular resistance.**
2) **Arterial pressure** depends on **cardiac output** and **total peripheral resistance.**

Autonomic and baroreceptor systems control arterial pressure in the short term by **reciprocal** changes in cardiac output and total peripheral resistance. But cardiac output and total peripheral resistance are determined by many factors that interact with each other. For long term control, the mechanisms that regulate the fluid balance of the body are of greatest importance. Finally, one must consider the circulating vasoactive and cardioactive substances that affect cardiac performance (eg, catecholamines) or vascular smooth muscle and peripheral circulation (eg, vasopressin, EDRF, angiotensin II, catecholamines).

The central and peripheral adjustments that operate to compensate or restore homeostasis will vary with the particular integrative physiologic response, such as dynamic exercise, circulatory or cardiogenic shock or orthostatic tolerance. For example, when assuming an upright posture from a horizontal position the baroreceptors interact with the heart and blood vessels to maintain sufficient arterial pressure to perfuse the brain; otherwise syncope will result. Patients with orthostatic intolerance usually are hypotensive and are prone to fainting when standing. Pooling of blood immediately occurs in the large veins of the dependent limbs, and this tends to take some of the central blood pool "out of circulation". This affects venous return, stroke volume and cardiac output. Muscular milking helps to return some of that pooled venous blood to the central blood volume.

Coupling between the Heart and the Peripheral Circulation

Heart rate and myocardial contractility (stroke volume) are cardiac factors that affect cardiac output. Preload and afterload depend on both cardiac and vascular systems, and they are also determinants of cardiac output. However, preload and afterload are determined by cardiac output and vascular status, because it is a coupled system. A graphing technique to learn cardiac and peripheral interactions has been used to analyze two functionally independent relationships between the cardiac output and the central venous pressure. These interactions are graphed as the **vascular function curve** and the **cardiac (ventricular) function curve** (Fig. 3-9).

The cardiac function curve is also an expression of the **Frank-Starling Mechanism.** Increased cardiac output is graphically depicted by a higher point on the cardiac function curve, demonstrating the Frank-Starling Effect (increased stroke volume due to increased preload). Cardiac function also improves with increased contractility (positive inotropic response) and is graphed by a parallel shift of the curve upward.

The vascular function curve depends upon peripheral resistance, arterial and venous compliance and blood volume. It graphs the results of an artificial heart-lung experiment using an "open loop" vascular system where venous return (or cardiac output) is the independent variable and central venous pressure (or right atrial pressure) is the dependent variable (Fig. 3-9A). When cardiac output increases, central venous pressure decreases; ie, central venous pressure varies inversely with cardiac output. The vascular function curve shows a parallel upward shift with an increase of blood volume and downward shift with decrease of blood volume. The curve shifts obliquely up (rotates; ordinate unchanged) for decreases in resistance and down for increases in resistance. The **systemic filling pressure** of 7 mm Hg represents the pressure in the vascular system at zero cardiac output.

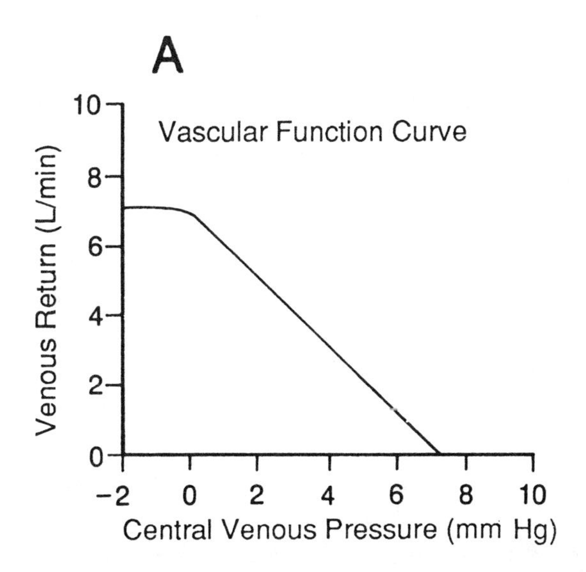

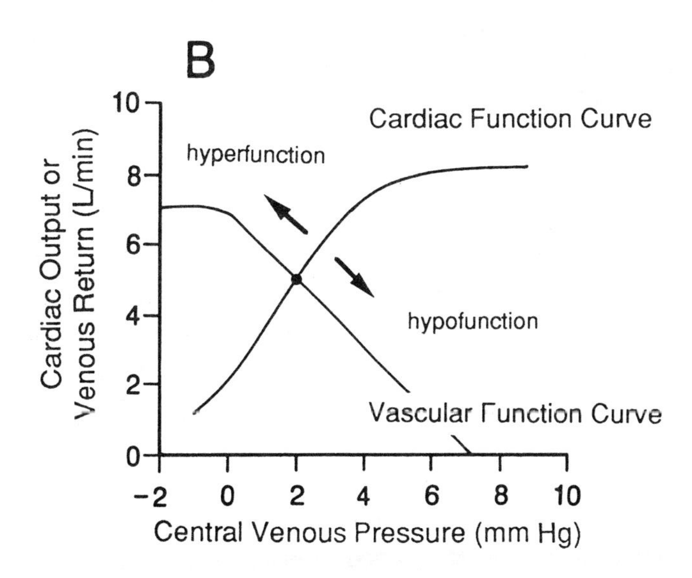

Figure 3-9. Vascular function curve (A) and interaction of cardiac function and vascular function curves (B).

These vascular changes can be coupled (closed loop system) to cardiac output by superimposing the vascular and cardiac function curves (Fig. 3-9B). The cardiac function curve may also shift due to physiological events, such as activity in the sympathetic nervous system. The intersection of both curves is called the **equilibrium point**, which represents the physiological status of the heart in the cardiovascular system. Movement of the equilibrium point must be commensurate with the physiologically-induced shifts in both the vascular and cardiac function curves. This graphing technique is a simple method for predicting cardiac performance with changing vascular and cardiac status. Myocardial ischemia may be caused by obstructive coronary artery disease, aortic value stenosis, severe aortic regurgitation hypertension or cardiomyopathy. Any of these would cause a downward shift of the cardiac function curve, representing decreased cardiac function (hypoeffective heart). Increases in contractility would shift the cardiac function curve upward. Related changes in blood volume would also shift the vascular function curve resulting in a new equilibrium point.

Aging of the Cardiovascular System

The arteries show reduced compliance with age. This causes reduction in the capacity of the aorta to "store" part of the stroke volume. Consequently, the load on the left ventricle increases, especially during exercise, increasing ventricular dimensions. The veins lose elasticity, and varicosities occur in veins under high pressure. Capillary basement membrane thickens, which threatens diffusion of nutrients into tissues.

There are also age-related changes in the heart. Time to peak isometric tension, duration of the relaxation phase, and the plateau phase of the action potential are prolonged, consistent with altered calcium movement into and out of the sarcoplasmic reticulum and other calcium storage pools. The aged myocyte is not as responsive to noradrenergic stimulation. There is a progressive loss of cells from the SA node with age. The SA and AV nodes, the His bundle, and the bundle branches become invaded by fibrous tissue.

Between 20 and 80 years of age, **cardiac output decreases** about 1% per year, while the **stroke volume decreases** at a slightly lower rate. The maximum heart rate that can be achieved changes in a linear fashion with age and may be predicted from the relationship: Maximum HR = 220 - age in years. The reduced chronotropic activity of the heart may be due to the loss of noradrenergic receptors, reduced release of neurotransmitter, or changes in the conductive properties of the heart due to fibrous invasion. Work of the heart tends to decrease slightly with age, while the total peripheral resistance increases steadily at a rate of approximately 1% per year from age 40 onward, decreasing tissue perfusion. Renal perfusion, as well as in the splanchnic and cutaneous beds, is reduced by 50%. Cerebral blood flow decreases by about 20% from age 40 to 80.

Longitudinal and cross-sectional studies have shown an **increase in systolic pressure** with age, with a lesser rate of increase in diastolic pressure. Physical conditioning can slow or reverse these changes. Both heart rate and blood pressure are sluggish to return to normal following increases with stress, like exercise.

Alterations in the cardiovascular and respiratory systems with age impair the conduction of oxygen from the atmosphere to active cells. However, three factors may be involved: 1) genuine physiological aging changes, 2) hypokinesis (reduced physical activity), and 3) undiagnosed pathology of the oxygen conduction system. Much of the reduction in oxygen conduction can be prevented or reversed with physical exercise extending into the 70s and 80s.

ANSWERS TO CARDIOVASCULAR PHYSIOLOGY QUESTIONS

1. B. The strength of contraction of <u>skeletal</u> muscle is graded by changing the number of fibers that contract (A). Extracellular Cl^- concentration does not affect strength of contraction (C). The initial fiber length before contraction begins has a direct effect on strength of contraction (D). Strength of contraction is independent of the amplitude of the action potential (E).

2. B. Actin, myosin, troponin, and tropomyosin, and transverse tubules are all present in cardiac as well as skeletal muscle (A, D). Sympathetic excitation modulates contraction strength (C). Entrance of "signal" Ca^{2+} is required during the plateau phase (E).

3. C. A and E are true statements but are unrelated to the all-or-nothing principle. Cardiac muscle does not depend on neuromuscular junctions to initiate contraction (B). Increasing the duration of the action potential would make it even more difficult to tetanize cardiac muscle (D).

4. A. Catecholamines act by stimulation of cardiac β-adrenergic receptors which increase intracellular cyclic AMP (B). Inhibiting the Na^+-K^+ ATPase increases the amount of Ca^{2+} available to myofibrils (A, C). Blocking or stimulating cardiac β-adrenergic receptors would modulate contractile performance but this is not the mechanism of glycoside action (D, E).

5. A. The delay allows atrial contraction to push an additional amount of blood into the ventricle before it contracts. The S-T segment is recorded after ventricular depolarization but before repolarization (B). The Q-T interval is a rough measure of the duration of systole (C). Slow propagation is associated with small rather than large diameter fibers (D). The bundle of His is composed of fast (1-4 m/sec) conducting fibers (E).

6. E. The bundle of His (and the Purkinje fibers) have the fastest conduction velocity of any portion of the heart.

7. D. Atrial repolarization is masked by the QRS complex (A). Atrial depolarization produces the P wave (B). Activity in the bundle of His is not recorded in a normal EKG (C). Ventricular depolarization produces the QRS complex (E).

8. D. The EKG does not provide information about patency or action of cardiac valves or about cardiac contractility.

9. E. The SA node is the pacemaker for the heart and initially stimulates the atrial musculature to contract.

10. D. See Fig. 3-4.

11. D. During isovolumic contraction atrial pressure actually decreases slightly from twisting of the ventricles (A, E). Aortic pressure is continuing to decrease slowly, because the aortic valve has not yet opened, and blood is running off to the peripheral arterial system (B, C).

12. D. The onset of atrial contraction does not close any valve, so there is no sound. Also see Fig. 3-4.

13. D. Positive inotropic agents increase maximum left ventricular dP/dt (A) and shift the ventricular function curve upward and to the left (B). The negative inotropic effect of ACh is mediated by muscarinic receptors (C). Positive inotropic agents tend to decrease left ventricular end-diastolic pressure (E).

14. E. In general a decrease in heart rate will result in an increase in stroke volume.

15. A. The indicator must be injected "upstream" from the sampling site, usually a vein. Indicator dilution techniques do not measure oxygen consumption as do Fick techniques (C) and they tend to be inaccurate at low cardiac outputs (A). Thermodilution uses cold saline (D). Stroke volume is calculated (E).

16. D. Stenosis of the aortic valve prevents adequate amounts of blood from being pumped out of the left ventricle into the aorta (D). A decrease in the number of β-receptors or poor cardiac muscle function would decrease cardiac performance, but the left ventricle would be unable to penetrate pressures of 190 mm Hg (A, E). Pulmonary ventilation is adequate if the arterial O_2 content is 19 mmO_2/100 ml of blood (B). Without an elevated left ventricular end-diastolic pressure this patient's cardiac performance would be even worse (C).

17. E. Administration of a α-blocking agent or a rapidly-acting diuretic would reduce the systemic arterial pressure which might further impair the cardiovascular status of the patient (A, B). A β-receptor blocker would reduce cardiac contractility and reduce the cardiac output (C). Blood oxygenation is adequate in this patient and does not need to be supplemented (D).

18. A. Flow increases 16 times when the radius is doubled (B); a decrease in viscosity will increase flow (C); flow decreases by a factor of 2 when the length is doubled (D), and flow in blood vessels is only an approximation of the flow equations for rigid tubes.

19. D. Maximal flow velocity occurs in the center of the stream where the erythrocytes tend to concentrate (A, B, D). The Law of Laplace relates to forces in the wall of the vessel rather than blood flow velocity (C). There is insufficient information to determine if turbulence is present (E).

20. A. See paragraph in text. Valve leaflets approach each other, because lateral pressure is reduced at a constriction (D). Distending pressure decreases as velocity increases (E).

21. C. Critical closing pressure increases if the vessel becomes less compliant (B) and decreases from smooth muscle relaxation induced by adrenergic β_2-receptor stimulation (A, D). Critical closing pressure has nothing to do with "critical" patient status (E).

22. D. Less pressure energy is absorbed by the wall of the aorta, hence pressure rises more rapidly. Choice E would have had the opposite effect as blood would enter the peripheral circulation with less resistance.

23. B. Beta-blocking drugs reduce heart rate and force of contraction (β_1 receptors), so the cardiac output would be lower. Therefore, arterial pressure would be lower without any change in vessel status. β_2 receptors on blood vessels are sparse and affect vasodilation minimally.

24. C. Since tonic vagal nerve activity partially restrains the nodal rate of depolarization, blocking the vagus increases heart rate.

25. B. The velocity of arterial blood flow is about 1/10th the pulse wave velocity (D). The pressure pulse propagates at about 7 m/sec and is palpable at the wrist 100 msec after ejection (A, C). Less compliant vessels propagate the wave at a greater velocity (E).

26. B. Pulse pressure decreases if compliance increases (B). Capillary pressure is always non-pulsatile (D). Systolic pressure increases in the arterial pressure wave form as the pressure pulse moves into peripheral arteries (C). Pulse pressure is systolic minus diastolic pressure (A). Aging is usually accompanied by decreased arterial compliance and elevated pulse pressure (E).

27. B. Arterial pressures above and below the heart are changed significantly between a supine and standing posture by gravitational effects (B, C, D). Mean arterial pressure at the level of the heart is closer to 100 mm Hg (A). Changes of emotional state affect the sympathetic nervous system (heart rate, cardiac output, vasoconstriction) and therefore arterial pressure (E).

28. B. Flow velocity is determined solely by blood flow and cross-sectional area of the vessel (D) so is lowest in the capillaries. Compliance changes will affect pulse wave velocity but not flow velocity (C). Pulse pressure is a function of aortic pressure which may be the same for various flow rates (A). As viscosity (hematocrit) increases, velocity decreases if all other things remain the same (E).

29. B. Arterioles up to and including the precapillary sphincter are controllable resistances because of the smooth muscle response to adrenergic agents.

30. D. The veins have little elastin or smooth muscle and can accommodate large volumes of blood.

31. C. Capillaries have the greatest cross-sectional area but are not all open at the same time, hence contain only 5% of the total volume.

32. E. Capillaries are in vasomotion in resting skeletal muscle (E) . Flow is continuous in arterioles (A), metarterioles (B) and discontinuous in true capillaries (D).

33. D. $F = (P_c + \pi_i) - (P_i + \pi_c)$; so $(20 + 5) - (2 + 25) = -2$

34. A. If capillary permeability decreases, fewer plasma proteins leak into the interstitial space, and interstitial fluid osmotic pressure decreases (B). Small molecules, such as glucose, have no effect on osmotic pressure (D). Electrolytes are not responsible for the osmotic pressure gradient (E). Lymphatic blockage would increase osmotic pressure (C).

35. E. Tissue metabolites are vasodilators that primarily affect metarterioles and precapillary sphincters but not capillaries (B). Sympathetic activity has its primary effect on arterioles (D). Some capillary beds are mainly controlled by local metabolic mechanisms (B, C). The tissue type (metabolic needs) determines relative capillary flow at rest (A)

36. A. Filtration is greatest in the proximal capillary (B) and increases when venules constrict (post capillary resistance) (C). A low plasma osmotic pressure would cause less reabsorption and thus greater net filtration (D). Interstitial osmotic pressure is affected by lymphatic drainage (E).

37. Blood flow to the calf and foot were reduced, probably because of peripheral vascular disease of the right femoral artery. This reduced blood flow to the foot (eg, atheromatous plaque reducing the radius of the artery).

38. This phenomenon is called intermittent claudication and is associated with reduced tissue blood flow. The mechanism activating pain fibers is unknown, but it may be due to accumulation of metabolites and changes in local pH. Therefore, when he rested and the metabolites were cleared, he was able to resume walking. The distance he could walk was related to blood flow relative to the intensity of the exercise work load. His right foot was colder, because heat from blood is normally dissipated as it leaves the warmer trunk. In this case decreased arterial flow reduced heat flow to the foot; reduced skin venous flow and less oxygenated blood both contributing to the skin pallor. Likewise, decreased metabolic support of the skin contributed to hair loss.

39. Resting blood flow is determined by perfusion pressure and resistance to flow. Arterioles peripheral to the obstruction will dilate because of metabolic autoregulation, so the total resistance is not remarkably different from the normal leg. During aerobic exercise (walking) the arterioles in both legs dilate and the leg/foot containing the atheromatous obstruction becomes the limiting factor to flow in the right leg.

40. Factors influencing oxygen transport include 1) the large surface area available for diffusion, which is dependent upon the number of open capillaries; 2) diffusion distance from capillary to muscle cells, which is also dependent upon the number of open capillaries; and 3) concentration difference across the capillary, which is usually determined by tissue uptake and blood flow into the capillary. Blood O_2 content is determined by pulmonary ventilation.

41. C. Metabolic products are vasodilators, not vasoconstrictors (B). Passive stretch of vascular smooth muscle promotes myogenic vasoconstriction (D). Metabolic theory proposes total metabolic control of blood flow (E). Metabolites produced in the tissue diffuse to dilate sphincters and increase blood flow to that local tissue site (A).

42. C. When peripheral arterial pressure is too high, the baroreceptors send information to the brainstem to reduce the amount of sympathoexcitation of vascular smooth muscle.

43. H. Vasomotor tone is the relative amount of vasoconstriction in vascular smooth muscle. Venomotor tone often refers to the same phenomenon in veins.

44. D. The Cushing Reflex is a clinical term for events related to head trauma and swelling. Cerebral blood flow is reduced; the brainstem is ischemic; and the vasomotor center is activated.

45. G. The hypothalamus is one higher center that stimulates vasomotor center activity, particularly during rage and anger emotions.

46. B. Renin release is elicited by decreased renal blood flow or pressure (A). The CNS ischemic response is elicited when arterial pressure falls below about 50 mm Hg but not by stretch receptors (C). Peripheral arteries are dilated (D). Urine production is increased by activating these "volume" receptors (E).

47. E. The baroreceptor reflex controls activity in sympathetic vasoconstrictor fibers (D). The aortic and carotid bodies contain chemoreceptors (A, B). Mild-tachycardia (sympathoexcitation) results from baroreflex "unloading" (C).

48. A tumor of the adrenal medulla called pheochromocytoma is characterized by hypertension, elevated heart rate and urinary catecholamines.

49. Elevated plasma catecholamines often produce anxiety. Since sweat glands are controlled by the sympathetic nervous system, they are activated by those catecholamines. Heart palpitation is the sensation of increased rate/vigor of cardiac contraction due to catecholamines. Cold extremities are due to peripheral cutaneous vasoconstriction, and pupillary vasodilation is caused by catecholamines.

50. The alpha$_1$-receptor blocker (eg, prazosin) blocks the vasoconstrictor effect of catecholamines, and the resulting peripheral vasodilation will lower arterial pressure. However, catecholamines have a beta-receptor function on the heart which is both chronotropic and inotropic.

51. She experienced tachycardia because norepinephrine and epinephrine act on the sinoatrial node pacemaker cells to increase the rate of depolarization. Mean arterial pressure is the product of cardiac output and total peripheral resistance, and total peripheral resistance is elevated because of peripheral vasoconstriction. Since cardiac output is elevated by the tachycardia and increased force of contraction, mean arterial pressure is increased. Stroke volume is also increased, because catecholamines shift the cardiac function curve upward (Fig. 3-9). The pulse pressure is

determined by two factors: 1) volume ejected from the left ventricle into the aorta, and 2) the distensibility or compliance of the aorta and large arteries. Stroke volume increases and compliance decreases (the aorta is stretched by the higher mean arterial pressure), so the stroke volume produces a greater pulse pressure.

52. Epinephrine increases both gluconeogenesis and glycogenolysis, thereby raising blood glucose levels.

53. C. The arteriovenous oxygen difference across the coronary circulation is greater than any other organ in the body.

54. C. Arterial pressure is one of the major determinants of coronary blood flow (A). Left circumflex coronary artery flow is greatest during diastole when the ventricular muscle is relaxed and extravascular compression is minimal (B). Coronary flow is about 5% of the cardiac output (D). Right coronary artery flow is greatest during systole when the driving pressure is greatest; extravascular compression is not a significant factor in reducing right coronary blood flow (E).

55. C. About 15% of cardiac output goes to the brain (A). Increases in intracranial pressure limit blood inflow, because of the fixed internal volume of the intracranial space (B, E). Blood flow is only minimally affected by the ANS (D).

56. C. Because it is a low resistance circuit, the pressures are lower. This is shown by the myocardial thickness in the right ventricle.

PULMONARY PHYSIOLOGY

Pulmonary physiology is concerned with the processes necessary for the exchange of oxygen (O_2) and carbon dioxide (CO_2) between cells of the body and the external environment. These processes include 1) gas transport properties of the blood, 2) mechanics of breathing, 3) O_2 and CO_2 exchange across the alveoli of the lungs, and 4) regulation of breathing.

GAS TRANSPORT PROPERTIES OF BLOOD

Transport of Oxygen

Most of the O_2 in blood is carried in chemical combination with **hemoglobin.** Normally, less than 2% of all the oxygen is physically dissolved, because O_2 solubility is only 0.003 ml O_2/ 100ml blood/ mm Hg. Blood from adult men typically contains 15 grams of hemoglobin/100 ml, and each gram of hemoglobin can bind 1.34-1.39 ml of O_2. This gives hemoglobin the **capacity** to carry 20.1-20.9 ml O_2/100 ml blood. The amount of O_2 carried by hemoglobin varies with the partial pressure of O_2 (P_{O_2}) in the blood. For a given P_{O_2} the amount of O_2 actually bound to hemoblogin divided by capacity (x 100) defines **% saturation** of hemoglobin with O_2. When either the % saturation, or O_2 content of blood, is plotted as a function of P_{O_2} (Fig. 4-1), the result is a sigmoid-shaped O_2 dissociation curve. A typical value for P_{O_2} in arterial blood at sea level is 90 mm Hg. Under these conditions the oxygen dissociation curve is nearly flat, indicating that the hemoglobin is almost saturated with O_2. At values of P_{O_2} found in tissue capillary blood (the actual value depends greatly on the tissue, but a "typical" value is 40 mm Hg) the dissociation curve is steep. On this lower portion of the dissociation curve large volumes of O_2 can be removed from hemoglobin with minimal reduction of P_{O_2}; this helps maintain an appropriate partial pressure gradient for O_2 diffusion into tissue (see p. 132). The oxygen dissociation curve can be shifted to the right by **increases** in the following: 1) partial pressure of CO_2 (P_{CO_2}), 2) hydrogen ion concentration, 3) temperature, and 4) 2,3-bisphosphoglycerate (2,3 BPG) concentration in red cells. This shift to the right allows removal of O_2 from the blood at higher P_{O_2} values. A shift to the left in the oxygen dissociation curve results from decreases in any of the preceding quantities. Shifts of the oxygen dissociation curve are sometimes described by changes in the P_{50} value for O_2 (the P_{O_2} value where % saturation is 50).

Abnormalities in Blood O_2 Transport

Carbon Monoxide (CO) Poisoning. Carbon monoxide strongly binds to sites on the hemogobin molecule which normally carry O_2. Therefore, CO reduces the total amount of O_2 that can be carried in the blood. Importantly, the O_2 dissociation curve is shifted to the left, and P_{50} will be reduced. This means that O_2 can be unloaded from the hemoglobin molecule only at a relatively low P_{O_2}, making it even more difficult to oxygenate body tissues.

Anemia. The carrying capacity of whole blood for O_2 is reduced when numbers of red cells are reduced, and/or there is inadequate hemoglobin in the red cells. Unlike CO poisoning, the O_2 dissociation curve is not shifted to the left, and may be shifted to the right under some circumstances by increases in 2,3 BPG.

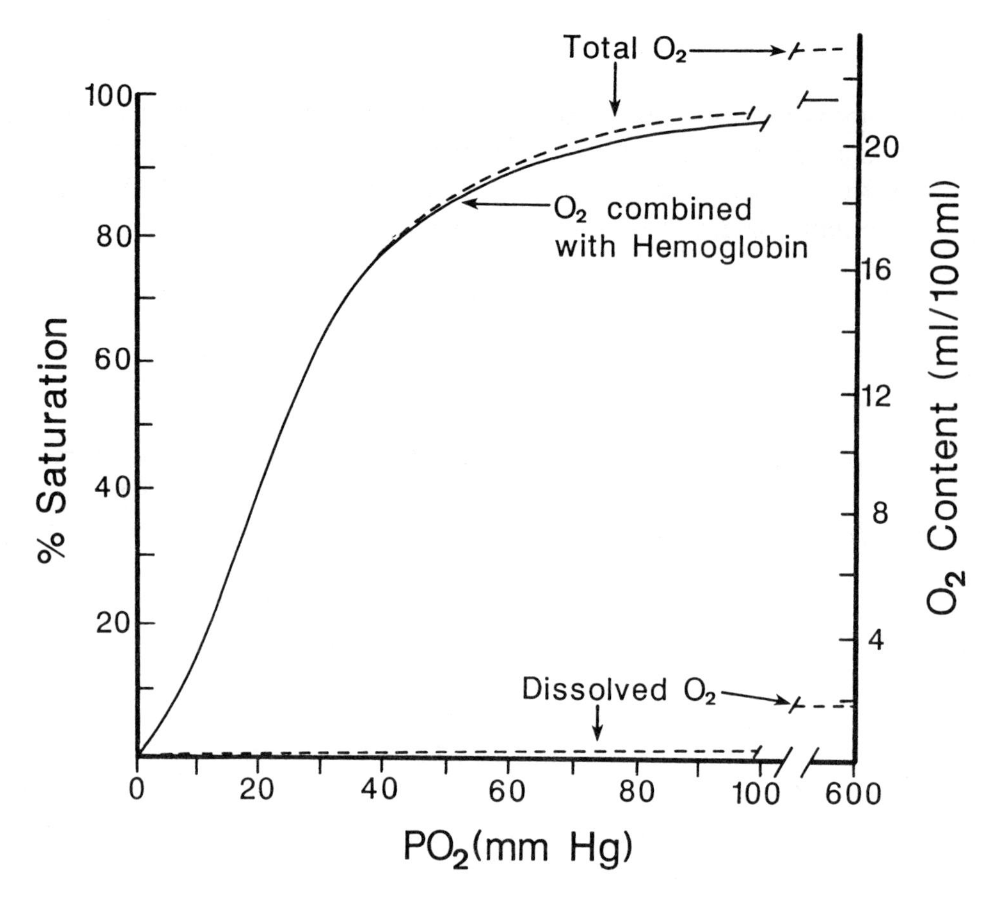

Figure 4-1. Oxygen dissociation curve for blood.

Transport of Carbon Dioxide

Significant quantities of CO_2 are carried in the blood as 1) physically dissolved CO_2 [because CO_2 is 20 times more soluble in blood than is O_2], 2) bicarbonate ion, and 3) carbamino CO_2. The relationship between CO_2 content in the blood and P_{CO_2}, the CO_2 dissociation curve, is **nearly linear** over P_{CO_2} values between normal venous blood (approximately 46 mm Hg) and normal arterial blood (approximately 40 mm Hg). In arterial blood about 90% of the CO_2 content is in the form of bicarbonate ion, while physically dissolved and carbamino-CO_2 account for approximately 5% each. As CO_2 is added to blood in the tissue capillaries, approximately 60% is converted into bicarbonate ion in the red cells by a carbonic anhydrase (CA) catalyzed reaction, as shown below:

$$CO_2 + H_2O \overset{CA}{\rightleftarrows} H_2CO_3 \rightleftarrows H^+ + HCO_3^-$$

The hydrogen ion produced is largely buffered by hemoglobin, and much of the bicarbonate diffuses into the plasma in exchange for chloride ion (the "chloride shift"). Deoxygenated hemoglobin is a better buffer than oxygenated hemoglobin. Therefore, removal of O_2 from the blood by the tissues increases the buffering of hydrogen ion in the blood and increases the amount of CO_2 that can be converted to bicarbonate ion at a given hydrogen ion concentration. Furthermore, deoxygenated hemoglobin binds considerably more carbamino-CO_2 than oxygenated hemoglobin, and about 30% of the CO_2 added to blood leaving the tissues is in the form of carbamino CO_2. Both these mechanisms enhance CO_2 uptake by the blood from the tissues.

Review Questions

1. As the saturation of hemoglobin with O_2 decreases,

    A. there is less bicarbonate [HCO_3^-] present for a given P_{CO_2}
    B. more CO_2 can be carried in the form of carbamino compounds
    C. the content of CO_2 in the blood at a given P_{CO_2} decreases
    D. the P_{CO_2} of the blood for a given content of CO_2 increases
    E. hemoglobin becomes a poorer buffer of hydrogen ion

2. An elderly person found unconscious after her gas house heater malfunctions. If one-half the hemoglobin in her blood is bound to carbon monoxide, then her

    A. P_{50} for O_2 will be decreased from normal
    B. O_2 content of blood at a P_{O_2} of 100 mm Hg will be normal
    C. percent saturation of hemoglobin with O_2 at a P_{O_2} of 100 mm Hg will be less than normal
    D. percent saturation of hemoglobin with O_2 at a P_{O_2} of 50 mm Hg will be less than normal
    E. O_2 carried in physically dissolved form at a P_{O_2} of 100 mm Hg will be less than normal

MECHANICS OF BREATHING

Elastic Properties of Lungs

Lungs contain smooth muscle and the connective tissue, elastin. Because of the elastic properties of these tissues, lungs behave like a balloon during changes in volume. When the lungs are nearly empty, the elastic elements are not stretched; this is the resting volume. To increase volume (inflate), the elastic elements must be stretched. This is done by increasing pressure on the inside surface above pressure on the outside surface (ie, like blowing up a balloon). Pressure difference between the inside and outside of the lungs under static conditions (conditions of no airflow) is the **elastic pressure (Pel)**. The Pel required to stretch lung elastic elements also reflects the potential energy stored within the elastic elements as the result of inflation. Therefore, inflated lungs will recoil to their nearly empty resting volume (again - like a balloon) unless prevented from doing so. Elastic pressures needed to inflate the lung are characterized by volume-pressure curves, such as those illustrated in Fig. 4-2. **Lung compliance**, the change in lung volume per unit change in Pel ($\Delta V/\Delta P$), can be obtained from the individual curves as shown. Note that lung compliance is reduced at high volumes and is changed with lung diseases.

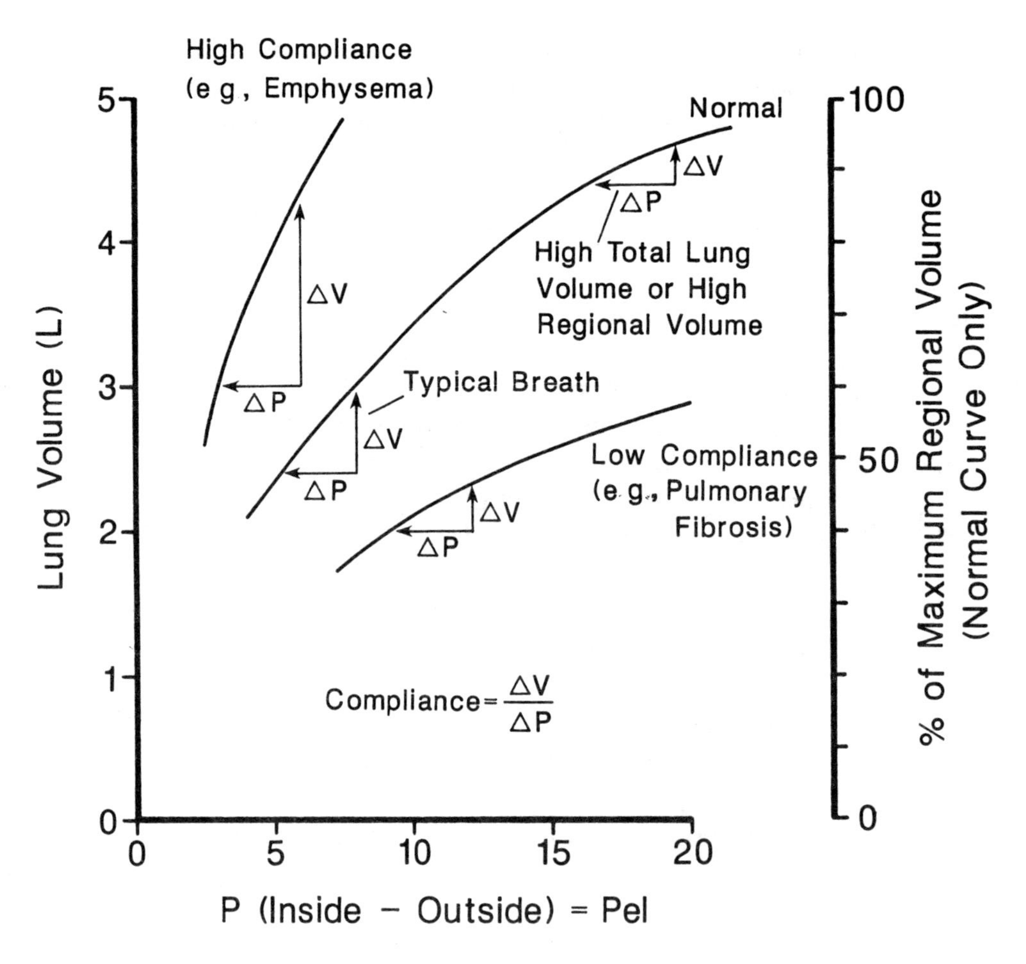

Figure 4-2. Volume-pressure curves for normal and abnormal lungs. Left ordinate shows lung volume; the abscissa represents the pressure difference between the inside and outside surface of the lungs (Pel). Each curve shows the change in volume (ΔV) for a given change in pressure (ΔP). To facilitate comparisons, ΔP in each case is 3 cm H_2O. Right ordinate shows percent of maximum regional lung volume and refers only to the curve for normal lungs. Lung tissue is relatively uniform, so volume-pressure characteristics of different parts or "regions" of the lung (eg, lung apices; lung bases) are similar.

Surface Tension in Alveoli

Besides **tissue elasticity, surface tension** also contributes to measured pressure differences between inside and outside surfaces of lungs. Surface tension results from the air-liquid interface in alveoli of lungs. The elevation in pressure on the inside of an alveolus with respect to the outside (P) due to surface tension (T) is inversely proportional to alveolar radius (r) according to the LaPlace relationship for a spherical surface (Equation 1):

EQUATION 1: $P = 2T/r$

For a constant surface tension the LaPlace relationship predicts that the elevation of pressure inside the alveolus will be greater for alveoli with **smaller** radii, and that small alveoli should collapse into the larger alveoli. This does not occur normally because a surface active agent, **pulmonary surfactant,** is secreted into alveoli and lowers surface tension. Most crucially, pulmonary surfactant has the property of lowering surface tension more in alveoli with small radii because of their smaller surface. The result is that **pressure due to surface tension does not differ** greatly in large and small alveoli. Failure of surfactant production and/or excessive surfactant breakdown occurs in infant and adult respiratory distress syndromes.

Lungs Volumes in the Intact Person

When we maximally inflate our lungs using our inspiratory muscles, the volume reached is called the **total lung capacity** (TLC). TLC is slightly larger in young adults than older individuals. The chest (thorax and diaphragm) is an elastic structure like the lungs. The chest is at its resting volume (the elastic elements are not stretched) when the lungs are at about 80% of TLC. As mentioned previously, the lungs are at their **resting volume** when nearly completely deflated. The outside surface of the lungs is **effectively** forced into contact with the inside surface of the chest (because of atmospheric pressure in the lungs), separated only by a very thin fluid-filled pleural space. When the airways are open after a passive expiration and all respiratory muscles are relaxed, the interaction between the elastic properties of the lungs and chest will cause the lungs to be inflated above their resting volume, while the chest is forced below its resting volume. The volume achieved by this interaction is the resting volume of the respiratory system and is called **functional residual capacity (FRC)**; for young adults in the seated position this volume is about 45% of TLC. The lungs cannot be deflated to their resting volume when the chest wall is intact; the volume of gas remaining in the lungs after a maximal expiration using our expiratory muscles (about 20% of TLC in young adults) is called the **residual volume (RV).** The residual volume increases with age.

Intrapleural Pressure. After a passive expiration to FRC the airways are normally open to the atmosphere, and pressure within alveoli is equal to barometric pressure. Because the lungs are inflated above their resting volume, the pressure in the pleural space must be lower than barometric pressure. Pressure in the pleural space minus barometric pressure is called **intrapleural pressure (Ppl).** Under the conditions described above for FRC, a typical value of Ppl is -5 cm H_2O and represents a (-) Pel of the lungs. The chest wall at FRC exerts an equal elastic pressure to expand.

Other Important Lung Volumes

- **Vital capacity (VC):** The volume between total lung capacity and residual volume (about 80% of TLC in young adults). VC is maximal in young adults and then decreases with age.
- **Tidal volume (V_T):** The volume of any breath (spontaneous or using a ventilator).
- **Inspiratory capacity (IC):** The difference between lung volume after expiration and TLC.
- **Inspiratory reserve volume (IRV):** The difference between lung volume after any tidal inspiration and TLC.
- **Expiratory reserve volume (ERV):** The difference between lung volume after any tidal expiration and residual volume.

Pressure Changes in the Respiratory System During Inspiration

Inspiration is usually an active process requiring contraction of the diaphragm; muscles of the chest wall assist the diaphragm to a variable degree. Changes in Ppl during inspiration are caused by resistive properties of airways and lung tissue as well as elastic properties of the lungs. In fact, Fig. 4-3, shows that Ppl is the algebraic sum of result of resistive pressures and (-) Pel. To overcome resistance of the airways to laminar and turbulent inspiratory airflow, pressure in the alveoli (representing the termination of the airways) must be lower than (barometric) pressure at the mouth. The major resistance to airflow occurs in medium-sized bronchi. Elastic pressure developed by the lungs as they increase in volume is measured as the pressure difference between the pleural space and alveoli; lung tissue resistance also makes a small contribution to this pressure difference. Note that Ppl is determined only by elastic properties of the lungs [(-)Pel] when there is no airflow and pressure in the alveoli is equal to atmospheric pressure at the mouth (eg, at the end of inspiration and at the end of expiration).

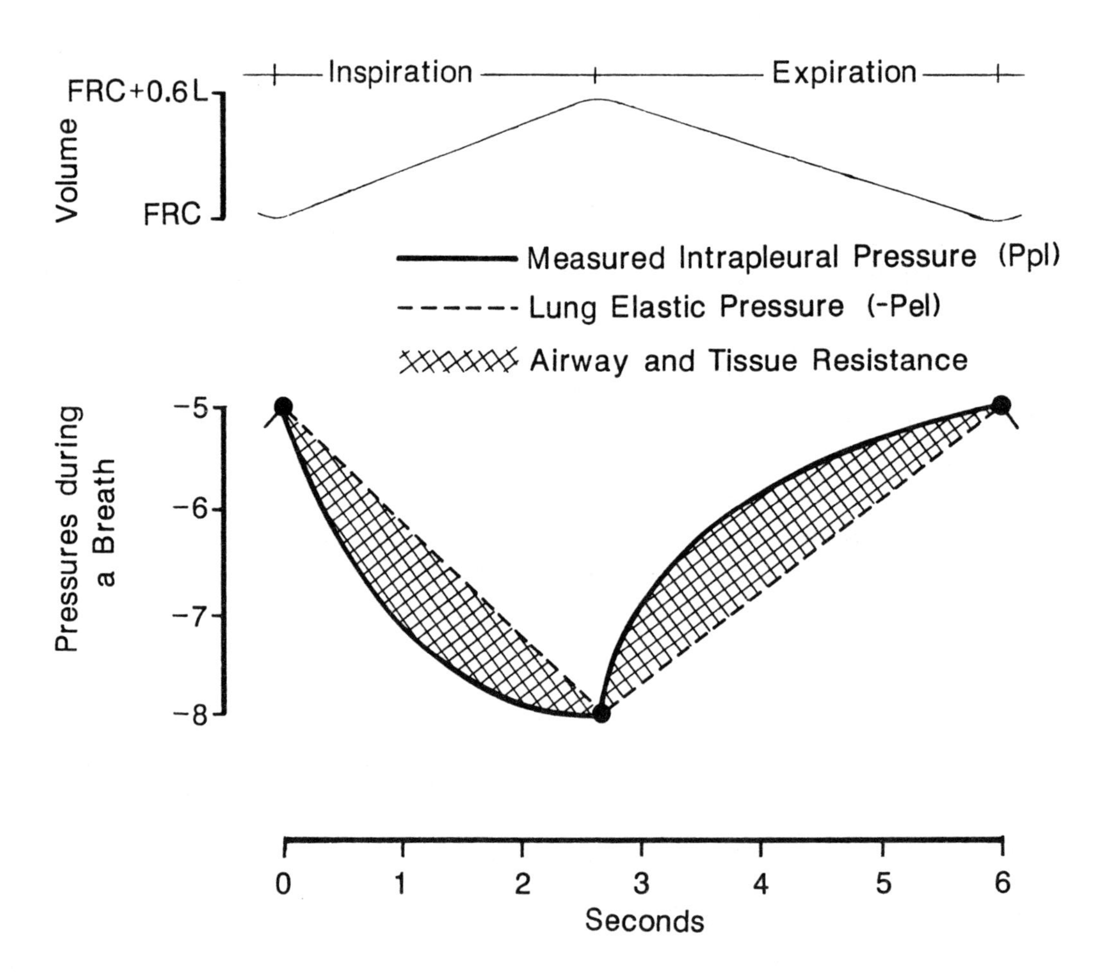

Figure 4-3. Changes in lung volume and intrapleural pressure (Ppl) during an inspired and expired breath. The components of Ppl due to elastic properties of the lungs and airway plus tissue resistance are shown.

Pressure Changes during Expiration

While inspiration normally requires muscular effort, the respiratory system can **passively recoil** back to FRC during expiration. To increase air flow during expiration, muscular effort (largely from abdominal muscles) can be used. During expiration pressure in the alveoli must be greater than pressure at the mouth because of airway resistance. This in turn affects Ppl, which becomes less negative than would be predicted on the basis of (-) Pel of the lungs. These pressure changes are summarized in Fig. 4-3. If expiratory effort is sufficiently high, Ppl will become "positive" (ie, pressure in the pleural space will rise above barometric pressure). If Ppl is "positive" during expiration, dynamic compression of airways may occur.

Dynamic compression of airways can be observed when expiration is performed with progressively increasing efforts. If a single lung volume is examined (within the lower 70% of vital capacity), airflow progressively increases with increasing expiratory effort at first. However, at some point increasing effort does not result in more airflow, because resistance of the airways is increased. This phenomenon, dynamic compression of the airways, is illustrated in Fig. 4-4. In this illustration pressure in the pleural space rises above atmospheric pressure because expiratory muscles are attempting to force air out of the lungs. Pressure within the alveoli is even higher than in the pleural space (because of lung elasticity), but resistance to expiratory airflow causes pressure to fall in the airways from the alveolus to the mouth. At some point, pressure inside the airway becomes less than in the pleural space (Fig. 4-4), and transmural pressure now squeezes the airways making them narrower. When lung volume is near the total lung capacity, **elastic pressure (Pel)** of the lungs is high (thereby producing a large reduction in pressure in the pleural space, with respect to the alveolar spaces and airways). Also, airway resistance is at a minimum value at high lung volumes because the airways tend to be stretched open. Both effects tend to keep pressure in the pleural space from rising above atmospheric pressure. Near total lung capacity, expiratory airflow is dependent on level of effort, and airway compression at high lung volumes does not occur in normal persons.

Other Aspects of Lung Mechanics

Forced Expired Vital Capacity (FVC). A forced expired vital capacity maneuver (an expiration from total lung capacity to residual volume using maximum effort) is often used to determine if the mechanical behavior of the respiratory system is normal. Patients with diseased lungs typically exhibit abnormal expiratory volumes and flows during a test of forced expired vital capacity. For example, if airway resistance is abnormally high or if lung compliance is high (Pel is low), dynamic compression of the airways and limitation of expiratory airflow will be emphasized. Reduced rigidity of airways may also augment compression. Diseases which decrease **rates** of airflow by increasing the resistance to airflow are termed **"obstructive"** (eg, emphysema, asthma). Diseases which decrease the compliance of either the lungs or chest wall will reduce the **volume** of the FVC, as will diseases that weaken respiratory muscles. These **"restrictive"** diseases (eg, pulmonary fibrosis, scoliosis) are not associated with reduced expiratory airflow (when airflow is normalized for the reduced expired volume) as long as airways are not obstructed. To differentiate between obstructive and restrictive disease states, two measurements are made; 1) the volume of the forced expired vital capacity expired in one second **(FEV_1)**, and 2) the FEV_1 as a percent of the FVC. For example, a purely obstructive disease would reduces both FEV_1 and FEV_1 as a percent of the FVC. On the other hand, a restrictive disease reduces

FVC and FEV_1 but does not reduce FEV_1 as a percent of the FVC. Normal adults can exhale at least 75% of their FVC in one second.

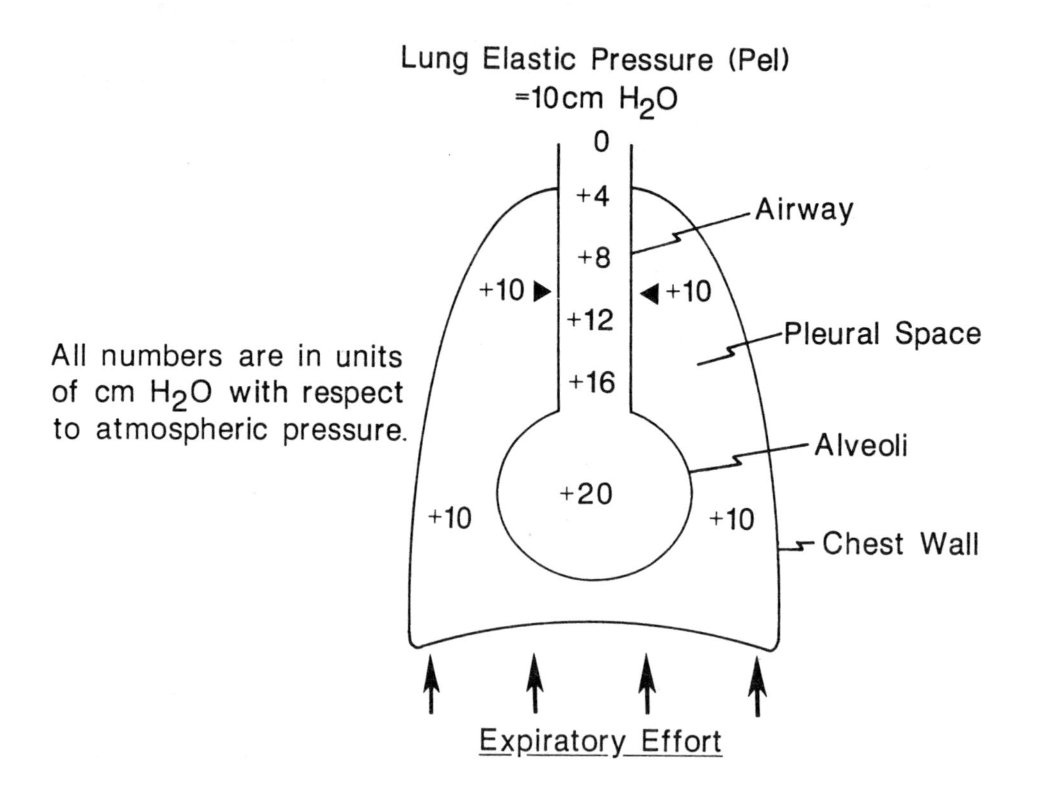

Figure 4-4. Pressures in the pleural space, alveoli, and airways during an expiration using muscular effort. At the arrows pressure in the airways begins to fall below pressure in the pleural space.

Distribution of Inspired Volumes Within the Lungs. When a breath is inhaled from the FRC, more of the breath will flow into the bases than apices of the lungs in a seated or standing subject. This occurs because gravity causes the weight of lung tissue below the apices to pull on the apical tissue. Since the lungs cannot normally separate from the inner surface of the chest, the apical elastic elements are stretched. At the FRC the gas volume contained in the apices will be relatively high in comparison with volumes at the bases of the lungs. Furthermore, because apical lung volume is high, elastic pressure is also high; this will cause the apices of the lungs to be on the relatively flat portions of their "regional" volume-pressure curves (Fig. 4-2). Since the lung bases are not as inflated at FRC, they are on a steeper (more compliant) portion of their "regional" volume-pressure curves (Fig. 4-2) and receive the larger portion of a breath taken from FRC.

The pulmonary circulation is described in Chapter 3 on p. 112.

Distribution of blood flow within the lungs. Like the inspired breath blood flow is not distributed uniformly throughout the lungs, and this effect is also due to gravity. For at least part of the cardiac cycle the hydrostatic column of blood may not reach the apices of the lungs in some seated or standing individuals, because pulmonary arterial pressure (eg, about 10 cm H_2O at diastole) is inadequate to pump blood up to that region. Lower down in the lungs arterial pressure is sufficient to perfuse capillaries, but pressure in alveoli is higher than pressure in the left atrium. This can occur when the left atrium is below the region being perfused. Consequently, driving pressure across the pulmonary capillary bed is the difference between arterial pressure (the amount in excess of that required to lift the blood above the heart) and alveolar pressure. Lower still in the lungs, the alveolar pressure is exceeded by hydrostatic pressure in the left atrium, and the driving pressure across the pulmonary capillary bed is from pulmonary artery to pulmonary veins. This region of the lungs receives the highest blood flow, because pulmonary arterial perfusion pressure is not dissipated by "lifting" the blood. Blood flow increases from the apices to bases of the lungs as a result of these mechanisms.

Review Questions

3. At the functional residual capacity (FRC)

   A. the chest wall exerts recoil toward lower volume
   B. the lungs exert no recoil and are therefore at their resting volume
   C. pressure outside the lungs (in the pleural space) is greater than pressure within the alveoli
   D. pressure in the pleural space is equal to pressure in the alveolar spaces
   E. the lungs exert recoil toward lower volume

4. In a premature infant without pulmonary surfactant

   A. surface tension of alveoli will be less than normal
   B. a greater than normal pressure difference between the inside and outside of the lungs will be required for lung inflation
   C. small alveoli will have higher surface tension than large alveoli
   D. pressures due to surface tension in large and small alveoli will be equal
   E. The lungs can be inflated with normal muscular effort

5. During inspiration

   A. barometric pressure (at the mouth) must be lower than pressure in alveoli
   B. intrapleural pressure reflects lung elasticity, airway resistance, and lung tissue resistance
   C. pressure in the alveoli must be greater than barometric pressure
   D. air movement is usually accomplished by the passive elastic properties of the respiratory system and not by muscular effort
   E. lung volumes must always be above FRC

6. During expiration

   A. airway compression may occur if pressure in the airways decreases to a value less than the pressure in the pleural space
   B. airway compression will most likely occur at high lung volumes
   C. pressures due to recoil properties of the lungs are greatest at low lung volumes
   D. airway resistance is highest at high lung volumes
   E. muscular effort is required for airflow

7. A patient with a normal vital capacity expires only 40% of his forced expired vital capacity in one second. This suggests

   A. abnormally low lung compliance
   B. restrictive disease
   C. that airflow limitation due to airway compression is less than normal
   D. weak inspiratory muscles
   E. obstructive disease

GAS EXCHANGE BETWEEN ALVEOLAR GAS AND ALVEOLAR CAPILLARY BLOOD

The transfer of O_2 and CO_2 between alveolar gas and alveolar capillary blood depends in part upon gas diffusion through lung tissue and blood. Rate of diffusion for a given gas is directly proportional to the surface area for diffusion and inversely proportional to the thickness of the diffusion barrier. In addition, the diffusion rate is proportional to the partial pressure difference and solubility coefficient for each particular gas. Finally, rate of diffusion is inversely related to the square root of the molecular weight of the gas. These factors are summarized in the relationship below.

$$\text{Rate of gas diffusion} \propto \frac{\text{Area x Solubility x } \Delta P}{\text{Thickness x } \sqrt{\text{Molecular Weight}}}$$

In the lungs the surface area for diffusion through tissue is very large (approximately 50-100 square meters), and the thickness of the alveolar epithelium and the alveolar capillary endothelium is very small (about one micron). Diffusion also occurs within the gas phase of the lungs, but it is about 10,000 times more rapid over a given distance than diffusion through tissue. It takes about **3/4 sec** for blood to pass through a typical alveolar capillary under resting conditions; the transfer of O_2 and CO_2 is completed in about **one-third** of this time. Consequently, partial pressures of O_2 and CO_2 in blood leaving an alveolus are equal to those in alveolar gas. During maximum exercise the blood spends much less time in alveolar capillaries, and equilibration of O_2 between blood and gas may **not** be achieved.

The rapid transfer by diffusion of CO_2 between blood and gas is aided by the **high solubility** coefficient for CO_2 in plasma and lung tissue (about 20 times higher than for O_2). Oxygen transfer from alveolar gas into capillary blood is aided by the characteristics of O_2-hemoglobin interactions in blood. Oxygen is loaded into venous blood on the steep part of the O_2 dissociation curve (Fig. 4-1). This maintains a high partial pressure difference to maximize O_2 diffusion into the blood.

Diffusing capacity is the volume of a given gas transferred per minute for each mm Hg partial pressure difference between alveolar gas and alveolar capillary blood. Its value depends upon three factors; 1) the characteristics of diffusion across the alveolar membrane and alveolar capillary, 2) the reaction rate of the gas with hemoglobin, and 3) capillary volume. Clinically, this measurement is most conveniently made using traces of carbon monoxide (CO), because the partial pressure of CO in blood can usually be assumed to equal zero (large amounts of CO are bound to hemoglobin at very low partial pressure), and the gradient for diffusion of CO from alveoli into blood can be approximated from alveolar partial pressures.

Distribution of Ventilation and Blood Flow

Both alveolar ventilation ($\dot{V}_A$) and alveolar capillary blood flow ($\dot{Q}_C$) increase from apices (top) to bases of the lungs in a normal seated or standing person. However, the ratio, $\dot{V}_A/\dot{Q}_C$, is not constant. Absolute levels (L/min) of $\dot{Q}_C$ are less than absolute levels (L/min) of $\dot{V}_A$ at the lung apices, while $\dot{Q}_C$ at the lung bases is greater than $\dot{V}_A$. Therefore, the ratio, $\dot{V}_A/\dot{Q}_C$, is high at the lung apices and low at the lung bases, as shown in the third column of Table 4-1. Differences in $\dot{V}_A/\dot{Q}_C$ affect gas composition and gas exchange in the lungs.

Table 4-1. Ventilation, blood flow, gas partial pressures, gas uptake, and respiratory exchange ratio as a function of lung region.

Lung Region	$\dot{V}_A$ (L/min)	$\dot{Q}_C$ (L/min)	$\dot{V}_A/\dot{Q}_C$	P_{O_2} (mm Hg)	P_{CO_2} (mm Hg)	$\dot{V}_{O_2}$ (ml/min)	$\dot{V}_{CO_2}$ (ml/min)	R
Apex	0.24	0.07	3.3	132	28	4	8	2.0
Base	0.82	1.29	0.64	89	42	60	39	0.65

Effects of $\dot{V}_A/\dot{Q}_C$ on Gas Composition

An alveolus with ventilation but no blood flow has an alveolar gas composition unchanged from that of the inspired air. The $\dot{V}_A/\dot{Q}_C$ of such an alveolus is infinity. The opposite extreme is an alveolus with $\dot{Q}_C$ but no $\dot{V}_A$; it equilibrates with venous blood entering the lungs. The $\dot{V}_A/\dot{Q}_C$ is zero. For alveoli with finite values of $\dot{V}_A/\dot{Q}_C$, those with high $\dot{V}_A/\dot{Q}_C$ have a composition closer to that of inspired air (ie, high P_{O_2} and low P_{CO_2}), and those with low $\dot{V}_A/\dot{Q}_C$ have a composition closer to that of venous blood (ie, low P_{O_2} and high P_{CO_2}). Since $\dot{V}_A/\dot{Q}_C$ is relatively high at the lung apices and low at the lung bases, alveolar P_{O_2} (or P_{O_2} of blood leaving alveolar capillaries) is high at the lung apices and lower at the lung bases. Similarly, because $\dot{V}_A/\dot{Q}_C$ is relatively low at the lung bases, higher values of alveolar P_{CO_2} are observed at the lung bases than at the lung apices. These effects are shown by the values listed in Table 4-1.

Effects of $\dot{V}_A/\dot{Q}_C$ on Uptake of O_2 and Elimination of CO_2

Lung regions with different $\dot{V}_A/\dot{Q}_C$ values do not exchange O_2 and CO_2 with equal efficiency. The **uptake of** O_2 ($\dot{V}_{O_2}$) by alveolar capillaries depends more upon $\dot{Q}_C$ than $\dot{V}_A$, but the **elimination of** CO_2 ($\dot{V}_{CO_2}$) from blood into alveolar gas depends more on $\dot{V}_A$ than $\dot{Q}_C$. These effects are caused by the different dissociation curves for O_2 and CO_2 in blood.

For O_2 all high $\dot{V}_A/\dot{Q}_C$ regions and all but the lowest $\dot{V}_A/\dot{Q}_C$ regions yield P_{O_2} levels (in capillary blood leaving those regions) on the flat part of the O_2 dissociation curve (Fig. 4-1).

$\dot{V}_{O_2}$ can be measured as

$$\text{EQUATION 2:} \quad \dot{V}_{O_2} = \dot{Q}_C \times (C_{CO_2} - C_{\bar{v}O_2})$$

where $(C_{CO_2} - C_{\bar{v}O_2})$ is the difference in O_2 content between blood entering the lungs ($C_{\bar{v}O_2}$) and the arterialized blood in the alveolar capillaries (C_{CO_2}). This is like the Fick equation to calculate blood flow described on p. 90. For alveolar capillary P_{O_2} values on the flat portion of the O_2 dissociation curve, C_{CO_2} is nearly constant as is $(C_{cCO_2} - C_{\bar{v}O_2})$, and $\dot{V}_{O_2}$ will vary directly with $\dot{Q}_C$.

In contrast, in the physiological range there are nearly proportional changes between P_{CO_2} and CO_2 content in blood. $\dot{V}_{CO_2}$ can be assessed as:

$$\text{EQUATION 3:} \quad \dot{V}_{CO_2} = \dot{Q}_C \times (C_{\bar{v}CO_2} - C_{cCO_2})$$

where, $(C_{\bar{v}CO_2} - C_{cCO_2})$ is the CO_2 content difference between venous blood entering the lungs ($C_{\bar{v}CO_2}$) and blood leaving alveolar capillaries (C_{cCO_2}). A high $\dot{V}_A/\dot{Q}_C$ is associated with a low P_{CO_2} and a corresponding low C_{cCO_2}. The reverse occurs for low $\dot{V}_A/\dot{Q}_C$ regions. Then (read this slowly and carefully) when $\dot{V}_A/\dot{Q}_C$ is raised by decreasing $\dot{Q}_C$, the decrease in $\dot{Q}_C$ will be (more or less) balanced by the fall in C_{cCO_2}, and the effect on $\dot{V}_{CO_2}$ will be small. However, if $\dot{V}_A/\dot{Q}_C$ is increased by raising $\dot{V}_A$ ($\dot{Q}_C$ is unchanged), then Equation 3 shows that $\dot{V}_{CO_2}$ will increase because $(C_{\bar{v}CO_2} - C_{cCO_2})$ has increased.

The relationship between $\dot{V}_{O_2}$ and $\dot{V}_{CO_2}$ for any portion of the lungs, or for the lungs as a whole, is characterized by the respiratory exchange ratio, R.

$$\text{EQUATION 4:} \quad R = \dot{V}_{CO_2} / \dot{V}_{O_2}$$

[Note that the similar term RQ, respiratory quotient, refers solely to metabolically consumed O_2 and metabolically produced CO_2]

Because $\dot{V}_{CO_2}$ depends more on $\dot{V}_A$, and $\dot{V}_{O_2}$ depends more on $\dot{Q}_C$, the high $\dot{V}_A/\dot{Q}_C$ at the lung apices is associated with relatively greater output of CO_2 than uptake of O_2; R is high. Conversely, the low $\dot{V}_A/\dot{Q}$ at the bases of the lungs represents a large blood flow relative to ventilation. Since $\dot{V}_{O_2}$ is greatly dependent on $\dot{Q}_C$, $\dot{V}_{O_2}$ will be larger than $\dot{V}_{CO_2}$, and R will be low. These effects are summarized in Table 4-1, which also shows that absolute levels of $\dot{V}_{O_2}$ and $\dot{V}_{CO_2}$ at the bases of lungs are greater than at the apices. This occurs because both $\dot{V}_A$ and $\dot{Q}_C$ are much greater at the bases of the lungs.

Factors Affecting the Efficiency of Pulmonary Gas Exchange

1. **Dead Space.** The conducting airways of the lungs do not participate in gas exchange. About 25-35% of each tidal volume under resting conditions remains in the airways and is not delivered to the alveoli of the lungs. The volume of the airways is called the **anatomical dead space.** Thus, the tidal volume (V_T) is divided into a volume ventilating alveoli (V_A) and a volume ventilating the dead space (V_D), so that

$$\text{EQUATION 5:} \quad V_T = V_A + V_D$$

Multiplying the above values by the number of breaths per minute (n) gives values for the **minute ventilation** of the lungs (V_T x n), **alveolar ventilation** (V_A x n), and **dead space ventilation** (V_D x n). Tidal volume can be increased 5 to 6-fold from resting levels, but anatomical dead space will increase by only about 75%. Therefore, increasing tidal volume puts most of the extra inspired gas into the alveoli. On the other hand, **increasing the number of breaths** per minute proportionally increases both dead space and alveolar ventilation (ie, doubling the number of breaths per minute will double the alveolar and dead space ventilation).

Alveoli that are ventilated but have little or no blood flow are another source of dead space. Air is inspired into these alveoli but little or no gas exchange occurs. This **alveolar dead space** volume is negligible in normal persons but may be appreciable in patients with diseased lungs. Anatomical dead space plus alveolar dead space is called **physiological dead space.**

There are several clinical approaches to the calculation or estimation of dead space. A common measurement is the ratio of physiological dead space to tidal volume (V_D/V_T). It requires measurement of P_{CO_2} in arterial blood (P_{aCO_2}) and in mixed expired gas (P_{ECO_2}).

$$\text{EQUATION 6:} \quad \frac{V_D}{V_T} = \frac{P_{aCO_2} - P_{ECO_2}}{P_{aCO_2}}$$

Equation 6 is based on the following assumptions: 1) P_{aCO_2} is the same as the CO_2 level from all alveoli of the lungs, except those alveoli constituting alveolar dead space, and 2) P_{ECO_2} reflects the dilution of CO_2 in alveolar gas by anatomical as well as alveolar dead space.

2. **Anatomic Shunt.** Even in normal persons about 2% of cardiac output bypasses alveolar gas exchange; this blood constitutes the anatomical shunt. Since shunted blood has the compositon of mixed venous blood, its P_{O_2} and O_2 content are lower than in blood from alveoli with gas exchange. Since P_{O_2} in blood from alveoli with gas exchange is considered to be in equilibrium with alveolar gas, the addition of shunted blood will reduce the P_{O_2} in systemic arterial blood to values below those found in alveolar gas (an alveolar to arterial P_{O_2} difference). In normal people this difference is about 6 mm Hg. If shunting is greater than normal, P_{O_2} and O_2 content in arterial blood is further reduced. Shunting also raises the systemic arterial P_{CO_2}, but the difference between arterial and venous P_{CO_2} is small, so the effect is minor. Clinically significant shunts can occur, for example, with cardiac septal defects, adult respiratory distress, and intrapulmonary arterial-venous anatomoses.

3. **Differences in $\dot{V}_A/\dot{Q}_C$.** A normal $\dot{V}_A/\dot{Q}_C$ for the lungs as a whole is about 0.85. Therefore, alveoli with $\dot{V}_A/\dot{Q}_C$ values higher than 0.85 contribute relatively more to total $\dot{V}_A$ than to total $\dot{Q}_C$, while alveoli with $\dot{V}_A/\dot{Q}_C$ values lower than 0.85 contribute relatively more to total $\dot{Q}_C$ than to $\dot{V}_A$. High $\dot{V}_A/\dot{Q}_C$ alveoli have high P_{O_2} values and low P_{CO_2} values, while low $\dot{V}_A/\dot{Q}_C$ alveoli have low P_{O_2} and high P_{CO_2} values. Therefore, P_{O_2} will be **lower in the blood** leaving the lungs than the mixed alveolar gas leaving the lungs. In addition, P_{CO_2} will be **lower in the alveolar gas** leaving the lungs than in the blood leaving the lungs. Figure 4-5 summarizes these effects in a two compartment model of the lungs; one compartment has a high $\dot{V}_A/\dot{Q}_C$, and the other has a low $\dot{V}_A/\dot{Q}_C$.

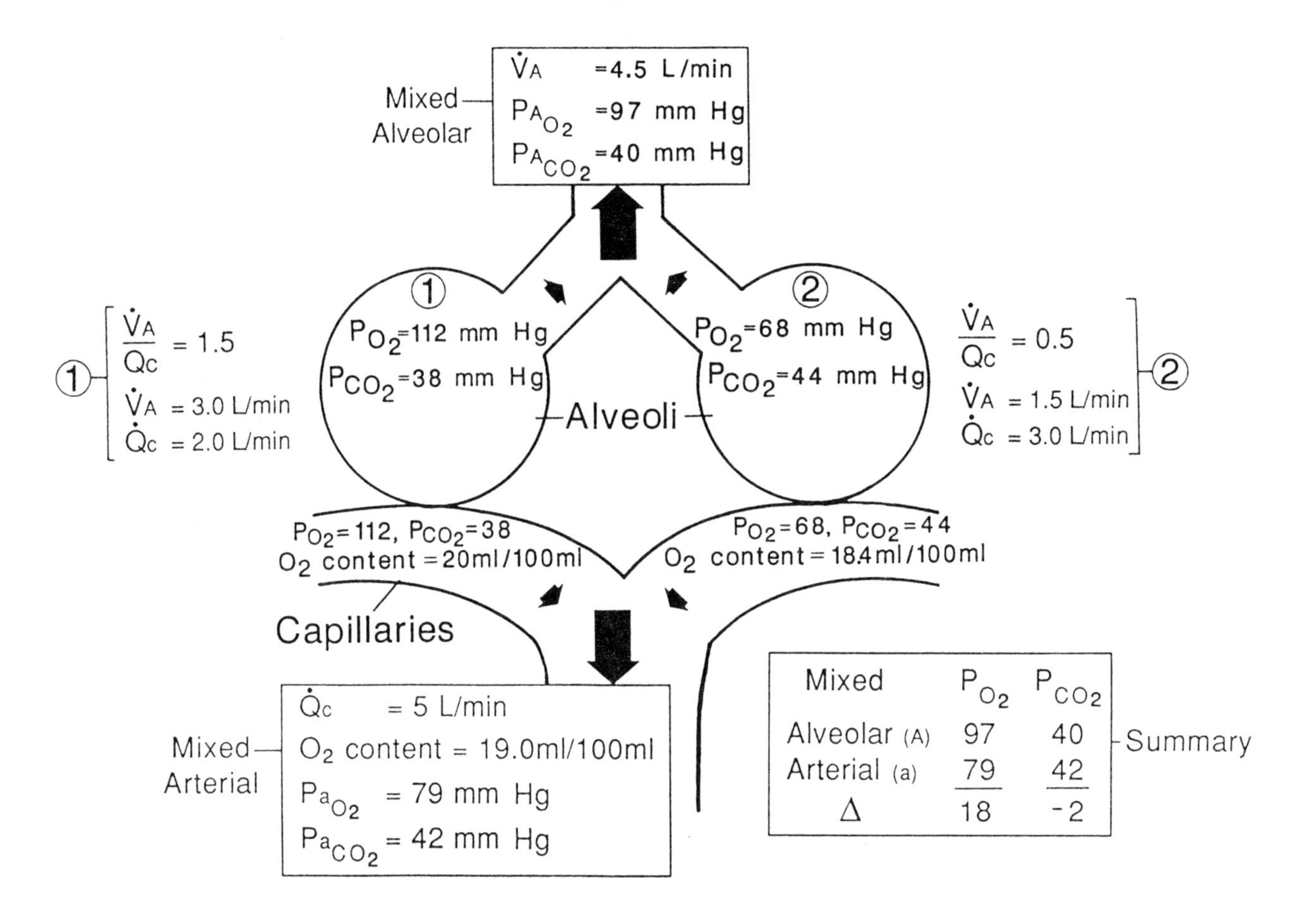

Mixed	P_{O_2}	P_{CO_2}
Alveolar (A)	97	40
Arterial (a)	79	42
Δ	18	-2

Figure 4-5. A two compartment model of lungs with abnormally uneven distribution of $\dot{V}_A/\dot{Q}_C$. Compartment 1 has a high $\dot{V}_A/\dot{Q}_C$ (1.5), while compartment 2 has a low $\dot{V}_A/\dot{Q}_C$ (0.5). Absolute values of $\dot{V}_A$ and $\dot{Q}_C$ for each compartment are shown on the figure. P_{O_2} and P_{CO_2} of alveolar gas in each compartment is in equilibrium with the values in blood leaving the alveolar capillaries. To obtain mixed alveolar P_{O_2} and P_{CO_2} (the mixture of alveolar gas expired from each compartment), a weighted mean of P_{O_2} and P_{CO_2} is calculated based on the amount that each compartment contributes to the mixed alveolar gas. For mixed arterial P_{O_2} and P_{CO_2} (the mixture of blood flowing from each compartment), the weighted means of O_2 and CO_2 content are calculated and then converted to partial pressures through their respective dissociation curves. The result is a relatively large alveolar to arterial P_{O_2} difference and a small alveolar to arterial P_{CO_2} difference.

In normal persons uneven $\dot{V}_A/\dot{Q}_C$ produces an alveolar to arterial P_{O_2} difference of about 4 mm Hg (combined with the effects of shunting, the normal total alveolar to arterial P_{O_2} difference is about 10 mm Hg). The normal difference between P_{CO_2} in alveolar gas and arterial blood is small and difficult to detect. In some lung diseases (eg, emphysema, chronic bronchitis) alveolar to arterial P_{O_2} differences due to uneven distribution of $\dot{V}_A/\dot{Q}_C$ throughout the lungs may be very large. In normal aging increased uneveness of gas exchange accounts in part for reductions in arterial P_{O_2} and increases in the alveolar to arterial P_{O_2} difference.

4. **Diffusion Defect.** In theory thickened alveolar membranes would slow diffusion of O_2 sufficiently to prevent equilibration of P_{O_2} in blood and gas. This would cause an alveolar to arterial P_{O_2} difference. However, under resting conditions in normal persons equilibration between blood and gas occurs in the first third of the transit of blood through the alveolar capillary, so there is a large **safety margin.** An alveolar to arterial P_{O_2} difference due to diffusion problems in diseased lungs is more likely at **high altitude** where inspired P_{O_2} is reduced, or during **exercise** where the transit time of O_2 through alveolar capillaries is reduced.

5. **Effects of O_2 Breathing.** One way that anatomical shunts can be differentiated from other causes of an alveolar to arterial P_{O_2} difference is to have patients breathe 100% O_2. Even in lungs with uneven $\dot{V}_A/\dot{Q}$or diffusion defect, P_{O_2} in end-capillary will rise to near inspired values (>600 mm Hg at sea level). With anatomical shunts, blood with venous composition mixes with end-capillary blood from alveoli, dramatically lower the P_{O_2} of arterial blood.

6. **Hypoventilation.** When $\dot{V}_A$ (for the lungs as a whole) is reduced from normal values, levels of CO_2 are increased in alveolar gas. This effect is shown by the following relationship:

$$\text{EQUATION 7:} \quad \dot{V}_{CO_2} = \frac{\dot{V}_A \times P_{ACO_2}}{K}$$

where $\dot{V}_{CO_2}$ is the output of CO_2 from the lungs; $\dot{V}_A$ is alveolar ventilation; P_{ACO_2} is the partial pressure of CO_2 in alveolar gas; and K is a constant. Equation 7 shows that if P_{ACO_2} is initially equal to 40 mm Hg (a normal value), then decreasing $\dot{V}_A$ to 1/2 of its initial value [with metabolism ($\dot{V}_{CO_2}$) remaining constant] will cause P_{ACO_2} to double (ie, P_{ACO_2} = 80 mm Hg). Besides increasing P_{ACO_2}, hypoventilation will reduce alveolar P_{O_2} (P_{AO_2}) as predicted by the alveolar gas equation, which can be approximated as follows:

$$\text{EQUATION 8:} \quad P_{AO_2} = P_{IO_2} - \frac{P_{ACO_2}}{R}$$

where P_{IO_2} is the inspired partial pressure of O_2. Alveolar levels of O_2 are largely determined by the amount of CO_2 which diffuses into the alveolar gas and the level of CO_2 achieved (P_{ACO_2}). It is also influenced by the relationship between the amount of CO_2 entering the lungs and the amount of O_2 leaving the lungs in the alveolar capillaries (R). Thus, with an R=1, a 40 mm Hg rise in CO_2 will exactly match a 40 mm Hg fall in the original inspired O_2, so that P_{AO_2} = 110 mm Hg. When R has a more typical value of 0.8, alveolar P_{O_2} is 100 mm Hg when P_{ACO_2} is 40 mm Hg. The above relationship is only approximate, because it does not account for differences in inspired and expired volumes when R is not equal to 1. The alveolar gas equation is more precisely expressed as

$$\text{EQUATION 9:} \quad P_{AO_2} = P_{IO_2} - \frac{P_{ACO_2}}{R} + \left[P_{ACO_2} \times F_{IO_2} \times \frac{(1-R)}{R} \right]$$

The last term [in brackets] corrects for changes in gas volume caused by R; F_{IO_2} is the fraction of inspired O_2. In summary, at a constant metabolic rate ($\dot{V}_{O_2}$, $\dot{V}_{CO_2}$) a reduction in $\dot{V}_A$ will cause a rise in P_{ACO_2} and a fall in P_{AO_2}. The relationship between these changes should be calculated for accuracy, but they are of similar magnitudes. Since arterial P_{O_2} and P_{CO_2} values are largely determined by alveolar levels, hypoventilation increases arterial P_{CO_2} while reducing arterial P_{O_2}. Hypoventilation by itself does not increase the alveolar to arterial P_{O_2} difference. Hypoventilation can occur with severe obstructive lung disease or as a result of drugs which depress breathing (eg, morphine-like substances).

Review Questions

8. If the respiratory exchange ratio (R) for alveolus A is greater than for alveolus B in the same lung, then

   A. the $\dot{V}_A/\dot{Q}_C$ of alveolus A is greater than $\dot{V}_A/\dot{Q}_C$ of alveolus B
   B. blood flow to alveolus A is greater than blood flow to alveolus B
   C. ventilation of alveolus A is greater than ventilation of alveolus B
   D. P_{O_2} in alveolus A is lower than P_{O_2} in alveolus B
   E. P_{CO_2} in alveolus A is higher than P_{CO_2} in alveolus B

9. Physiological dead space volume will be increased by each of the following **EXCEPT**

   A. increases in volume of the alveolar dead space
   B. large increases in tidal volume
   C. increases in breathing frequency
   D. increases in volume of the anatomical dead space
   E. breathing through a tube (eg, a snorkel)

10. A decrease in arterial P_{O_2} with chronic bronchitis is most commonly associated with

   A. an increase in the $\dot{V}_A/\dot{Q}_C$ for all regions of the lungs
   B. abnormally uneven $\dot{V}_A/\dot{Q}_C$
   C. a decrease in shunting of blood from the right to left side of the circulation
   D. a higher P_{O_2} in arterial blood than in alveolar gas

11. Two regions of the lungs (A and B) have the same alveolar ventilation, but region A has twice the blood flow of region B. These data suggest which of the following?

A. Oxygen uptake and CO_2 elimination from the two regions are similar.
B. Region A may have nearly twice the O_2 uptake as region B, but only one half the CO_2 elimination.
C. Region A may have nearly twice the O_2 uptake as region B, and a lower respiratory exchange ratio.
D. Region A may have about twice the CO_2 elimination as region B, but a similar O_2 uptake.

12. An alveolar to arterial P_{O_2} difference will be most likely to increase when

A. alveolar ventilation is reduced by a drug overdose
B. shunting of blood from the right to left side of the circulation is increased when a lobe of the lung fills with fluid (pneumonia)
C. diffusing capacity decreases because one lobe of a lung is removed
D. breathing through a tracheostomy
E. $\dot{V}_A$ and $\dot{Q}_C$ in all parts of the lung are doubled

13. For a "typical" alveolar capillary

A. P_{O_2} increases to the level in the alveolar gas
B. P_{CO_2} normally remains at the same level as blood entering the alveolar capillary
C. the partial pressure differences for O_2 and CO_2 between blood entering and leaving the capillary are nearly the same
D. diffusion rate for O_2 (per mm Hg partial pressure difference) is about 20 times greater than for CO_2

CONTROL OF BREATHING

Neural Organization

Medullary Centers. Breathing, the rhythmic activity of respiratory muscles, originates from neural discharge patterns in the **medullary reticular formation.** Neurons that discharge in phase with breathing are found in two major groupings. The dorsal respiratory group of cells is in the region of the nucleus tractus solitarius, while the ventral respiratory group is a long column of cells in the ventrolateral medulla. This column starts rostral to nucleus ambiguus and continues caudally through and behind this nucleus. Other concentrations of respiratory neurons have been described in the rostral portion of the ventrolateral medulla. The details of how these neuronal groups generate a rhythmic breathing pattern is not completely understood, but some aspects of their function can be suggested. 1) Some groups of cells that discharge during inspiration or expiration are organized into **self-exciting networks** (ie, stimulation of one inspiratory-phased neuron causes excitation of other medullary inspiratory neurons). 2) Stimulation of inspiratory-phased neurons can result in **inhibition** of expiratory-phased cells, and less often, the activation of expiratory-phased neurons can result in inhibition of inspiratory cells. 3) The termination of discharge in the inspiratory neural network may involve a reduction in neuronal excitability and/or specific neurons which act as an inspiration **off-switch.** Whether pacemaker-like cells in the medulla (eg, as in the SA node of the heart) participate in generating breathing rhythm is disputed.

Pontine Centers. Two regions in the pons modify the respiratory rhythm set by the medulla. Some investigators believe that these pontine regions may be required for production of relatively normal breathing patterns. In the caudal pons there is a region classically called the **apneustic center** which functions to **prolong** the act of inspiration. The **pneumotaxic center**, located in the rostral pons (associated with the medial portions of the parabrachial nuclei), can act to **terminate** inspiration. The vagus nerves also carry information to the central nervous system to terminate inspiration. An experimental animal with the brainstem transected between the apneustic and pneumotaxic centers and the vagi cut will show prolonged inspiratory efforts, only occasionally interrupted by expiration (apneustic breathing).

Receptors in the Lungs and Airways

Three types of sensory receptors have afferent fibers in vagus nerves.

1. **Slowly Adapting Airway Receptors.** These receptors are concentrated in medium to large size airways, but some are found in more peripheral sites including lung parenchyma; their afferent fibers are myelinated. They increase their discharge rates as lung volume is increased, and their discharge shows little adaptation if lung volume is maintained constant. Increasing rates of lung inflation cause a further transient stimulation of these receptors. Their effects on the control of breathing include **decreasing duration** and **volume** of inspired breaths, as well as **prolonging** expired breaths. These responses may be significant in adults only with elevated tidal volumes (eg, during exercise) and possibly during sleep and anesthesia. These reflexes are much more prominent in newborn babies.

2. **Rapidly Adapting Airway Receptors.** These receptors are associated with epithelium in all but the smallest conducting airways; their afferent fibers are myelinated. They increase their discharge rate **during** changes in lung volume and are sensitive to several chemical irritants. They are also stimulated by lung congestion. These receptors may influence breathing patterns under normal conditions, but they may have more important effects in cardiopulmonary diseases. Rapid breathing rates have been associated with stimulation of these receptors. They may also help to elicit spontaneous deep breaths.

3. **J-receptors.** These receptors are associated with unmyelinated fibers and are sensitive to a variety of irritants as well as congestion of the lungs. J-receptor stimulation may contribute to breathing patterns observed in cardiopulmonary diseases.

Chemical Regulation of Breathing

Levels of arterial P_{CO_2} are remarkably stable. This stability is due in part to receptor mechanisms that detect small alterations in blood CO_2 levels and produce an appropriate ventilatory response. For example, raising arterial P_{CO_2} causes an increase in ventilation that brings P_{CO_2} down toward normal values. The left side of Figure 4-6 shows how breathing increases when CO_2 levels are artificially raised in a normal subject by inhaling gas mixtures containing CO_2. People also breathe more when arterial P_{O_2} values are markedly reduced from normal values near sea level by inhaling mixtures with low levels of O_2 (right side of Fig. 4-6). However, little change in breathing is observed with small decreases from normal values of P_{O_2} in arterial blood. Consequently, small changes from normal CO_2 levels are much more important for regulation of breathing than are small changes from normal O_2 levels.

Peripheral Arterial Chemoreceptors. The primary receptors for detection of reduced levels of O_2 in arterial blood are the **carotid bodies**. The **aortic bodies** also respond to reductions in blood O_2 but have little effect on breathing. The carotid bodies respond to reductions in P_{O_2} rather than reductions in O_2 content. In addition, these receptors can be stimulated by increases in arterial P_{CO_2} or concentration of H^+ ions.

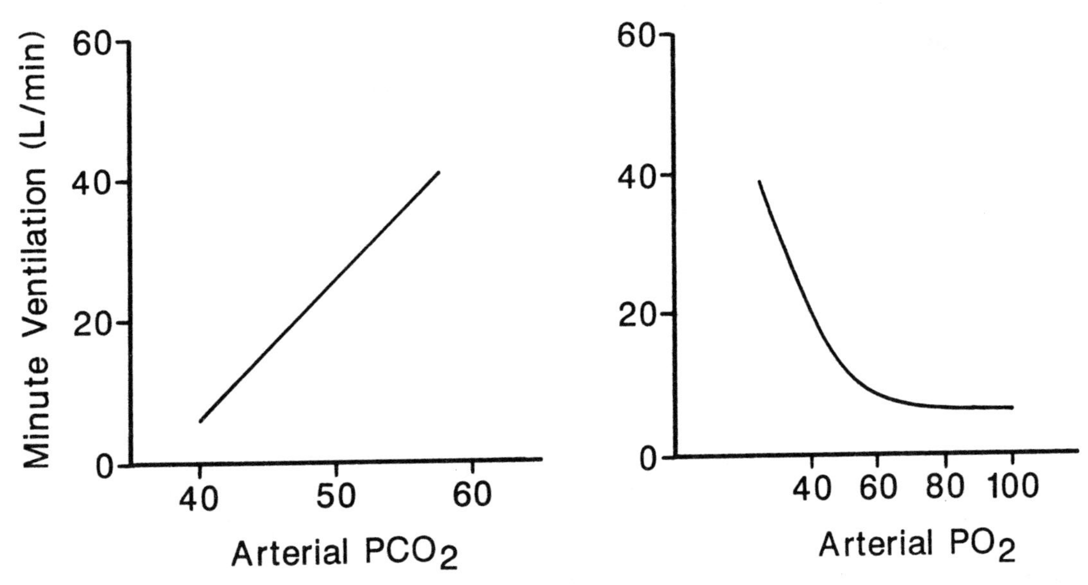

Figure 4-6. Effects on breathing when arterial P_{CO_2} is artificially raised (left panel) or arterial P_{O_2} is artifically lowered (right panel).

The highest discharge rates of carotid body afferent fibers occur when arterial P_{O_2} is reduced substantially below normal levels. Their discharge as a function of P_{O_2} has the same form as the relationship between ventilation and P_{O_2} (Fig. 4-6). There is little change in carotid body discharge when arterial P_{O_2} rises or falls by small amounts from the normal arterial P_{O_2}. Raising blood P_{CO_2} or the concentration of H^+ ions increases sensitivity to hypoxia. In addition, increased acidity or P_{CO_2} in arterial blood will increase carotid body discharge, even if P_{O_2} is normal.

Central Chemoreceptors. While the peripheral arterial chemoreceptors can induce an increase in breathing in response to a rise in arterial P_{CO_2}, stimulation of **medullary chemoreceptors** elicits a larger portion of the ventilatory response to elevated levels of CO_2. These chemoreceptors are probably located near the ventral surface of the medulla and are exposed to brain interstitial fluid. They are separated from the circulation by the blood-brain barrier. When P_{CO_2} is increased in cerebral capillaries, molecular CO_2

crosses the blood-brain barrier and increases the concentration of physically-dissolved CO_2 in interstitial fluid. A small portion of the dissolved CO_2 is converted into H^+ and HCO_3^-. This increase in **hydrogen ion concentration** stimulates the chemoreceptors, leading to an increase in breathing. Since the blood-brain barrier is poorly permeable to hydrogen and bicarbonate ions, the central chemoreceptors are **not** directly affected by rapid changes of the levels of these ions in the blood. Nevertheless, important effects on central chemoreceptors accompany metabolic **acidosis** or metabolic **alkalosis**. For example, during metabolic acidosis stimulation of peripheral arterial chemoreceptors (and possibly other receptors on the "blood" side of the blood-brain barrier) causes ventilation to rise and blood P_{CO_2} to fall. CO_2 diffuses out of the interstitial fluid into the blood. The removal of CO_2 from the medullary tissue interstitial fluid reduces the H^+ concentration and makes it alkaline compared with initial conditions. This decreases the activity of medullary chemoreceptors and limits the ventilatory increase accompanying metabolic acidosis. Over many hours (up to one or two days) bicarbonate leaves the brain extracellular fluid. Reduction in bicarbonate makes the brain extracellular fluid slightly acidic compared to initial conditions. Once their environment becomes more acidic, medullary chemoreceptors contribute to the stimulation of breathing. In fact, the **increase in breathing** with acute metabolic acidosis is less than that for chronic conditions. Metabolic alkalosis is associated with an acute acidosis in brain interstitial fluid, followed by slightly alkalotic conditions in the chronic state.

Review Questions

14. The apneustic center

    A. is located in the hypothalamus
    B. is located in the rostral portion of the pons
    C. must be intact for breathing to occur
    D. requires an intact vagal innervation to elicit apneustic breathing
    E. can under appropriate circumstance elicit prolonged inspiratory efforts

15. Prolongation of expiration is associated with

    A. activity from the apneustic center
    B. activity from slowly adapting lung stretch receptors
    C. activity from rapidly adapting airway receptors
    D. shortened period of activity for medullary expiratory neurons
    E. None of the above is correct

16. Compared with acute metabolic acidosis, chronic metabolic acidosis will be associated with

    A. a decrease in arterial P_{CO_2}
    B. increased levels of bicarbonate in the medullary tissue interstitial fluid
    C. a more alkaline pH in the medullary tissue interstitial fluid
    D. reduced stimulation of medullary chemoreceptors
    E. more acidic arterial blood

For Questions 17-23. Match one of the lettered choices with each question.

A. Alveolar to arterial P_{O_2} difference
B. Arterial O_2 content
C. Arterial P_{O_2}
D. Arterial P_{CO_2}
E. End capillary blood in alveoli
F. Extracellular pH around medullary chemoreceptors
G. FEV_1
H. FVC
I. Intrapleural pressure
J. Lung surfactant
K. Respiratory exchange ration (R)
L. $\dot{V}_A/\dot{Q}_C$

17. ____ is decreased when a patient loses 1.5L of blood through a stomach ulcer.

18. ____ when decreased, can depress the activity of medullary chemoreceptors.

19. ____ is elevated after a narcotic overdose which depresses breathing.

20. ____ tends to equilibrate transmural pressure in large and small alveoli.

21. ____ is determined by lung elasticity at the end of inspiration.

22. ____ is reduced if airway resistance during expiration is increased.

23. ____ is increased from normal when a lobe in the lung is filled with fluid during pneumonia.

For Questions 24-29. Travelling to high altitude produces a variety of profound responses and adaptions of respiration. Effects on gas exchange, gas transport in the blood, and control of breathing are among the most important. The reduction of barometric pressure with increased altitude leads to a decrease in inspired partial pressure of O_2 (P_{IO_2}) according to the following relationship:

EQUATION 10: $P_{IO_2} = F_{IO_2} \times (P_B - 47)$ mm Hg

Inspired air is 20.9% O_2 which is a fractional concentration (F_{IO_2}) of 0.209. Multiplying this value by barometric pressure (PB) corrected for water vapor pressure at body temperature (47mm Hg) yields P_{IO_2}. For example, at sea level (PB=760 mm Hg) P_{IO_2} is 149 mm Hg. At 18,000 ft, barometric pressure is about one half the value at sea level (PB=380), and P_{IO_2} is only 70 mm Hg.

Lowering inspired O_2 levels leads to a reduction in P_{AO_2}. For example, at 18,000 ft, if P_{ACO_2} is 40 mm Hg and R=0.8, the alveolar gas equation (EQUATION 8) predicts that P_{AO_2} will be 20 mm Hg.

24. Why must arterial P_{O_2} be even lower than 20 mm Hg?

25. Why are large reductions in arterial P_{O_2} expected to stimulate breathing?

26. Using Equations 7 and 8 predict what will happen to $P_{A_{CO_2}}$ and $P_{A_{O_2}}$ if $\dot{V}_A$ increases to three times its value at sea level.

27. Although breathing is stimulated by acute exposure to high altitude, what would be the expected effect on central chemoreceptors?

Over time at high altitude, more red blood cells are produced, thereby increasing the hematocrit and hemoglobin concentration (polycythemia). In addition, levels of 2,3 BPG in red cells can be elevated.

28. How could these adaptations aid delivery of O_2 to body tissue?

In addition to the preceding adjustments, ventilation rises over a period of several days at high altitude.

29. How would this rise in ventilation influence delivery of O_2 to body tissues?

ANSWERS TO PULMONARY PHYSIOLOGY QUESTIONS

1. B. Reduced hemoglobin can bind more CO_2 in the carbamino form than oxygenated hemoglobin. Choices A, C, and D are opposite to the correct direction. Reduced hemoglobin is a better buffer than oxygenated hemoglobin (E).

2. A. Poisoning of hemoglobin with carbon monoxide causes the oxygen dissociation curve to shift to the left, so the partial pressure where the available hemoglobin is 50% saturated with O_2 is reduced (A). Since carbon monoxide competes with O_2 for sites on the hemoglobin molecule, less O_2 will be carried at a given P_{O_2} level (B). Nevertheless, saturation of the hemoglobin sites available for O_2 will be approached at P_{O_2} levels consistent with arterialized blood (ie, 90 mm Hg) (C). Since the oxygen-hemoglobin dissociation curve is shifted to the left, more of the available O_2 will remain bound to hemoglobin as P_{O_2} is decreased (ie, to 50 mm Hg) (D). Presence of CO has no effect on O_2 solubility in blood (E).

3. E. At FRC the lungs recoil toward a smaller volume, while the chest wall exerts a recoil toward larger volume that is equal and opposite to that of the lungs (A, B). Pressure in the pleural space has a lower absolute value (more negative) than pressure in the alveolar spaces (C, D).

4. B. Lack of surfactant will increase surface tension (A). Since $P = 2T/r$, a greater pressure difference is required to inflate alveoli to overcome surface forces (B). Surfactant makes pressures similar in large and small alveoli (D). More effort would be needed to inspire (E).

5. B. During inspiration and expiration pressure in the pleural space minus barometric pressure (intrapleural pressure) represents not only lung elastic pressure (-Pel), but also resistive pressures due to airflow and tissue movement. Choices A and C are opposite to what is correct. Inspiration requires muscular work under most conditions (D). Inspiration can occur at any volume between the residual volume and the total lung capacity (E).

6. A. If pressure in the pleural space is greater than that within the airways, the pressure difference can compress the airways (A). Airway compression is least likely to occur at high lung volumes (B), because elastic recoil of the lungs is maximal (C), and airway resistance is minimal (D). Expiratory airflow can be produced by potential energy stored in the lungs from the previous inspiration (E).

7. E. In restrictive disease FEV_1 as a percent of forced vital capacity is not reduced (A, B, D).

8. A. A higher R means that CO_2 output (which depends primarily on ventilation) is greater than the O_2 uptake (which depends primarily on blood flow). Therefore, the $\dot{V}_A/\dot{Q}_C$ is high (A). Absolute levels of either ventilation or blood flow cannot be determined from the data (B, C). Alveoli with higher R values will have higher P_{O_2} and lower P_{CO_2} levels (D, E).

9. C. Increasing the frequency of breathing will increase the ventilation of the dead space per minute, but it will not significantly affect the volume of the dead space for each breath.

10. B. Chronic bronchitis is associated with uneven $\dot{V}_A/\dot{Q}_C$.

11. C. Doubling blood flow with the same alveolar ventilation will nearly double the O_2 uptake, as long as P_{O_2} remains on the (nearly) flat portion of the oxygen-hemoglobin dissociation curve. Since CO_2 elimination is largely dependent on alveolar ventilation, which is the same for these two regions, the respiratory exchange ratio ($\dot{V}_{CO_2}/\dot{V}_{O_2}$) will be lower for region A.

12. B. Hypoventilation by itself or a change in dead space is not associated with an increased alveolar to arterial P_{O_2} difference (A, D). A decrease in diffusing capacity might cause such an effect under extreme conditions (C), but shunt will clearly produce this result (B). Uniformly doubling the $\dot{V}_A$ and $\dot{Q}_C$ in all parts of the lungs will not change the ratio, $\dot{V}_A/\dot{Q}_C$ (E).

13. A. Both O_2 and CO_2 are typically equilibrated between alveolar gas and blood leaving the alveolar capillary (A); P_{CO_2} is reduced between the beginning and end of the alveolar capillaries (B). There is a much greater partial pressure difference between venous and arterial P_{O_2} than for P_{CO_2} (C); solubility of CO_2 is about 20 times greater than of O_2 in plasma (D).

14. E. The apneustic center is located in the caudal pons (A, B). An animal can breathe with only the medulla intact (C). The pneumotaxic centers must be eliminated and the vagi cut to produce apneustic breathing (D).

15. B. Choices A, C and D are better associated with a shortened expired breath.

16. A. A decrease in HCO_3^- levels of medullary interstitial fluid when acidosis is maintained (B) will cause pH to fall (C) and increase the excitation of medullary chemoreceptors (D). The subsequent increase in breathing will cause arterial P_{CO_2} to decrease (A) and acidity of arterial blood to decrease (E).

17. B. Arterial O_2 content

18. D. Arterial P_{CO_2}

19. D. Arterial P_{CO_2}

20. J. Lung surfactant

21. I. Intrapleural pressure

22. G. FEV_1

23. A. Alveolar to arterial P_{O_2} difference

24. Arterial O_2 is always lower than alveolar P_{O_2} because of shunted venous blood and unevenness in $\dot{V}_A/\dot{Q}_C$.

25. A large reduction in arterial P_{O_2} will stimulate the peripheral chemoreceptors (carotid bodies).

26 A 3-fold rise in alveolar ventilation ($\dot{V}_A$) will cause P_{ACO_2} to decrease to 1/3 of its initial value; with $\dot{V}_{CO_2}$ remaining constant, 40 mm Hg x $\dot{V}_A$ = P_{ACO_2} (altitude) x 3 x $\dot{V}_A$, so P_{ACO_2} (altitude) = 13.3 mm Hg and P_{AO_2} = 70-13.3/0.8 = 53.4 mm Hg.

27. Stimulation of breathing at high altitude tends to depress central chemoreceptors because of reductions in arterial P_{CO_2} and alkalinization of brain extracellular fluid. A complicating factor, when P_{O_2} is sufficiently reduced, is production of metabolic acid(s) by brain tissue.

28. Increasing the number of red blood cells will increase the amount of O_2 transported to the tissues in the arterial blood. Shifting the O_2 dissociation curve to the right by 2,3 BPG will cause O_2 to be unloaded at the tissues at a higher P_{O_2} (the actual shift in the dissociation curve will also be influenced by the reduced P_{CO_2} and acid-base adaptations). There is no advantage to such an adaptation at very high altitudes, since both arterial and venous blood are on the steepest portion of the O_2 dissociation curve.

29. Further increases in ventilation would decrease alveolar and arterial CO_2 and raise alveolar and arterial P_{O_2} towards the inspired value.

RENAL PHYSIOLOGY

Renal physiology seeks to understand the importance of the water content and the solute composition of **body fluid compartments** and the mechanisms used by the **kidneys** to regulate them. All substances in the body fluids come from either intake or metabolism and are eliminated by either excretion or metabolic consumption. To maintain relatively constant concentrations of these substances in the body, the total amount taken in and produced must equal the total amount excreted and consumed. Therefore, the regulation of body fluid composition is accomplished by **adjusting** the **output** of **water** and **electrolytes** to match their input, such that both **water and electrolyte balances** are maintained.

The intake of water (amounts drunk, eaten, and metabolically produced) normally equals its output (evaporation from skin and lungs, losses via urine, sweat, and feces). The amount of water intake is influenced partly by sociological and habitual factors and controlled primarily by **thirst** mechanisms in the hypothalamus. Body water is balanced on the output side primarily by precise control of water loss by the kidney. Electrolytes are affected only by ingestion and excretion, and the electrolyte output via urinary excretion is tightly regulated. The kidneys also rid the body of waste substances and foreign chemicals such as drugs and pesticides.

In addition to homeostatic and excretory functions, the kidneys also perform **hormonal function.** The kidneys produce 1) **renin,** which helps regulate total body Na^+, extracellular fluid volume, and blood pressure; 2) vasoactive substances (**prostaglandins** and **kinins**) which are autocoids that act on renal hemodynamics, Na^+ and water excretion, and renin release, and 3) the activated form of **Vitamin D_3,** which maintains calcium levels by stimulating calcium absorption from gut, bone, and renal tubular fluid.

BODY FLUID COMPARTMENTS AND THEIR COMPOSITION

The major component of body fluid compartments is **water**, which accounts for about 60% of body weight (**total body water, TBW**). This percentage declines with increased age and increased amount of body fat. TBW is distributed among two major compartments within the body: **intracellular fluid compartment (ICF)** (40% of body weight) and **extracellular fluid compartment (ECF**; 20% of body weight). ECF is composed of **plasma volume** (about 4% of body weight), **interstitial fluid volume** (about 15% of body weight), and **transcellular fluid volume** (cerebrospinal fluid, peritoneal fluid, etc.; about 1-3% of body weight).

The volumes of fluid compartments can be measured indirectly by **indicator dilution** methods. The volume of the compartment in question is found by determining the final concentration of a known quantity of a substance which has been added to the compartment, or $V = Q/C$, where V is the compartmental volume in which the substance X is uniformly distributed; C is the measured final concentration of X; and Q is the quantity of X added to the compartment minus the amount lost from the compartment by excretion or metabolism during the measurement.

Total body water is measured using **tritiated water** (THO), deuterium oxide (D_2O), or antipyrine. **ECF** is measured using saccharides (eg, **inulin**, sucrose, or mannitol) or ions (eg, thiosulfate, thiocyanate, or radioactive chloride). **Interstitial fluid volume** is not measured directly but is calculated as the **difference between ECF and plasma volume. Plasma volume** is determined using substances that neither leave the vasculature nor enter red blood cells, such as Evans blue dye or radioactive **serum albumin.** Another method is to label red blood cells with ^{32}P or ^{51}Cr and reinject them back into the circulation. The dilution of tagged red blood cells and the hematocrit are used to determine red cell volume and plasma volume. **ICF** volume is difficult to measure; it is calculated as the **difference between TBW and ECF volume.**

Electrolytes, comprising up to 95% of total solutes, are the most abundant constituents of body fluids next to water; organic solutes (glucose, amino acids, urea, etc.) constitute only a small portion. Within the **ICF, K^+ and Mg^{2+}** are the major cations, and **proteins and phosphates** are the major anions. **Na^+, Cl^- and HCO_3^-** are the major ions of the **ECF. Interstitial fluid,** unlike plasma, is relatively **free of proteins.** The presence of electrolytes exerts a variety of effects on body fluids: contributing to the osmotic pressure of the fluids, functioning as substrates for membrane transport, and acting as determinants of membrane potential and pH of body fluids.

Normally, the **osmolality** (ie, the concentration of osmotically-active solute particles) of the ICF and ECF compartments is the same, about 285 mosmol/kg of body water. The osmotic content of ICF is determined by the concentrations of K^+ and charged proteins and associated ions, and the osmolality of ECF is primarily determined by its NaCl content. Osmotic equilibration is normally maintained between ECF and ICF compartments. Any alteration in ECF osmolality will result in water movement between the two compartments (water flows from hyposmotic to hyperosmotic compartments).

Review Questions

1. Which of the following statements regarding fluid compartments is correct?

   A. The extracellular compartment is the largest fluid compartment.
   B. Total body water can be determined using inulin.
   C. The volume of the plasma compartment can be determined using labeled serum albumin and the hematocrit.
   D. The volume of the plasma compartment equals the volume of the interstitial compartment.
   E. The volume of the transcellular compartment is greater than the volume of the plasma compartment.

2. Use the following values to determine the volume of the intracellular fluid compartment.

Inulin infused	= 1.6g	Tritiated water infused	= 2.5g
Inulin excreted	= 0.2g	Tritiated water excreted	= 0.4g
Inulin concentration	= 0.1 mg/ml	Tritiated water concentration	= 0.05mg/ml
		Hematocrit	= 45%

   A. 14 L
   B. 16 L
   C. 28 L
   D. 34 L
   E. 42 L

3. If a patient hemorrhages one liter of blood in 5 minutes, you would restore their body fluid balance by giving one liter of which of the following?

   A. 5% glucose intravenously
   B. 5% glucose intraperitoneally
   C. 0.9% NaCl intravenously
   D. 4.5% NaCl intravenously
   E. Distilled water intravenously

4. Interstitial fluid is different from plasma because of its

   A. concentration of small solutes
   B. osmolality
   C. sodium ion concentration
   D. plasma protein concentration
   E. pH

A SYNOPSIS OF RENAL FUNCTION

The kidney accomplishes its function through functional units, the **nephrons**, by means of three basic processes: **ultrafiltration, tubular reabsorption,** and **tubular secretion**. As blood passes through the kidneys, substances are first removed by glomerular filtration. Substances that the body needs are returned to the blood by tubular reabsorptive mechanisms. Some substances can also be added to the filtrate by renal tubular secretion; these secreted substances are then excreted in the urine.

Arterial blood is delivered to glomerular capillaries via **afferent arterioles**. Glomerular filtration is the process of plasma ultrafiltration through glomerular membranes. The plasma filtrate passes into Bowman's capsule, and the unfiltered blood leaves the glomerulus via **efferent arterioles** and flows into **peritubular capillaries** surrounding the nephrons. **Glomerular filtration rate (GFR)** is the volume of plasma that is filtered each minute by all glomeruli in the kidneys. The average GFR for a healthy 70 kg male is 125 ml/min. This value is lower in children and females and greater in larger persons. **Renal plasma flow (RPF)** is about 600 ml/min. Only 20% of this RPF is filtered at the glomerulus; this fraction is known as **filtration fraction**. Since plasma represents about 55% of whole blood, **renal blood flow (RBF)** is about 1 L/min, or about 20% of cardiac output.

Glomerular capillaries and basement membranes are freely permeable only to small solutes; the glomerular filtrate thus contains the same solute concentrations as the plasma **with the exception of proteins**. However, urine is quite different from glomerular filtrate, because tubular reabsorption and secretion processes alter the composition and volume of the filtrate as it flows down renal tubules. **Tubular reabsorption** is the movement of solutes and water from the filtrate in tubule lumen to the blood in peritubular capillaries. **Tubular secretion** is the movement of solutes from peritubular capillaries into the tubule lumen. Therefore, **urinary excretion** is a result of the combined functions of the nephron, namely filtration, reabsorption, and secretion.

GLOMERULAR FILTRATION

Determinants of GFR

Filtration Forces. Ultrafiltration of plasma occurs as plasma moves from glomerular capillaries into Bowman's capsule under the influence of net filtration pressures. Glomerular filtration is essentially the same phenomenon as systemic capillary filtration; the balance between hydrostatic and oncotic forces across the glomerular membrane determines the direction of fluid movement. Net filtration pressure driving water and solutes across the glomerular membrane is the sum of 1) the **glomerular capillary hydrostatic pressure** (P_c, 45 mm Hg) in an outward direction minus 2) the **hydrostatic pressure in Bowman's capsule** (P_t, 10 mm Hg) and 3) the **colloid osmotic pressure** of plasma in glomerular capillaries (π_p, 28 mm Hg). Therefore, **net glomerular filtration pressure = P_c - P_t - π_p**, with a normal mean value of about 7 mm Hg. The glomerular capillaries are much more permeable than average systemic capillaries; approximately 180 L/day of fluid are filtered across glomerular capillaries. If these were systemic capillaries, only a net amount of 4 L/day of fluid could have been filtered.

Permselectivity of Glomerular Membranes. The permeability of glomerular membranes is dictated by the permselectivity of the glomerular filtration barrier. The barrier is composed of capillary endothelial cells, endothelial basement membrane, and epithelial cells of Bowman's capsule. Two factors, molecular size and electrical charge, determine the barrier's permeability to a given substance. In general, molecules larger than about 10,000 mw do not pass through the filtration barrier. The "pores" or "channels" that substances pass through are 7.5-10 nm in diameter and are surrounded by negative charges. Therefore, the barrier is most permeable to small, neutral or positively charged molecules and is relatively impermeable to large negatively charged molecules like proteins.

In addition to the three aforementioned determinants of filtration rate (glomerular capillary pressure, plasma colloid oncotic pressure and glomerular membrane permeability), the rate of glomerular filtration is also directly influenced by the **rate of renal plasma flow.**

Hemodynamics of GFR

Both glomerular hydrostatic pressure and RPF are major determinants of GFR. The magnitudes of RPF and GFR are influenced by renal autoregulation, autonomic innervation, and factors affecting arteriolar resistance.

Renal Autoregulation. RPF and GFR remain almost constant over a wide range of arterial blood pressure (80-180 mm Hg). As blood pressure increases over this range, resistance in afferent arterioles increases proportionately to prevent large increases in RPF and GFR. Autoregulation is an intrinsic property of the kidney, independent of neural influence and extrarenal humoral stimulation. The two intrarenal mechanisms responsible for renal autoregulation are 1) the **myogenic mechanism**, involving an intrinsic property of the afferent arteriolar smooth muscle, and 2) the **tubulo-glomerular feedback mechanism**, involving a flow-sensing feedback loop between the macula densa of the distal tubule and the afferent arteriole of the same nephron. Autoregulation helps to decouple the renal regulation of salts and water excretion from fluctuations in arterial blood pressure.

Even in the face of autoregulation, changes in RPF and GFR can occur by **local changes in vascular resistance of afferent and efferent arterioles.** These resistance changes can be caused by actions of the **autonomic nervous system** and various **vasoactive humoral agents.** An increase in sympathetic activity to the kidney will result in afferent and efferent arteriolar vasoconstriction, increased renal vascular resistance, and decreased GFR. Similarly, decreased sympathetic tone results in decreased renal vascular resistance and increased GFR. Norepinephrine, epinephrine, acetylcholine, angiotensin, vasopressin, prostaglandins, and kinins are vasoactive in the kidneys. When resistance is altered in **afferent arterioles only**, then **RPF and GFR change in the same directions.** When resistance is altered in **efferent arterioles only**, then **RPF and GFR change in opposite directions.** Therefore, GFR tends to decrease less than RPF when sympathetic tone is increased, since both afferent and efferent arterioles are constricted.

Renal Oxygen Consumption and Metabolism

Per unit of tissue weight, the kidneys are perfused by more blood, and they consume more **oxygen** than does almost any other organ except the heart. Yet, the **renal arteriovenous (a-v) oxygen content difference is lower** than that of other organs. This unique feature reflects the high filtering capacity of the kidney, consequently its blood flow is far in excess of the basal oxygen requirements. The high renal oxygen consumption reflects the amount of energy required for reabsorption of the filtered sodium. Energy is derived from **aerobic oxidative metabolism** (mostly of fatty acids) **in the renal cortex.** In contrast, **renal medullary structures** derive energy from **anaerobic metabolism** of glucose.

Review Questions

5. Which of the following opposes glomerular filtration?

   A. Colloid osmotic pressure of plasma in glomerular capillaries
   B. Colloid osmotic pressure of plasma in peritubular capillaries
   C. Hydrostatic pressure in peritubular capillaries
   D. Hydrostatic pressure in glomerular capillaries
   E. Hydrostatic pressure in efferent arterioles

6. Renal autoregulation of blood flow and glomerular filtration rate requires

   A. autonomic innervation
   B. epinephrine
   C. aldosterone
   D. antidiuretic hormone
   E. no action of extrarenal factors

7. A drug causes a decrease in glomerular filtration rate. That drug might be

   A. dilating afferent arterioles
   B. causing an obstruction in the urinary system
   C. constricting efferent arterioles
   D. decreasing plasma albumin concentration
   E. decreasing renal sympathetic nerve activity

RENAL TUBULAR TRANSPORT

Classification

Renal tubular transport, regardless of its direction, can be either a passive process (requiring no energy expenditure) or an active process (requiring energy). For substances that are actively transported by renal tubule cells, their renal transport can be characterized by the **capacity** of the renal tubules to transport them at any one time, ie, there is an upper limit for the rate of transport either in the reabsorptive or secretory direction. The highest attainable rate of tubular transport of any given solute is its **maximum tubular transport capacity (Tm)**, and transport systems exhibiting tubular transport maxima are known as **Tm-limited** transport processes. The existence of the Tm phenomenon can be explained in terms of saturation of the transport carriers and/or sites for the particular substance.

Substances with a reabsorptive Tm include phosphate and sulfate ions, glucose and other monosaccharides, many amino acids, and Krebs cycle intermediates. The plasma concentration at which a reabsorbed solute reaches its Tm and begins to appear in urine is its **threshold** concentration and is characteristic for that substance. For example, glucose is not normally excreted, because all filtered glucose is reabsorbed. However, glucose will be excreted at high plasma concentrations (above 300 mg/100 ml). Assuming that GFR remains constant, the filtered load of glucose will be proportional to plasma glucose concentration. As plasma glucose and consequently filtered load increase, the renal glucose transport sites become saturated, and the maximum transport rate of glucose is reached. Therefore, the amount of glucose not being reabsorbed will start spilling into the urine. Further increases in plasma glucose will be followed by increases in the amount of glucose excreted. The maximum reabsorptive rate (Tm) for glucose is 375 mg/min.

Some substances can also be secreted by Tm mechanisms: organic acids (eg, uric acid, p-aminohippurate (PAH)), organic bases (eg, creatinine, histamine), and other compounds not normally found in man (eg, penicillin, morphine). These secretory transport systems are important for the elimination of drugs and other foreign chemicals from the body. Secretory rates of these substances will increase as their arterial concentrations increase until their secretory Tm's are reached at threshold concentrations. At above threshold concentrations secretory rates reach plateau, and the contribution of the secretion process to total urinary excretion will decrease even though the amount excreted continues to increase.

For some solutes there is no definite upper limit for the rate of renal tubular transport, ie, no Tm. Instead, their rates of transport are limited by the solute concentration gradient differences between the filtrate and the peritubular blood. This type of transport is known as a **gradient-limited** transport process. Sodium reabsorption along the nephron is representative of this type of transport.

Quantitation of Renal Function: Renal Clearance

Renal Clearance (C) measures the efficiency of kidneys in removing a substance from plasma. It is a useful concept in renal physiology, because it can be used to quantitatively measure the intensity of several aspects of renal function, ie, filtration, reabsorption, and secretion. **Renal clearance is defined as the theoretical volume of plasma from which a given substance is completely cleared**

by the kidneys per unit time. Each substance has a specific renal clearance value. In general, for a given substance X, renal clearance of X is the ratio of its excretion rate to its concentration in plasma. The formula for calculating renal clearance of substance X is:

$$C_x = \frac{U_x \cdot \dot{V}}{P_x}$$

where C_x is the renal clearance of the substance in ml/min, U_x and P_x are the concentrations of X (mg/ml) in urine and plasma, respectively, and $\dot{V}$ is urine output per minute or urine flow rate (ml/min). In a renal clearance measurement, $\dot{V}$ is obtained by measuring the volume of urine produced per unit time, and the concentrations of X in the urine sample (U_x) and in the plasma (P_x) are measured. Since filtration, reabsorption, and secretion mechanisms all contribute to urinary excretion (or plasma clearance), the values for renal clearance of various substances give information about renal function and the manners by which the substances are handled by the kidneys.

The direction of net renal tubular transport and the rate of tubular transport of a substance can be quantitatively determined using the renal clearance measurement as follows: at any one time the total renal excretion of a substance must equal the algebraic sum of the following three processes:

Total amount excreted = Amount filtered + Amount secreted - Amount reabsorbed

Total amount excreted (Excretion rate, mg/min) = $U_x \cdot \dot{V}$
Amount filtered (Filtered load, mg/min) = $P_x \cdot GFR$
Net amount secreted (mg/min) = Amount excreted - Amount filtered
Net amount reabsorbed (mg/min) = Amount filtered - Amount excreted

where U_x = concentration of substance X in urine (mg/ml)
$\dot{V}$ = urine flow rate (ml/min)
P_x = concentration of substance X in plasma (mg/ml)
GFR = glomerular filtration rate (ml/min)

For a substance undergoing net tubular reabsorption:

Amount excreted < Amount filtered

For a substance undergoing net tubular secretion:

Amount excreted > Amount filtered

The renal clearance method can also be used to determine whether or not renal transport of a substance is a Tm-limited transport process. This is done by constructing a **renal titration curve**, a combined plot of the filtered load, the urinary excretion rate and the transport rate of substance X against the increasing plasma concentrations of X. For a Tm-limited transport process, the transport rate will become constant at high plasma concentrations of X.

Renal Clearance of Various Solutes

Inulin is a non-toxic polysaccharide that is not bound to plasma proteins, is freely-filtered at the glomerulus, and is neither reabsorbed nor secreted by renal tubules. Therefore, renal clearance of inulin is a **measure of GFR**, because the amount of inulin excreted in the urine represents the amount that is only filtered at the glomerulus. Consequently, the volume of plasma cleared of inulin per minute must equal the volume of plasma filtered per minute, ie, GFR.

Creatinine, an end-product of skeletal muscle creatine metabolism, exists at a fairly constant concentration in plasma under normal conditions. **The 24 hr creatinine clearance is used clinically as an estimate of GFR**. Renal clearance of creatinine is slightly greater than GFR estimated with inulin, because creatinine is secreted in small amounts in addition to being filtered and not reabsorbed. Creatinine is used rather than inulin, because it is **endogenous** and does not need to be administered. Since plasma creatinine levels are stable, creatinine production normally equals creatinine clearance by the kidney. If GFR decreases to half of normal, creatinine production will exceed renal clearance, and serum creatinine will double. Similarly, if GFR decreases to one-fourth of normal, serum creatinine will increase four times. Therefore, there is an excellent inverse relationship between plasma creatinine level and the magnitude of GFR, such that **an increase in plasma creatinine concentration can be used as an indicator of a decrease in GFR** of similar magnitude.

The decline in GFR is the most clinically significant deficit in renal function occurring with age. This decline is due to declines in renal plasma flow, cardiac output, and renal tissue mass. Creatinine clearance, the index of GFR, also decreases with age whereas plasma creatinine concentration remains constant. Decreased creatinine production as a result of a reduction in muscle mass occurring with age is matched by decreased renal creatinine excretion. Therefore, serum creatinine concentration can be unreliable as an estimate of renal function in elderly individuals.

Urea is filtered and reabsorbed. Under conditions when urea reabsorption is approximately a constant fraction of its filtered load, urea clearance could be used to estimate GFR. The serum level of urea is also used to estimate renal function in the same way as serum creatinine. This level is expressed as **blood urea nitrogen** concentration (**BUN**). When GFR falls, BUN usually rises in parallel to serum creatinine. However, urea clearance or BUN is usually not a reliable indicator of the magnitude of GFR. Plasma urea concentration varies widely, depending on protein intake, protein catabolism, and variable renal reabsorption of urea under different states of hydration.

Glucose. The renal clearance of glucose is zero at normal plasma glucose concentration (80 mg/100 ml) and up to 300 mg/100 ml, because all filtered glucose is reabsorbed by renal tubules. If plasma glucose levels increase above three times normal, the renal reabsorptive rate of glucose will reach its Tm, and glucose excretion will increase until its clearance approaches GFR. That is, at high plasma glucose concentrations, the majority of excreted glucose comes from its unreabsorbed filtered load.

p-Aminohippuric acid (PAH). The renal clearance of PAH is greater than GFR. In addition to being filtered, PAH is also secreted into renal tubules, therefore more PAH is excreted than the amount originally filtered. Since PAH is both filtered and secreted, below its secretory Tm virtually all plasma supplying nephrons can be cleared of PAH. In other words, PAH will be completely cleared from the plasma by renal excretion during a single circuit of plasma flow through the kidney. Consequently, **renal clearance of PAH can be used to estimate the magnitude of renal plasma flow.** Normally 85-90% of the total plasma flowing through the kidney is cleared of PAH, therefore PAH clearance is a measure of **"effective" renal plasma flow (ERPF).** The **effective renal blood flow (ERBF)** is calculated from the ERPF and hematocrit (Hct): $ERBF = \frac{ERPF}{1 - Hct}$. At higher plasma PAH concentrations, renal secretory transport of PAH reaches its Tm; therefore, the contribution of secretion to renal clearance of PAH decreases. Consequently, as plasma PAH levels increase, renal clearance of PAH decreases towards the value of GFR.

Review Questions

8. Use the data below to calculate renal blood flow.

Arterial plasma PAH conc.	= 0.04 mg/ml	Urine PAH conc.	= 10 mg/ml
Venous plasma PAH conc.	= 0.004 mg/ml	Urine inulin conc.	= 20 mg/ml
Arterial plasma inulin conc.	= 0.4 mg/ml	Urine flow rate	= 2 ml/min
		Hematocrit	= 55%

A. 100 ml/min
B. 250 ml/min
C. 556 ml/min
D. 1000 ml/min
E. 1235 ml/min

9. Use the data from Question 8 above to calculate GFR.

A. 50 ml/min
B. 100 ml/min
C. 125 ml/min
D. 200 ml/min
E. 500 ml/min

10. Which of the substances below would produce the following measurements during a renal clearance study?

Plasma concentration	=	2 mg/100 ml
Urine concentration	=	12 mg/ml
Urine flow rate	=	1 ml/min
Glomerular filtration rate	=	125 ml/min

A. Glucose
B. Alanine
C. Inulin
D. Para-aminohippuric acid (PAH)
E. Sodium ions

11. If plasma concentration of para-aminohippuric acid (PAH) is increased to two times threshold, then all of the following are correct **EXCEPT**

    A. the transport system for PAH will be saturated
    B. the amount of PAH excreted in the urine will be less than two times the amount excreted at threshold
    C. the tubular maximum for transport will be exceeded
    D. the amount of PAH excreted will increase as the concentration increases past threshold
    E. all of the PAH in renal arteries will be secreted by the renal tubules

12. Use the data below to calculate the rate of excretion of X in the urine. Substance X is freely filtered across the glomerular membrane.

Glomerular filtration rate	=	125 ml/min
Plasma concentration of X	=	2 mg/ml
Tubular reabsorption of X	=	30 mg/min
Tubular secretion of X	=	60 mg/min

    A. 160 mg/min
    B. 220 mg/min
    C. 250 mg/min
    D. 280 mg/min
    E. 340 mg/min

13. A patient submits a 12-hour urine collection with 400 mg of creatinine. The serum creatinine is 0.8 mg/dl. Calculate the creatinine clearance in ml/min.

14. The same patient in question 13 now has a new renal transplant. He has had a stable serum creatinine of 0.6 mg/dl. Six months after the transplant the level rises to 1.2 mg/dl. This change is most likely due to

    A. a significant loss of GFR
    B. the normal daily fluctuation of serum creatinine within the normal range
    C. a rise in creatinine production
    D. impairment of protein synthesis due to corticosteroids (for immunosuppression)

Mechanisms of Renal Tubular Transport

Renal tubular transport of solutes involves transepithelial movement of solutes across two different membranes, the **luminal** and **basolateral** membranes. Renal tubular transport thus requires the transport systems at both membranes to work together in series.

Glucose reabsorption in proximal tubules is an example of Na^+-dependent **secondary active transport.** Na^+, K^+-ATPase in the basolateral membranes of renal tubule cells establishes an electrochemical gradient for Na^+ by extruding Na^+ from cells and pumping K^+ in across the basolateral side. The electrochemical gradient for Na^+ provides the energy source for the uphill glucose transport into the cells across the luminal membrane. Once glucose accumulates within the cells, it leaves across the basolateral membrane by **facilitated diffusion.** The transport systems for glucose on both membranes are specific for the D-forms of sugars and are inhibited by D-sugar analogues and specific inhibitors.

Other organic solutes (amino acids, Krebs cycle intermediates, metabolic intermediates) including phosphate are also reabsorbed in proximal tubules by specific secondary active transport processes, co-transported with Na^+ across the luminal membrane, and by facilitated diffusion across the basolateral membrane.

PAH is taken up into the proximal tubule cells from peritubular capillary blood across the basolateral membrane against its electrochemical gradient by an active transport mechanism specific for organic anions. PAH accumulates within proximal tubule cells and is secreted into luminal fluid by facilitated diffusion.

Transport of Ions and water. Sodium is actively reabsorbed along the whole length of the renal tubule. Na^+ moves from the lumen into cells down its electrochemical gradient by several mechanisms; 1) co-transport with organic solutes (eg, glucose), 2) counter-transport (exchange) with H^+, and 3) simple diffusion via Na^+ channels. Once Na^+ enters cells, it is actively transported across the basolateral membrane by the Na^+, K^+-ATPase system. **Chloride** is mostly passively reabsorbed, since its electrochemical gradient favors the movement of Cl^- from lumen to peritubular blood. **Filtered potassium** is generally reabsorbed by renal tubules via active reabsorption at luminal membranes and passive outflux at peritubular membranes. **Hydrogen ions** are generated within renal cells by cellular metabolism and then are secreted into tubular fluid by secondary active transport coupled to Na^+ entry (Na^+-H^+ exchange) as well as by primary active transport (H^+-ATPase). Hydrogen ions are generated from carbonic acid which is formed by hydration of CO_2 within renal cells in a reaction catalyzed by the enzyme **carbonic anhydrase.** In the proximal tubule lumen, secreted H^+ react with filtered HCO_3^-, forming carbonic acid with the aid of carbonic anhydrase present on the external luminal membranes of proximal tubule cells. This carbonic acid in proximal tubule fluid dissociates into CO_2 and H_2O. CO_2 diffuses back into proximal tubule cells and is rehydrated in the cells to carbonic acid, which in turn dissociates to HCO_3^- and H^+. HCO_3^- crosses basolateral membranes into peritubular blood by an active transport system, and H^+ are re-secreted. The net results are the conversion of secreted H^+ in the proximal segment into H_2O and the transfer of filtered HCO_3^- from proximal tubule fluid to peritubular blood, ie, **HCO_3^- reabsorption.** In the lumen of the distal nephron secreted H^+ combine with other urinary buffers (HPO_4^{2-}, NH_3), and the products are then excreted in the urine, ie, in the distal segment secreted H^+ are excreted. **Water** is passively reabsorbed by osmosis in response to solute osmotic gradients.

Transport of Major Solutes and Water in Various Segments of the Nephron

Proximal Tubule. Sixty-five to ninety percent of the glomerular filtrate is normally reabsorbed in this segment, and the **reabsorption is isosmotic.** Reabsorption of Na^+ and other solutes tends to transiently decrease osmolality of tubular fluid and raise osmolality of surrounding interstitial fluid. Since the proximal tubule has relatively high permeability to water, water is reabsorbed in response to this osmotic gradient in the same proportion as solutes. The reabsorbed fluid then moves from interstitial space into peritubular capillaries by bulk flow caused by the net balance of hydrostatic and oncotic pressures acting across the capillaries. In the early part of the proximal tubule the transepithelial potential difference is about 4 mV lumen negative; in the late part of the proximal tubule it becomes slightly positive (+3 mV). Na^+, K^+, Ca^{2+}, Cl^-, HCO_3^-, organic solutes and water are reabsorbed, and H^+ are actively secreted against a concentration gradient of 25:1. Organic acids and bases can be secreted in this segment. Under steady-state conditions a relatively constant fraction of the filtered Na^+ is reabsorbed in the proximal tubule despite variations in GFR **(glomerulo-tubular balance)**. Therefore, the absolute rate of Na^+ reabsorption in the proximal tubule will increase proportionately with the increase in GFR or Na^+ filtered load; this helps blunt changes in Na^+ excretion which might accompany changes in GFR.

Loop of Henle. The descending limb of the loop of Henle is relatively permeable to water but poorly permeable to solutes like Na^+, Cl^-, and urea. In contrast, the **ascending limb of loop of Henle is impermeable to water**. Very little reabsorption of water occurs in the ascending limb while salts are reabsorbed; therefore, this is the primary site for dilution of urine. In the **thick ascending limb of the loop of Henle,** where the transepithelial potential difference is 10 mV lumen positive, there is net transepithelial reabsorption of Na^+, Cl^-, and K^+. Na^+, K^+, and 2 Cl^- enter cells together by a secondary active transport process **(cotransport of K^+ and Cl^- with Na^+)**. Na^+ is transported out of cells at the peritubular side by the same primary active process as in the proximal tubule (involving Na^+, K^+-ATPase). **Cl^- transport across the luminal membrane of the ascending limb is an active process,** and Cl^- leaves cells on the peritubular side by a passive mechanism. Net K^+ reabsorption in this segment is very small compared to net reabsorption of Na^+ and Cl^-. In addition, the **thick ascending loop of Henle is the major site for Mg^{2+} reabsorption** which occurs intercellulary through tight junctions. The characteristics of reabsorption of ions and water in this segment are such that **high ion and osmolar concentration gradients are established between the lumen of the ascending limb of loop of Henle and the peritubular fluid (medullary interstitium).**

Distal Tubule and Collecting Duct. The transepithelial potential difference across the distal nephron is about 50 mV, lumen negative. In this segment water reabsorption is dissociated from salt reabsorption, because the **permeability to water of the distal nephron is under the control of antidiuretic hormone.** Na^+ is reabsorbed in a similar manner to that in the proximal tubule, except that in the distal nephron Na^+ concentration in the tubular fluid can be reduced to zero. **Na^+ reabsorption in this segment is regulated by the hormone aldosterone,** which enhances Na^+ reabsorption. Cl^- reabsorption is mostly passive, although there is evidence for some active transport. **K^+** is actively reabsorbed in early distal tubules but **secreted in late distal tubules and collecting ducts.** K^+ secretion occurs by passive entry of K^+ from cells into the lumen. Since most filtered K^+ is reabsorbed in the proximal tubule, the rate of K^+ excretion is proportional to its secretory rate in the distal nephron. **The rate of K^+ secretion is controlled by cell K^+ content** (depending on K^+ intake, acid-base balance), **tubular fluid flow rate** (depending on Na^+ excretion rate), and **transepithelial potential difference** (influenced by K^+ intake and Na^+ excretion rate). These factors are under the control of **aldosterone** which **stimulates K^+ secretion.** The distal tubule is the **major site where Ca^{2+} reabsorption is regulated by parathyroid hormone and the activated form of vitamin D.** The distal nephron also actively secretes H^+ against a concentration gradient of 1000:1; consequently, the tubular fluid can be significantly acidified in this segment.

Fig. 5-1 summarizes the transport of substances by various parts of the nephron.

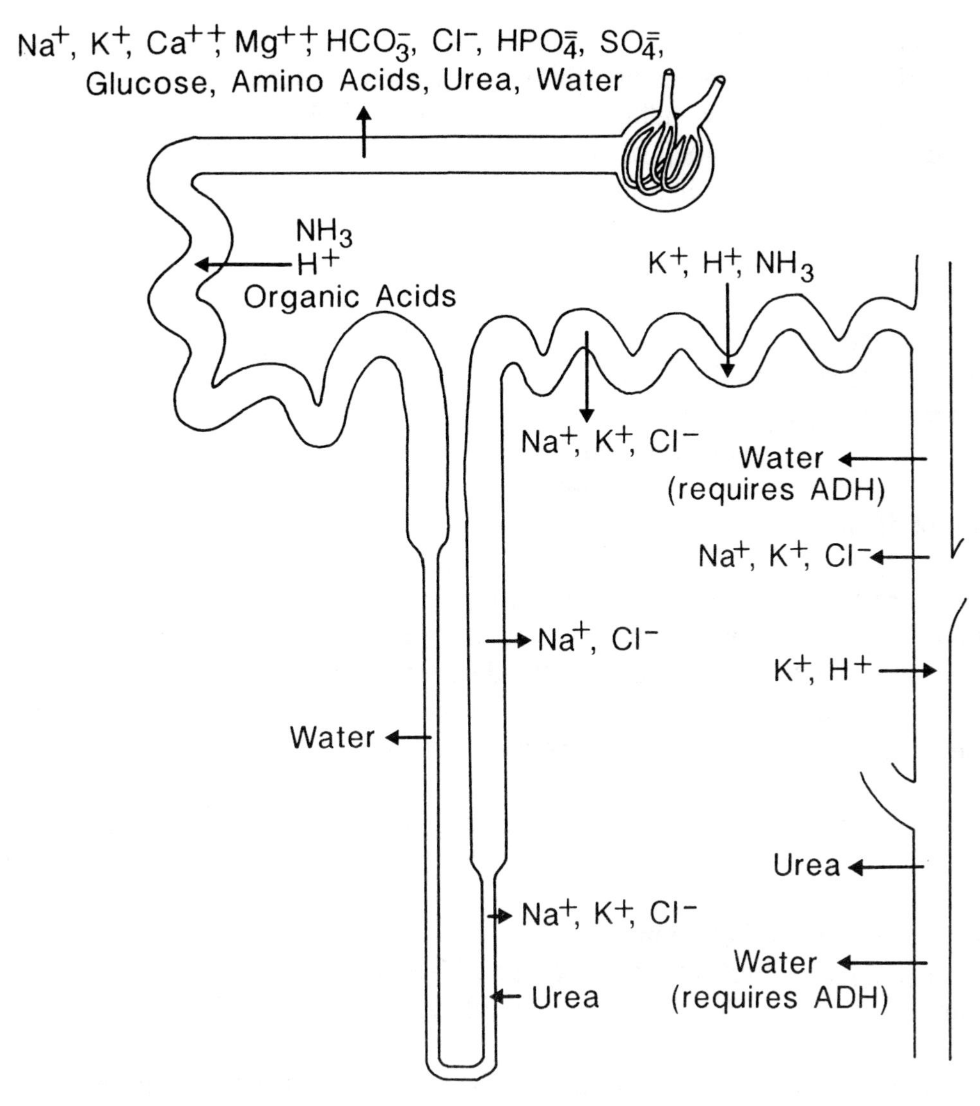

Figure 5-1. Transport of substances at various sites in the nephron. (Modified with permission from Sheng, HP, "Renal Tubular Transport", in *Contemporary Medical Physiology*, Vick, RL [ed]. Copyright 1984 by Addison-Wesley, Reading, MA).

Review Questions

15. Na^+ can be transported across the luminal membrane of renal tubule cells by each of the following mechanisms **EXCEPT**

    A. Na^+ channels
    B. cotransport with organic solutes
    C. counter-transport with H^+
    D. Na^+, K^+-ATPase system
    E. cotransport with K^+ and Cl^-

16. Filtered Cl^- is reabsorbed by secondary active transport in

A. proximal tubules
B. proximal tubules and descending limbs of the loop of Henle
C. proximal tubules and ascending limbs of the loop of Henle
D. ascending limbs of the loop of Henle
E. distal tubules and collecting ducts

17. The kidney "handles" K^+ by

A. filtration only
B. filtration and reabsorption only
C. filtration and secretion only
D. filtration, reabsorption and secretion

18. In the loop of Henle

A. water is reabsorbed from the ascending limb
B. little or no water is reabsorbed from the ascending limb
C. active solute reabsorption occurs in the descending limb
D. both solute and water are reabsorbed from the ascending limb
E. water is added to tubular fluid in the ascending limb, making it hyposmotic

19. A 49-year-old woman saw her physician because of weakness, easy fatigability, and loss of appetite. During the past month she has lost 7 kg. The following information was obtained.

Blood pressure = 80/50 mm Hg
Serum $[Na^+]$ = 130 mEq/L (normal: 135-145)
Serum $[K^+]$ = 6.5 mEq/L (normal: 3.5-50)

From the symptoms and laboratory data, the physician concluded that this patient has Addison's disease (decreased levels of adrenal cortical steroids). The possible causes for her hyperkalemia include all of the following **EXCEPT**

A. decreased K^+ secretion by the distal tubule and collecting duct
B. decreased Na^+ reabsorption by the collecting duct
C. decreased levels of serum aldosterone
D. decreased volume flow rate of tubular fluid along the nephron
E. increased K^+ reabsorption by the proximal tubule

CONCENTRATION AND DILUTION OF URINE

The kidneys are able to produce urine that is either more concentrated or more dilute than plasma; the range of urine concentrations is 50-1400 mosmols/kg H_2O. The ability of the kidneys to concentrate urine makes it possible for a person to survive with minimal water intake. This water conservation task is accomplished through the operation of **countercurrent multiplication** in the loop of Henle. The hairpin turn and close apposition of descending and ascending limbs of the loop of Henle in the medulla of the kidney provide the proper environment (counter-flow) for the operation of a countercurrent multiplier. In addition, the epithelia of the loop of Henle have special permeability characteristics: the descending limb of the loop of Henle is highly permeable to water but poorly permeable to solutes, whereas the ascending limb of the loop is permeable to Na^+ and Cl^- but relatively impermeable to water. In addition, Na^+ and Cl^- are actively reabsorbed in the thick ascending limb of the loop of Henle. These permeability characteristics allow the ascending limb to separate its solute transport from water transport (**"single effect"**), creating a horizontal osmotic gradient between tubular fluid in the ascending limb and that in the descending limb. This horizontal osmotic gradient is then multiplied vertically along the length of the descending loop of Henle, generating an osmotic gradient within the tubular fluid of the descending limb from 300 mosmols/kg H_2O to 1400 mosmols/kg H_2O at the bend of the loop. The **medullary interstitium** is equilibrated with the fluid in the descending limb since this nephron segment is highly permeable to water. A **concentration gradient is thereby established within the medullary interstitium from renal cortex to inner renal medulla, with the highest concentration occurring at the tip of the papilla.**

Formation of Hyposmotic (dilute) Urine [Water Diuresis]

Isosmotic tubular fluid from the proximal tubule enters the descending limb of the loop of Henle. It becomes progressively more concentrated as it moves towards the bend of the loop, because fluid in the descending limb equilibrates osmotically with fluid within the medullary interstitium. The concentrated fluid at the bend of the loop then becomes progressively more diluted as it flows through the ascending loop of Henle, because NaCl can be reabsorbed without water following. When the body is well hydrated, there is a very low circulating level of **antidiuretic hormone (ADH or vasopressin). Without ADH, the permeability to water of the distal tubule and collecting duct is very low.** Consequently, no water will be reabsorbed in the distal nephron even though salts continue to be reabsorbed. Therefore, the diluted tubular fluid that emerges from the ascending loop of Henle will remain hyposmotic as it flows through the distal nephron, producing dilute or hyposmotic urine.

Formation of Hyperosmotic (concentrated) Urine [Antidiuresis]

When there is a **high circulating level of ADH in the blood, the epithelia of the distal nephron are highly permeable to water.** The dilute tubular fluid from the ascending loop of Henle can equilibrate with the osmotic gradient of the medullary interstitium fluid as it flows down the distal nephron, becoming progressively more concentrated. A hyperosmotic or concentrated urine, at the same osmotic concentration as that of the medullary fluid at the tip of the papilla, is produced.

Roles of Vasa Recta and Urea in Renal Concentrating Mechanisms

The **vasa recta** are hairpin capillary beds in renal medulla formed from efferent arterioles of juxtamedullary glomeruli in apposition to loops of Henle and collecting ducts. They are permeable to solutes and water similar to other systemic capillaries. Because of their unique countercurrent arrangement, the vasa recta acts as a **passive counter-current exchanger** system. As solutes are transported out of the ascending loop of Henle, they diffuse down their concentration gradients into the descending vasa recta. Thus the blood in the descending vasa recta becomes progressively more concentrated as it equilibrates with the corticomedullary osmotic gradient. In the ascending vasa recta, these solutes diffuse back into the medullary interstitium and the descending vasa recta, so blood leaving the renal medulla becomes progressively less concentrated as solutes return to the inner medulla. In this manner solutes recirculate within the renal medulla, keeping the solute concentration high within the medullary interstitium. The passive equilibration of blood within each limb of vasa recta with the pre-existing medullary osmotic gradient at each horizontal level **helps maintain the medullary osmotic gradient** necessary for the production of hyperosmotic urine.

Urea is the other major solute within the medullary interstitium besides NaCl. Filtered urea undergoes net reabsorption passively in the proximal tubule. However, because of water reabsorption urea concentration at the end of the proximal tubule is approximately twice that of plasma. Due to the **low permeability of the loop of Henle and distal tubule to urea**, urea concentration in the tubule fluid remains high until the fluid flows through the medullary collecting duct which is more permeable to urea. Urea diffuses out of the collecting duct, entering the interstitium and vasa recta as well as re-entering the descending loop of Henle. Therefore, there is a **medullary recycling of urea**. In the presence of ADH (antidiuresis), urea constitutes about 40% of papillary osmolality, because **ADH increases the permeability of medullary collecting ducts to urea as well as to water**. In the absence of ADH (water diuresis), less than 10% of medullary interstitial osmolality is due to urea. The medullary recycling of urea thus helps establish an osmotic gradient within the medulla with less energy expenditure (urea transport is passive) and enhances water conservation.

Measurement of Concentrating and Diluting Ability of the Kidney

The concentrating and diluting ability of the kidney can be assessed quantitatively. The simplest approach is to measure maximum and minimum urine osmolality. The more widely-used method is to quantitate water excretion. The principle of this quantitation is based on the concept that urine flow is divisible into two components; 1) the urine volume needed to excrete solutes at the same concentration as that in plasma (volume of isosmotic urine = **osmolal clearance or C_{osm}**), and 2) the volume of water that is free of solutes or **free water**. The kidney generates solute-free water in the ascending limb of the loop of Henle by reabsorption of NaCl without water. By excreting dilute or concentrated urine, solute-free water can either be removed or added back to plasma. **Free water clearance (C_{H_2O})** is defined as the amount of distilled water that must be subtracted from or added to the urine (per unit time) in order to make that urine isosmotic with plasma.

$$C_{H_2O} = \dot{V} - C_{osm}$$

$$C_{H_2O} = \dot{V} - \frac{U_{osm} \cdot \dot{V}}{P_{osm}}$$

where

C_{osm} = Osmolal clearance
U_{osm} = Urine osmolality
P_{osm} = Plasma osmolality
$\dot{V}$ = Urine flow rate

For hyposmotic urine: C_{H_2O} is positive (ie, solute-free water is removed from body fluid).

For hyperosmotic urine: C_{H_2O} is negative (also written as $T^{C}_{H_2O}$, ie, solutes are removed without water from the tubular fluid or free water is added to body fluid).

For isosmotic urine: C_{H_2O} is zero.

Review Questions

20. The most significant contribution of the loop of Henle to the process of urine concentration and dilution is

    A. the production of hyposmotic tubular fluid
    B. the generation of high osmotic gradient for fluid within the loop
    C. acting as a countercurrent exchanger that helps maintain the medullary osmotic gradient
    D. the generation of high osmotic gradient within the medullary interstitium
    E. being the site for urea secretion and participating in medullary urea recycling

21. A person with normal plasma osmolality (300 mOsm/kg) excretes 2 liters of urine per day at an osmolality of 600 mOsm/kg. The effect of this urine excretion on the body fluid osmolality would be identical to that of

    A. subtracting 2 liters of pure water
    B. subtracting 2 liters of pure water and adding 1 liter of 300 mOsm/kg fluid
    C. adding 2 liters of 300 mOsm/kg fluid
    D. adding 2 liters of pure water
    E. maintaining body fluid osmolality constant

REGULATION OF WATER BALANCE

The regulation of body water depends upon the dynamic balance between rates of water movement into and out of the body. The two major mechanisms responsible for water balance are **thirst** and **ADH regulation of urinary water excretion.** The circulating level of ADH regulates the amount of water reabsorption from distal tubules and collecting ducts. Therefore, the regulation of renal excretion of water is ultimately determined by factors which influence the rates of synthesis and release of ADH into the blood and its renal action. ADH is a peptide hormone synthesized in specialized hypothalamic neurons and transported to the posterior pituitary where it is stored until release. Since water gain or deficit significantly affects total solute concentration within body fluids, **plasma osmolality** is the **most**

important regulator of ADH release. The amount of ADH released increases with a rise in plasma osmolality via stimulation of **hypothalamic osmoreceptors.** A decrease in **blood volume** also stimulates ADH release, because the resulting decrease in atrial pressure relieves the inhibitory effect of atrial baroreceptors on ADH release. ADH binds to specific receptors on **peritubular membranes** of epithelial cells of the distal nephron and, via an **activation of the adenylate cyclase** enzyme system, **increases the permeability of luminal membranes of the epithelial cells to water.**

Water intake is regulated through a **thirst center** located in the hypothalamus. Thirst is also **stimulated by both an increase in plasma osmolality and a reduced extracellular fluid volume,** thus working in concert with the ADH mechanism to maintain water balance. The hormone **angiotensin** of the renin-angiotensin system is an important stimulus of the thirst mechanism.

REGULATION OF SODIUM BALANCE

Sodium is the most abundant solute in extracellular fluid. Consequently, the status of **Na^+ balance critically determines the size (volume) of the extracellular fluid compartment** and the **long-term regulation of blood pressure.** The kidney regulates Na^+ balance by adjusting the amount of Na^+ excretion according to Na^+ intake. Na^+ excretion is the result of two processes: glomerular filtration and tubular reabsorption. The kidneys conserve Na^+ by normally **reabsorbing 99.4% of filtered Na^+.** Autoregulation of GFR automatically prevents excessive changes in the rate of Na^+ excretion in response to spontaneous changes in blood pressure. In addition, glomerulotubular balance compensates for changes in the filtered load of Na^+ due to acute changes in GFR under normal Na^+ and volume status. When there is a chronic change in GFR due to a change in the size of the extracellular fluid compartment, glomerulotubular balance is abolished, and Na^+ excretion varies more directly with GFR, such that Na^+ balance and extracellular fluid volume are restored.

The final and most important regulation of Na^+ excretion is at the level of **distal tubules and collecting ducts,** where the steroid hormone from the adrenal cortex, **aldosterone,** acts to **stimulate Na^+ reabsorption.** Aldosterone secretion is increased in response to **decreased plasma Na^+ concentration, increased plasma K^+ concentration, and increased plasma angiotensin II concentration.** Therefore, the **renin-angiotensin-aldosterone** system is the most important regulator of Na^+ balance. Since the amount of renin release determines the level of angiotensin II, the factors that govern renin release ultimately influence renal excretion of Na^+. The most important stimulus for renin release is depletion of the extracellular fluid compartment volume, the response to which is mediated by **sympathetic nervous system** and sensor mechanism(s) within the **juxtaglomerular apparatus complex** and **renal afferent arterioles.** Increased renal sympathetic nerve activity will also directly increase Na^+ reabsorption. The **Na^+-retaining system** thus consists of the renin-angiotensin-aldosterone system and the sympathetic nervous system. There is also a humoral **Na^+-losing system,** the **natriuretic hormones** which increase renal excretion of Na^+. One such hormone is **atrial natriuretic peptide (ANP),** which is released from the atria in response to an expansion of the extracellular fluid compartment. ANP increases Na^+ excretion, in part, by raising GFR and by reducing tubular reabsorption of Na^+ in collecting ducts. ANP also inhibits renin and aldosterone secretion. In addition, **renal interstitial hydraulic pressure** can also influence Na^+ reabsorption in an inverse manner.

In disease states when there is a decrease in the effective extracellular fluid volume (eg, cirrhosis, heart failure), renal tubular Na^+ reabsorption increases resulting in Na^+ retention **(generalized edema)**.

Review Questions

22. Antidiuretic hormone (ADH) conserves water by

    A. constricting afferent arterioles, thereby reducing GFR
    B. increasing water reabsorption by the proximal tubule
    C. stimulating active reabsorption of solutes in the ascending limb of the loop of Henle
    D. increasing water permeability of the distal tubule and collecting duct
    E. stimulating urea synthesis in the collecting duct

23. The volume of the extracellular fluid (ECF) compartment is determined by the _____ of the ECF.

    A. sodium concentration
    B. sodium content
    C. potassium concentration
    D. potassium content
    E. chloride concentration

24. During severe exertion in a hot environment a person may lose 4 liters of hypotonic sweat per hour. This would result in

    A. decreased plasma volume
    B. decreased plasma osmolality
    C. decreased circulating levels of ADH
    D. decreased circulating levels of aldosterone
    E. decreased circulating levels of renin

25. Which of the following actions of a drug is possible if treatment of a patient causes formation of a large volume of urine with an osmolarity of 100 mOsm/L.? (Consider each action separately).

    A. Inhibition of renin secretion
    B. Increase of ADH secretion
    C. Increased permeability to water in distal tubules and collecting ducts
    D. Decreased active chloride reabsorption by the ascending limb of the loop of Henle
    E. Inhibition of aldosterone secretion

For Questions 26-28. A 45-year-old man who has had a blood pressure of 150/90 mm Hg for the past six months consents to physiologic studies for research purposes prior to treatment. He is stabilized on a daily 200 mEq Na^+ diet for three days prior to provocative testing. He remains on this diet throughout the study period.

	24-Hour urinary Na^+	Plasma Volume	Plasma Renin	Blood Pressure
Baseline measurements	200 mEq	3400 ml	2 ng/ ml/hr	150/90 mm Hg

For each pharmacologic maneuver below, select the letter that most closely approximates the laboratory findings.

	24-Hour urinary Na^+(mEq)	Plasma Volume (ml)	Plasma Renin (ng/ml/hr)	Blood Pressure (mm Hg)
A.	200	3400	1.5	140/85
B.	50	2900	5	140/85
C.	100	3800	5	140/85
D.	200	3400	5	155/95
E.	100	3800	1.5	155/95

26. **One day after** one dose of a short-acting powerful diuretic drug

27. During administration of a direct arteriolar vasodilator drug

28. During administration of aldosterone

For Questions 29-31. A 60-year-old man who was found unconscious is brought to the emergency room by ambulance. A neighbor states that the patient has not felt well over the past week and looks as though he has lost weight lately. After an initial physical examination, laboratory studies are obtained.

For each set of laboratory findings below, select the most appropriate diagnosis.

A. Diabetes insipidus
B. Dehydration
C. Excessive production/release of antidiuretic hormone

	Serum Na^+ (mEq/L)	Plasma Osmolality (mOsm/Kg)	Urine Osmolality (mOsm/Kg)
29.	150	310	200
30.	150	310	850
31.	125	260	500

REGULATION OF ACID-BASE BALANCE

Acid-base balance or the concentration of H^+ in the extracellular fluid is tightly regulated, such that the pH of the arterial blood is maintained within a small range, **pH 7.37-7.42 or $[H^+]$ = 40 nmols/L.** This delicate balance is threatened continuously by additions of extra acids or bases to body fluids from metabolic processes. Cellular respiration produces some 20,000 mmoles of CO_2 (or H_2CO_3, **volatile acid**) daily; this is continuously eliminated by the lungs so that no pH change occurs under normal conditions. Metabolism of foodstuffs produces **non-volatile or fixed acids** from protein diets (eg, H_2SO_4, H_3PO_4) and bases from vegetarian diets (eg, lactate, citrate). In addition, fixed acid concentrations may rise during exercise or many pathological conditions. Most of these acids/bases are buffered by **buffering systems** within the body. These buffers act within seconds, therefore chemical buffering is the **first line of defense** against changes in H^+ concentration. The remaining acids are eliminated through the **lungs** by appropriate adjustments in alveolar ventilation within seconds or minutes (**second line of defense**). Finally, acid-base balance is adjusted by the **kidneys** through renal excretion of H^+ and renal reabsorption of HCO_3^-, which take days to complete (**third line of defense**).

Buffering Systems

Phosphate buffer ($HPO_4^{2-}/H_2PO_4^-$), with a pK of 6.8, contributes little to the buffering capacity of the extracellular fluid because of its low concentration. Organic phosphate is an important chemical buffer within intracellular fluid.

Protein buffers include plasma proteins (extracellular), hemoglobin (Hb) within red blood cells, and other intracellular proteins. The abundance of plasma proteins and hemoglobin together in blood and their broad-ranged pK values make them strong buffers. Hb, as protein buffer, helps buffer H^+ generated from CO_2 during CO_2 transport in blood from tissues to the lungs. In addition, the buffering capacity of Hb is further enhanced during the process of deoxygenation. Deoxygenated Hb is less acidic than oxygenated Hb, so it can act as a base and accepts extra H^+ formed from CO_2 within red cells during the passage of CO_2 from tissues to the lungs. The enhanced buffering capacity of deoxygenated Hb prevents significant pH changes between arterial and venous blood during CO_2 transport.

Bicarbonate buffer (HCO_3^-/H_2CO_3 or HCO_3^-/P_{CO_2}) is the **most important physiological buffer of extracellular fluid**, because 1) it exists in high concentration in plasma (24 mM), and 2) **the buffer pair can be tightly regulated: CO_2 by the lungs and HCO_3^- by the kidneys**. Since all buffer pairs in plasma are in equilibrium with the same concentration of H^+, a change in the buffering capacity of the entire blood buffer system will be reflected by a change in the buffering capacity of only one buffer pair (**isohydric principle**). Therefore, the acid-base status or pH of extracellular fluid can be evaluated by examining only the bicarbonate buffer system. The pH of extracellular fluid containing HCO_3^- buffer can be expressed as functions of the buffer pair concentrations using the Henderson-Hasselbalch equation.

$$CO_2 + H_2O \rightleftarrows H_2CO_3 \rightleftarrows H^+ + HCO_3^-$$

$$pH = 6.1 + \log \frac{[HCO_3^-]}{[H_2CO_3]} \quad \text{(pK for this system is 6.1)}$$

$$pH = 6.1 + \log \frac{[HCO_3^-]}{[0.03 \times P_{CO_2}]}$$

(the proportionality constant between dissolved CO_2 and P_{CO_2} is equal to 0.03)

$$pH = 6.1 + \log \frac{24 \text{ mmoles/L}}{0.03 \times 40 \text{ mm Hg}}$$

$$pH = 6.1 + \log \frac{20}{1}$$

Thus, the maintenance of a normal plasma pH depends on the preservation of the ratio of HCO_3^- concentration to CO_2 concentration in plasma at approximately 20:1.

Utilization of Various Buffers. If the acid-base disturbance is derived from the addition of the bicarbonate buffer pair to the extracellular fluid, then more than 95% of the buffering will be done by proteins and phosphates within cells, because the HCO_3^- buffer system cannot buffer itself. If the disturbance is derived from the addition of fixed acids or bases, extracellular buffering by HCO_3^- buffer will account for nearly half of the total chemical buffering occurring in the body fluids.

Respiratory Regulation of Acid-Base Balance

When arterial pH is made more acidic or alkaline, the rate of alveolar ventilation is altered via changes in signals from respiratory chemoreceptors. The resulting hyper- or hypoventilation will change arterial P_{CO_2} in the direction that will return arterial pH towards normal.

$$pH \downarrow \rightarrow \text{alveolar ventilation} \uparrow \rightarrow P_{CO_2} \downarrow \rightarrow \frac{[HCO_3^-]}{CO_2} \uparrow \rightarrow pH \uparrow$$

$$pH \uparrow \rightarrow \text{alveolar ventilation} \downarrow \rightarrow P_{CO_2} \uparrow \rightarrow \frac{[HCO_3^-]}{CO_2} \downarrow \rightarrow pH \downarrow$$

The respiratory system cannot by itself restore pH back to normal. As a result of chemical buffering, the body store of HCO_3^- buffer will be depleted, and the buffered acids or bases in the body fluids need to be eliminated. Only the renal system can perform these two tasks and finally restore acid-base balance.

Renal Regulation of Acid-Base Balance

The kidneys regulate acid-base balance in three ways; 1) **Conservation of filtered HCO_3^-**; more than 99.9% of filtered HCO_3^- is reabsorbed by renal tubules. This process prevents the development of acidosis due to HCO_3^- loss. 2) **Replenishment of depleted HCO_3^- store (formation of new HCO_3^-)**; and 3) **Excretion of excess H^+.** The latter two processes involve combining secreted H^+ with non-HCO_3^- buffers in the urine, simultaneously forming new HCO_3^- moieties and excreting acids, thus alkalinizing body fluids.

Bicarbonate Reabsorption. Under normal conditions urine is almost totally free of HCO_3^-. Most of the HCO_3^- reabsorption (80-90% of the filtered load) occurs in proximal tubules. Another 2% of the filtered load is reabsorbed in the loops of Henle, and the rest (8%) is reabsorbed in distal tubules and collecting ducts. In the proximal tubule net HCO_3^- reabsorption occurs as a result of the combination of secreted H^+ with filtered HCO_3^- in tubular fluid catalysed by luminal carbonic anhydrase. The secreted H^+ are consumed by reactions with HCO_3^-; this keeps H^+ concentration in the tubular fluid low, resulting in only a small pH change. The slight pH change favors more H^+ secretion as well as preventing backflux; therefore, essentially all filtered HCO_3^- is reabsorbed. In the distal nephron some secreted H^+ can combine with HCO_3^- without the action of luminal carbonic anhydrase, and further HCO_3^- reabsorption occurs.

The rate of reabsorption of HCO_3^- by renal tubules is influenced by many factors. 1) **Filtered load of HCO_3^- can vary over a wide range**, and the rate of HCO_3^- reabsorption will increase accordingly such that HCO_3^- reabsorption remains complete. It is possible that high concentrations of HCO_3^- in the filtrate raise tubular fluid pH, favoring more H^+ secretion and thus more HCO_3^- reabsorption. 2) The **status of the extracellular fluid compartment volume** influences the rate of HCO_3^- reabsorption. An expansion of extracellular fluid volume will result in decreased Na^+ reabsorption, therefore Na^+-coupled H^+ secretion and consequently HCO_3^- reabsorption will decrease. 3) High **arterial P_{CO_2}** will increase the rate of HCO_3^- reabsorption, possibly through the effects of P_{CO_2} on HCO_3^- formation in blood and on the cellular production and secretion of H^+. The dependence of HCO_3^- reabsorption on P_{CO_2} allows the kidney to respond to acid-base disturbances originating from respiratory causes. 4) The rate of HCO_3^- reabsorption depends on the **concentrations of other ions in the extracellular fluid.** The rate of HCO_3^- reabsorption is inversely correlated with plasma concentration of Cl^-. Increasing plasma K^+ level will lead to a reduction in H^+ secretion and HCO_3^- reabsorption; decreasing plasma K^+ has the opposite effect. 5) HCO_3^- reabsorption is affected by certain **hormones.** Corticosteroids enhance HCO_3^- reabsorption, and parathyroid hormone has the opposite effect.

Formation of New Bicarbonate. Hydrogen ions are produced within renal cells from carbonic acid and then secreted into tubular fluid. The secreted H^+ can also combine with other urinary non-HCO_3^- buffers in tubular fluid, namely phosphates and ammonia. In this case H^+ remain in tubular fluid as parts of the buffer pairs and later are excreted. The HCO_3^- that is formed within renal cells at the same time is transported across the basolateral membrane by an active mechanism. Therefore, for every H^+ that is secreted and excreted with non-HCO_3^- buffers, a new moiety of HCO_3^- is formed within renal cells and added to body fluids. Urinary phosphate buffer that combines with secreted H^+ (ie, $H_2PO_4^-$) is called **titratable acid.** It can be measured as the amount of strong base required to titrate 1 ml of urine back to the pH of the glomerular filtrate or plasma. The amount of titratable acid formed is limited by the supply of urinary phosphate buffer (HPO_4^{2-}), since about 75% of filtered phosphate is reabsorbed. Therefore, another urinary buffer such as ammonia is more effectively used to buffer the secreted H^+, so a large amount of H^+ can be excreted without urine pH dropping to a very low level. H^+ secretion is a gradient-limited transport process; the distal nephron cannot transport H^+ against a concentration gradient exceeding 1000:1 (ie, when urine pH equals 4.4). If the urine pH drops below pH 4.4, the body will not be able to get rid of any more excess H^+.

The NH_3/NH_4^+ buffer system has a pK of 9.2, so it is a relatively poor buffer in the pH range found in tubular fluid. However, there is a plentiful supply of NH_3 from renal cells. NH_3 is produced within renal cells by transamination reactions of amino acids, glutamine being the principal source. Ammonia, being uncharged and lipid-soluble, freely diffuses across cell membrane down its concentration gradient. In the lumen it combines with secreted H^+ to form NH_4^+. Ammonium ion (NH_4^+), being charged and relatively impermeable, is "trapped" in the tubular fluid and excreted in urine in the form of neutral salts, such as $(NH_4)_2SO_4$ or NH_4Cl. The transport process of NH_3 is known as **diffusion-trapping or non-ionic diffusion**, which is a special case of passive diffusion. Additionally, in the proximal tubule NH_4^+ may enter the lumen by substituting for H^+ on the Na^+-H^+ exchanger. The effectiveness of the NH_3 buffer system is enhanced during an acid load. The rate of synthesis of NH_3 is regulated according to the acid-base status of the individual; more NH_3 is synthesized during acidosis, permitting more excretion of excess acid.

Under normal conditions, all filtered HCO_3^- is reabsorbed, and an additional 40-60 mmoles of acid is secreted, contributing 40-60 mmoles of new HCO_3^- added to blood to replenish the HCO_3^- used to buffer the acid produced from metabolism. The secreted acid is excreted with HPO_4^{2-} (25%) or NH_3 (75%). Urine is usually slightly acidic, pH ~ 6. During acidosis the kidneys compensate by excreting more acidic urine, still completing HCO_3^- reabsorption and increasing the excretion of titratable acid and NH_4^+. During alkalosis cell pH rises, providing less driving force for H^+ secretion, so less HCO_3^- is reabsorbed. The unreabsorbed HCO_3^- will alkalinize the urine, less NH_3 will be trapped and less acid excreted. Overall, more HCO_3^- will be eliminated from the body, and the body fluid will become more acidic.

Quantitation of Renal Tubular Acid Secretion and Excretion

Total rate of H^+ secretion = Rate of HCO_3^- reabsorption + rate of titratable acid excretion + rate of NH_4^+ excretion

Total rate of H^+ excretion = Rate of titratable acid excretion + rate of NH_4^+ excretion
= Total rate at which new HCO_3^- is added to the blood

Acid-Base Disturbances: Defect, Consequence, and Compensation

Disturbances of acid-base balance result in deviations from normal values of the variables that describe the acid-base status of extracellular fluid These are arterial pH of 7.4, [HCO_3^-] of 24 mM, P_{CO_2} of 40 mm Hg, and [HCO_3^-]/P_{CO_2} ratio of 20:1. **Acidosis** is the state where arterial pH is below 7.36, and **alkalosis** is the state where arterial pH is above 7.44. Since P_{CO_2} is regulated by the rate of alveolar ventilation, any disturbance in H^+ concentration that results from a primary change in P_{CO_2} is called a **respiratory** disturbance. Changes in the concentration of HCO_3^- are most commonly caused by the addition or loss of fixed acids or bases derived from metabolic processes. Therefore, any abnormality of pH that results primarily from a change in HCO_3^- concentration is called a **metabolic** disturbance. There are **four** primary acid-base disturbances: respiratory acidosis/alkalosis and metabolic acidosis/alkalosis. In most cases a primary disturbance of one origin is accompanied by a secondary or **compensatory response of the opposite origin.** The compensatory response shifts pH towards its normal value. Compensations for metabolic disturbances are almost instantaneous but those for primary respiratory disturbances require several days. The efficiency of compensatory responses is indicated by how close arterial pH is brought back to 7.4. The acid-base patterns during the primary acid-base disturbances and their compensations are graphically represented on the pH-bicarbonate diagram in Fig. 5-2.

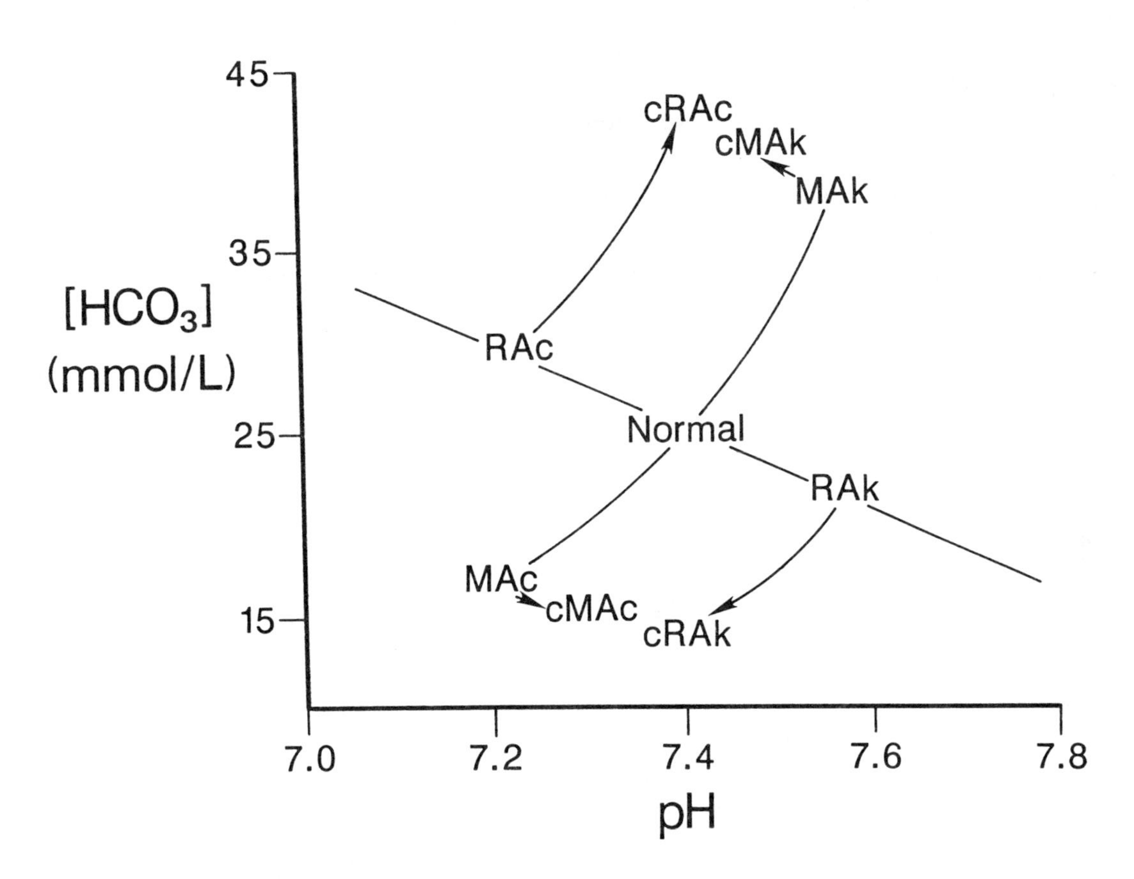

Figure 5-2. Changes of pH and $[HCO_3^-]$ with acid-base disturbances and their compensations. (Modified with permission from Davenport, HW, *The ABC of Acid-base Chemistry*, 5th ed. Copyright 1969 by Univ. of Chicago Press).

Respiratory acidosis (RAc) results from high P_{CO_2} due to failure of the lungs to excrete CO_2 adequately (eg, due to pulmonary diseases). The increase in carbonic acid will increase HCO_3^- concentration, changing the $[HCO_3^-]/P_{CO_2}$ ratio to less than 20:1 (change in P_{CO_2} is greater) and causing a fall in plasma pH. As a compensation, increased P_{CO_2} will increase H^+ production and secretion from

ONE REASON FOR STONE FORMATION ?

renal tubular cells, HCO_3^- reabsorption will be increased, more H^+ will be excreted, and new HCO_3^- will be produced, returning the $[HCO_3^-]/P_{CO_2}$ ratio closer to 20:1. An elevated P_{CO_2} and a normal pH indicate compensated respiratory acidosis (cRAc).

Respiratory alkalosis (RAk) results from low P_{CO_2} due to excessive loss of CO_2 (eg, hyperventilation). There will be a decrease in HCO_3^- concentration, raising the $[HCO_3^-]/P_{CO_2}$ ratio to greater than 20:1 (change in P_{CO_2} is greater) and elevating pH. Renal compensation includes reduction in H^+ secretion and increased excretion of HCO_3^-, further reducing HCO_3^- concentration and returning the $[HCO_3^-]/P_{CO_2}$ ratio nearer to 20:1. Compensated respiratory alkalosis (cRAk) is indicated by normal pH but a still-depleted HCO_3^- store.

Metabolic acidosis (MAc) results from abnormal retention of fixed acids which are buffered by HCO_3^-. There will be a decrease in HCO_3^- concentration, decreasing the $[HCO_3^-]/P_{CO_2}$ ratio to less than 20:1 and decreasing pH. The respiratory center will be stimulated to eliminate more CO_2, thus reducing H_2CO_3 concentration. Renal compensation also occurs unless acidosis is due to renal failure. There will be virtually complete reabsorption of HCO_3^- and increased excretion of titratable acid and NH_4^+. The compensation will restore the $[HCO_3^-]/P_{CO_2}$ ratio nearer to 20:1 and the pH closer to 7.4 (cMAc).

Metabolic alkalosis (MAk) results from excessive loss of H^+ (eg, loss of HCl from vomiting) or excessive intake or retention of bases (eg, $NaHCO_3$, lactate). There will be an increase in HCO_3^- concentration, increasing the $[HCO_3^-]/P_{CO_2}$ ratio to greater than 20:1, thus raising pH. Compensatory decrease in respiration will lead to retention of CO_2 and increased H_2CO_3 concentration. Renal compensation involves increased renal excretion of bicarbonate. The $[HCO_3^-]/P_{CO_2}$ ratio thus returns towards 20:1, and the arterial pH is restored towards 7.4 (cMAk). However, the compensatory retention of CO_2 will tend to increase H^+ secretion and HCO_3^- reabsorption, thereby limiting the effectiveness of respiratory compensation.

Review Questions

32. The HCO_3^-/CO_2 buffer system is the most important physiological buffer, because

    A. it is the most effective buffer for preventing pH changes when volatile acid is added to body fluids
    B. its pK is very close to the normal pH of body fluids
    C. P_{CO_2} and $[HCO_3^-]$ can be altered by the lungs and the kidneys, respectively
    D. its concentration is high in the intracellular fluid compartment
    E. it is the only buffer system in the body

33. A state of alkalosis will lead to

    A. a decreased rate of ventilation
    B. increased renal excretion of ammonium
    C. increased renal reabsorption of bicarbonate
    D. increased renal excretion of titratable acid
    E. excretion of an acidic urine

34. The kidney regulates acid-base balance by all of the following mechanisms **EXCEPT**

A. reabsorbing filtered bicarbonate
B. excreting titratable acid
C. active reabsorption of H^+
D. acidifying the urine via H^+ secretion
E. excreting ammonium salts

35. Which of the following would tend to increase excretion of H^+?

A. Increased renal reabsorption of HCO_3^-
B. Increased renal synthesis of NH_3
C. Decreased concentration of urinary buffers (eg, phosphate)
D. Increased carbonic anhydrase activity in the lumen of proximal tubules
E. Decreased P_{CO_2}

36. Which of the following will cause a decrease in H^+ secretion by the proximal tubule?

A. Inhibition of carbonic anhydrase
B. Increasing filtered HCO_3^-
C. Increasing P_{CO_2}
D. Decreasing intracellular Na^+
E. Increasing filtered Na^+

For Questions 37-38. Match one line of values from the table below with the description for each question. Normal P_{CO_2} = 40 mm Hg and normal [HCO_3^-] = 24 mM.

	P_{CO_2} (mm Hg)	[HCO_3^-] (mM)	pH
A.	60	37	7.4
B.	66	28	7.25
C.	29	22	7.51
D.	60	57	7.60
E.	35	15	7.25

37. ___ Uncompensated respiratory alkalosis

38. ___ Partially compensated metabolic acidosis

For Questions 39-41. A patient who has been suffering from chronic lung disease is found to have an arterial pH of 7.37 and an arterial P_{CO_2} of 54 mm Hg.

39. One would classify his condition as a primary

A. respiratory acidosis
B. respiratory alkalosis
C. metabolic acidosis
D. metabolic alkalosis
E. normal acid-base status

40. One would expect his arterial

A. plasma bicarbonate concentration to be higher than normal
B. plasma bicarbonate concentration to be lower than normal
C. plasma hydrogen ion concentration to be lower than normal
D. plasma dissolved CO_2 concentration to be higher than normal
E. blood buffer base to be higher than normal

He later develops severe diarrhea. Plasma tests indicate the following laboratory values: pH = 6.92, P_{CO_2} = 40 mm Hg and HCO_3^- concentration = 8 mEq/L

41. His condition is now classified as

A. metabolic alkalosis and respiratory acidosis
B. metabolic acidosis and respiratory acidosis
C. metabolic acidosis and respiratory alkalosis
D. metabolic acidosis
E. respiratory acidosis

ANSWERS TO RENAL PHYSIOLOGY QUESTIONS

1. C. The volumes of the intracellular, extracellular, interstitial, plasma and transcellular compartments represent about 40%, 20%, 15%, 4%, and 1% of body weight, respectively. The inulin space measures ECF volume.

2. C. Total body water (TBW) (Tritiated water) = (2.5 - 0.4)/0.05 = 42 L. Extracellular fluid (ECF) (inulin) = (1.6 - 0.2)/0.1 = 14 L. Intracellular fluid = TBW - ECF = 28 L.

3. C. Hemorrhage is isotonic volume depletion. Isotonic NaCL (0.9%) restores both water and ions (C). Glucose is metabolized, so it does not restore ion concentrations (A,B).

4. D. Interstitial fluid is relatively protein-free (D); both it and plasma are extracellular.

5. A. Only the Starling forces in the glomerular capillaries are involved in glomerular filtration. Glomerular capillary hydrostatic pressure forces fluid out of the capillaries (D).

6. E. Renal autoregulation is an intrinsic property of the kidney.

7. B. Obstructing the urinary tract increases the hydrostatic pressure in Bowman's capsule, decreasing GFR (B). Dilating afferent arterioles increases renal plasma flow, increasing GFR (A). Constricting efferent arterioles increases hydrostatic pressure in glomerular capillaries, increasing GFR (C). Decreasing plasma albumin decreases plasma osmotic pressure, leading to increased GFR (D). Decreased renal sympathetic nerve activity would increase GFR (E).

8. E. RPF = (10 mg/ml x 2 ml/min)/0.04 mg/ml - 0.004 mg/ml) = 556 ml/min; RBF = 556/1 - 0.55 = 1235 ml/min.

9. B. GFR = (20 mg/ml x 2 ml/min)/0.4 mg/ml = 100 ml/min.

10. D. Clearance of X = (1 ml/min x 12 mg/ml)/0.02 mg/ml = 600 ml/min. Since the volume of plasma being cleared of X is greater than the GFR, the substance must have been secreted into the tubule and so could be PAH (D). Inulin is only filtered (C); others are reabsorbed (A, B, E).

11. E. Excreted PAH derives from both filtration and secretion. Increasing the plasma concentration of PAH past threshold will saturate the transport system for PAH (A), and the tubular maximum will be exceeded (C). The amount of PAH secreted will remain constant once threshold has been exceeded. Nevertheless, the amount of PAH excreted will increase as more PAH is filtered (D). Since the amount of secreted is not increasing, the amount of PAH excreted will be less than two times the amount excreted at threshold (B).

12. D. (125 ml/min x 2 mg/ml) - 30 mg/min + 60 mg/min = 280 mg/min.

13. $$C_{creatinine} = \frac{\text{Creatinine excreted/day}}{\text{Serum creatinine}} = \frac{400 \text{ mg/12h x 24h/day}}{8 \text{ mg/L}} = 100 \text{ L/day or } 69 \text{ ml/min}$$

14. A. Serum creatinine has been stable for 6 months, and creatinine production will only rise if there is an increase in muscle mass (C). For serum creatinine to double, the person must have doubled his muscle mass, an unlikely situation. Serum creatinine is not related to protein synthesis (D).

15. D. Na^+, K^+-ATPase is located in the the basolateral membrane.

16. D. Na^+-dependent secondary active transport of Cl^- is restricted to the ascending limb of the loop of Henle (D).

17. D. K^+ is freely filtered, and then reabsorbed mainly in the proximal tubule and secreted in the distal tubule and collecting duct (D).

18. B. The ascending limb is relatively impermeable to water but actively reabsorbs NaCl (A, B, D, E). The descending limb is relatively impermeable to solutes and does not actively reabsorb solutes (C).

19. E. Urinary K^+ excretion is determined mostly by the amount of K^+ secreted into tubular fluid by the distal tubule and collecting duct, not by obligatory proximal K^+ reabsorption. K^+ secretion at these nephron sites is reduced (A) by a decrease in serum [aldosterone] (C) as well as in [Na^+] (B) and flow rate of tubular fluid (D). Therefore, K^+ excretion by the distal tubule and collecting duct will be reduced in this woman, and she will be in positive K^+ balance (intake > excretion).

20. D. The loop of Henle acts as a countercurrent multiplier. Although it does generate a high osmotic gradient within the loop (B), produces hyposmotic tubular fluid (A), and participates in urea recycling (E), its most significant contribution is generation of the medullary osmotic gradient (D). Without the gradient, concentrated urine cannot be produced.

21. D. $C_{H_2O} = 2 \text{ L/day} - \dfrac{600 \text{ mOsm/kg} \times 2 \text{ L/day}}{300 \text{ mOsm/kg}} = -2 \text{ L/day}$. A negative value for free water clearance means free water has been added back to body fluids (D).

22. D. ADH increases permeability of the distal tubule and collecting duct to water (D).

23. B. Potassium is the major intracellular electrolyte (C, D). Chloride is distributed close to equilibrium between intra- and extracellular compartments (E). Concentration has no effect on volume (A). Content = volume x concentration, therefore for a given Na^+ concentration, ECF volume is determined by Na^+ content (B).

24. A. Sweat contains K^+ as well as Na^+ and Cl^-. Therefore, there is a net loss of water, K^+, Na^+, and Cl^-. Plasma volume would be smaller (A). But the loss of water is relatively greater than that of Na^+ and Cl^-, so there would be an increase in plasma osmolality stimulating the release of ADH (B, C). Net Na^+ loss will result in an increased circulating levels of renin and aldosterone (D, E).

25. D. Inhibition of renin and aldosterone secretion inhibits Na^+ reabsorption; this increases urine osmolarity (A,E). Increase of ADH secretion increases water reabsorption in distal tubules and collecting ducts, producing a concentrated urine (B). Increased permeability to water produces a concentrated urine (C). Decreased chloride reabsorption in the loop of Henle decreases the corticomedullary osmotic gradient and produces a dilute urine (D).

26. B. Following administration of a diuretic, plasma volume is contracted. So the kidney reabsorbs more Na^+, and renin level rises.

27. C. A direct arteriolar vasodilator lowers blood pressure and reduces filtered Na^+. Therefore, the kidney retains Na^+, and plasma renin level reflexly rises. Plasma volume increases as a result of Na^+ retention.

28. E. Aldosterone causes retention of Na^+ and expansion of plasma volume. Blood pressure rises. Renin output would be mildly suppressed by the hypervolemia.

29. A. Diabetes insipidus occurs when the antidiuretic hormone system is non-functional. This results in excretion of large volumes of very dilute urine and subsequent loss of body water, causing increases in plasma Na^+ and osmolality.

30. B. Dehydration causes increases in both serum Na^+ and plasma osmolality, and the kidneys excrete very concentrated urine.

31. C. Too much ADH results in production of concentrated urine and retention of excess body water, lowering serum Na^+ and plasma osmolality.

32. C. The HCO_3^-/CO_2 buffer system is important, because the amounts of acid and base in the buffer pair are regulated by physiological processes.

33. A. Alkalosis results from effective loss of H^+. The kidney will compensate by decreasing H^+ secretion and excretion (E), decreasing excretion of titratable acid and ammonium (B, D), and decreasing HCO_3^- reabsorption (C). The respiratory response is hypoventilation (A).

34. C. H^+ are not reabsorbed by renal tubules (C).

35. B. Increased buffering capacity of tubular fluid by increasing phosphate (C) and NH_3 content (B) will enhance H^+ excretion. The process of HCO_3^- reabsorption in proximal tubules does not result in H^+ excretion (A, D). Decreased P_{CO_2} will reduce H^+ secretion and excretion (E).

36. A. Inhibition of carbonic anhydrase and decreased P_{CO_2} reduce formation of H_2CO_3, which limits the amount of H^+ secretion. The electrochemical gradient for H^+ secretion is enhanced when the amount of filtered HCO_3^- is increased (B). Either decreasing intracellular Na^+ (D) or increasing filtered Na^+ (E) will enhances the driving force for Na^+-H^+ counter-transport.

37. C. Reduced P_{CO_2} and HCO_3^- with an alkaline pH.

38. E. Reduced P_{CO_2}, reduced HCO_3^- and acidic pH.

39. A. Lung disease causes elevation of P_{CO_2}, thereby inducing respiratory acidosis.

40. A. Elevation of P_{CO_2} causes an increase in plasma bicarbonate concentration.

41. B. A person with diarrhea eliminates a fluid rich in $NaHCO_3^-$, inducing a metabolic acidosis. The appropriate respiratory compensation for metabolic acidosis would reduce pCO_2 below normal range. The failure to do this is because of respiratory acidosis due to respiratory insufficiency.

GASTROINTESTINAL PHYSIOLOGY

The gastrointestinal (GI) system consists of the mouth, pharynx, esophagus, stomach, small and large intestines, rectum and associated organs and glands (ie, salivary glands, exocrine pancreas, and liver). Its overall function is to obtain nutrients from the external environment. With the exception of oxygen, all required nutrients are obtained through the GI system.

To accomplish this overall function, the GI system must perform a number of subsidiary functions. These include 1) **motility**, the contractile activities responsible for mixing GI contents and for controlled propulsion of the contents distally; 2) **secretion**, both exocrine secretion of digestive and lubricative substances and endocrine secretion of hormones; 3) **digestion**, the breakdown of macromolecules into smaller components; 4) **absorption**, the transfer of nutrients from the lumen to the portal circulation or lymph, and 5) **excretion**, the elimination of indigestible and nonabsorbable residue.

REGULATION OF GASTROINTESTINAL ACTIVITIES

Neural Regulation

Most GI activities are under both neural and hormonal control. The GI tract is richly innervated by both intrinsic and extrinsic nerves. The intrinsic neural supply consists of two major plexi, 1) the **myenteric** (Auerbach's) **plexus** located between longitudinal and circular layers of smooth muscle and 2) the **submucosal** (Meissner's) **plexus** located in the submucosa. These plexi, collectively referred to as the **enteric nerve plexus**, are an independent integrative system and not merely distribution centers relaying information from extrinsic nerves. Thus, the enteric nervous system possesses all the elements necessary for **short reflex** regulation of GI functions, ie, modification of motor and secretory activity by afferent and efferent nerves entirely within the GI tract.

The GI system is also innervated by **extrinsic autonomic nerves**. The **parasympathetic** nerve supply comes from the cranial **vagus nerve** except for the distal colon, which is supplied by pelvic nerves from the sacral spinal cord. **Sympathetic** postganglionic efferent nerves innervate the GI tract from the celiac, superior mesenteric, and superior and inferior hypogastric plexi. Most preganglionic parasympathetic fibers and postganglionic sympathetic fibers synapse with neurons in the enteric nerve plexus. These extrinsic nerves provide for **long reflexes**, which coordinate activities at widely separated sites along the GI tract. **Vagovagal** reflexes are long reflexes in which both the efferent and afferent limbs are mediated by the vagus nerve. "In general", the parasympathetics are stimulatory to GI functions while the sympathetics are inhibitory.

Endocrine Regulation

The mucosa of certain regions of the GI tract, particularly the gastric antrum and upper small intestine, contains endocrine cells. The major physiologically active hormones secreted by the GI tract are **gastrin, secretin, cholecystokinin** (CCK), **motilin** and **gastric inhibitory peptide** (GIP). GIP is also known as glucose-dependent insulinotrophic peptide. The source, stimuli for secretion, and major physiological actions of the above hormones are summarized in Table 6-1. Other actions have also been proposed for these hormones.

Table 6-1. Major actions of gastrointestinal hormones.

HORMONE	SOURCE	STIMULUS	ACTIONS*
Gastrin	gastric antrum and duodenum	peptides, Ca^{2+} and neural mechanisms	S acid S antral motility S mucosal growth
CCK	duodenum and jejunum	fats and amino acids	S pancreatic enzyme secretion S gallbladder contraction P pancreatic HCO_3 secretion I gastric emptying
Secretin	duodenum and jejunum	duodenal pH below 4.5	S pancreatic HCO_3 secretion S biliary HCO_3 secretion P pancreatic enzyme secretion
GIP	duodenum and jejunum	carbohydrate, fats and amino acids	S pancreatic insulin release I acid secretion ?
Motilin	upper small intestine	uncertain	S migrating myoelectric complex

*Key: S = stimulates; I = inhibits; P = potentiates; ? = uncertain significance

Paracrine Regulation

Paracrine messengers are secreted into interstitial fluid and diffuse to neighboring cells where they exert their effects. Important paracrines of the GI tract are **histamine** (which stimulates parietal cell secretion of HCl) and **somatostatin** (which inhibits secretion of gastrin by G cells).

Review Question

1. Most vagal preganglionic efferent nerves to the GI tract synapse with

   A. smooth muscle cells
   B. endocrine cells
   C. the enteric nerve plexus
   D. ganglia outside the GI tract
   E. exocrine glands

MOTILITY

Oral, Pharyngeal, and Esophageal Motility

Mouth and Pharynx. When large chunks of food are taken into the oral cavity, the teeth grind the food into small pieces which are mixed with **saliva** to form a **bolus**. The tongue propels the bolus of food to the back of the oral cavity, initiating a complex **swallowing reflex**. Pressure receptors in the walls of the pharynx are stimulated and send afferent signals to the **swallowing center** in the medulla. From the swallowing center efferent signals are sent to various **skeletal** muscles of the pharynx, larynx, and the upper third of the esophagus, and the **smooth** muscles of the lower two-thirds of the esophagus and the stomach. As food moves into the pharynx, the soft palate is elevated to seal off the nasal cavity. Respiration is momentarily inhibited, and the larynx is raised. The glottis closes, and the epiglottis swings back over it so that food cannot enter the trachea. Sequential contraction of the skeletal muscles of the pharynx forces the bolus through the pharynx toward the esophagus.

Esophagus. To enter the esophagus, the bolus of food must pass through the hypopharyngeal or **upper esophageal sphincter** (UES). The UES is composed of skeletal muscle and is normally closed due to elasticity of the sphincter and tonic neural excitation. During the oropharyngeal phase of swallowing the sphincter muscle relaxes reflexly, allowing the bolus to pass. The UES then closes to prevent regurgitation. Pressure in the resting esophagus parallels intrapleural pressure, and thus is subatmospheric during inspiration. Therefore, closure of the UES also prevents entry of air into the esophagus during inspiration.

In an upright posture swallowed liquids can simply flow down the esophagus due to the force of gravity. The passage of semi-solid food down the esophagus requires **peristalsis**. A ring of muscle contracts immediately above the bolus while muscles distal to the bolus are inhibited. This creates a pressure gradient that forces the bolus to move distally. The ring of contraction moves aborally and propels the bolus toward the stomach. The peristalsis that follows the oropharyngeal phase of swallowing is known as **primary peristalsis**. **Secondary peristalsis** is triggered by local distention of the esophagus and occurs in the absence of an oropharyngeal phase. It removes food left behind by inefficient primary peristalsis and returns material refluxed up from the stomach.

Peristalsis in the esophagus is coordinated centrally and peripherally. In the upper esophagus (skeletal muscle) orderly propagation of the peristaltic wave requires intact extrinsic innervation. Thus, bilateral section of the vagus paralyses the upper esophagus. In contrast, secondary peristalsis can occur in the lower esophagus (smooth muscle) without extrinsic innervation.

The cardiac sphincter or **lower esophageal sphincter** (LES), at the lower end of the esophagus, is closed under resting conditions due in part to tonic myogenic contraction. As the peristaltic wave approaches the LES, the sphincter relaxes due to vagal (non-cholenergic) inhibition, allowing food to pass into the stomach. The LES closes behind each bolus to prevent reflux of gastric contents into the esophagus. Tone of the LES is increased by many factors including elevated intra-abdominal and intra-gastric pressures, cholinergic agents and gastrin. LES tone is reduced by prostaglandin E and progesterone. Elevated progesterone levels contribute to the increased incidence of gastro-esophageal reflux, and consequent **heartburn**, seen in late pregnancy.

Review Questions

2. Cutting the extrinsic nerves to the esophagus and its sphincters will

   A. abolish primary peristalsis in the upper esophagus
   B. abolish secondary peristalsis in the distal esophagus
   C. cause increased constriction of the UES (upper esophageal sphincter)
   D. cause complete relaxation of the LES (lower esophageal sphincter)
   E. not affect esophageal function

3. The upper esophageal sphincter (UES)

   A. prevents entrance of liquids into the trachea
   B. reduces the respiratory dead space
   C. prevents regurgitation of gastric acid into the esophagus
   D. facilitates passage of material from the esophagus into the pharynx
   E. remains open between swallows

Gastric Motility

The primary function of the stomach is to store food for a variable length of time and to release it in a slow, controlled fashion into the duodenum. In the stomach food is converted to a thick soup called **chyme**. The stomach is separated into two major motor regions. The **proximal stomach** is the fundus and the oral 1/3 of the corpus. The **distal stomach** is the aboral 2/3 of the corpus, the antrum and the **pyloric sphincter** or **gastroduodenal junction**. Before the swallowed bolus reaches the stomach, the muscles of the proximal stomach relax, decreasing intragastric pressure and facillitating entrance of the bolus. This **receptive relaxation** is mediated reflexly by vagal nerves. The proximal stomach also exhibits the property of **accommodation**, ie, the volume of the stomach can increase greatly with little increase in intragastric pressure. Once **deglutition** (swallowing) has ceased, the proximal stomach exhibits two types of contractions; 1) **slow** sustained contractions of 1-3 min duration and 2) **rapid** phasic contractions lasting 10-15 sec superimposed on the sustained contractions. The sustained contractions exert steady pressure to gradually move gastric contents toward the distal stomach. These contractions may also play an important role in gastric emptying of liquids and small solids.

Contractions of the **proximal** stomach can be either excited or inhibited by vagal nerves. Inhibition predominates during deglutition resulting in receptive relaxation and accommodation. Once the stomach becomes distended, **vagovagal** and **short** reflexes stimulate contractions in the proximal stomach. On the other hand, **CCK** inhibits proximal sustained contractions.

The **distal** stomach exhibits **peristaltic contractions**. The maximum frequency, direction and speed of propagation of peristaltic contractions are determined by myogenic **slow wave depolarizations**. The smooth muscle cells exhibit slow wave depolarizations (about 3/min) consisting of a rapid upstroke followed by a plateau potential and return to resting levels. Slow wave depolarizations are not sufficient to cause significant contractions, since they occur even in the resting stomach. When the stomach is stimulated to contract, the plateau potential is elevated above threshold, and action potentials may be posed on the plateau potential. Slow wave depolarizations and peristaltic contractions usually n the corpus and spread distally, forcing chyme toward the **gastroduodenal junction** (GDJ). ractions are weak in the body of the stomach where smooth muscle layers are relatively thin. Contractions are more forceful in the more muscular antrum. As the antrum begins to contract, some chyme may pass through the GDJ into the duodenal bulb. As antral contractions continue (**antral systole**), the GDJ closes and the bulk of the chyme is retropeled back into the body of the stomach; the process then repeats. Peristaltic contractions and antral systole function to mix gastric contents and to break up digestible solids, allowing them to be suspended in and emptied along with the liquid. Gastric peristalsis is stimulated by 1) long and short **cholinergic reflexes** initiated by distention and 2) elevated serum **gastrin** (see Regulation of Gastric Secretion, p. 182). Elevated serum gastrin not only stimulates contractions but also increases the frequency of slow wave depolarizations.

Chyme that leaves the stomach enters the duodenum as fluid and small solids (less than 0.25mm in diameter). While peristaltic contractions of the distal stomach aid in gastric emptying, elevation of intragastric pressure due to sustained contractions of the proximal stomach also contribute to emptying of gastric contents. Thus, **CCK** reduces gastric emptying by inhibiting proximal contractions.

The stomach empties different components of a meal at different rates. Liquids empty more readily than digestible solids; small solids empty more readily than large solids. The rate of gastric emptying of liquids increases with increased volume of the gastric contents. A meal consisting primarily of carbohydrate will empty more rapidly than one consisting primarily of protein, which will empty more readily than a fatty meal. Increased acidity or osmolarity slows emptying. Isotonic NaCl is an exception, since it empties more readily than pure water. The effects of chemical composition on the rate of gastric emptying are mediated by specific receptors in the upper small intestine through both neural (**enterogastric reflex**) and hormonal mechanisms (**CCK** and unidentified hormones).

Contractions of the **duodenal bulb** are coordinated with gastric contractions. When the pylorus is open, the duodenal bulb is relaxed, facilitating gastric emptying and minimizing reflux of duodenal contents into the stomach. After antral systole and closure of the pylorus, the duodenal bulb contracts, moving the contents distally. This coordination is mediated via the **enteric** nerve plexus.

During fasting the stomach exhibits contractile activity called the **migrating myoelectric complex** (MMC), the migrating motility complex or the interdigestive myoelectric complex. The MMC lasts for 90-120 min and can be divided into four phases. During **Phase 1** the stomach is quiescent while **Phase 2** is characterized by intermittent action potentials and contractions increasing toward the end of the phase. **Phase 3** lasts 5-10 min and is characterized by intense bursts of action potentials and powerful gastric contractions. **Phase 4** is a short transition from Phase 3 to Phase 1. Contractions in late Phase 2 and Phase 3 are strong peristaltic contractions of the distal stomach that almost occlude the lumen. These contractions remove large (>1mm) non-digestible solids left behind after emptying of a meal. The timing of Phase 3 of the MMC appears to be controlled by the hormone **motilin**. Each phase of the MMC starts in the stomach and migrates to the distal ileum. In the **small intestine** the MMC functions during fasting to remove mucus, sloughed cells and bacteria and help prevent bacterial overgrowth.

Vomiting is a mechanism for the rapid evacuation of gastric and duodenal contents. Vomiting is usually preceded by nausea, increased salivation and retching. During retching the duodenum and antrum contract, forcing their contents into the relaxed body of the stomach. Inspiration occurs against a closed glottis, lowering intraesophageal pressure. The abdominal muscles contract, increasing intra-abdominal pressure and creating a pressure gradient between the stomach and esophagus that forces gastric contents into the esophagus. The abdominal muscles then relax, the esophageal contents drain back into the stomach, and the cycle may begin again. Vomiting is similar to retching. The abdominal muscles contract more forcefully. In addition, the larynx and hyoid bone are drawn forward, decreasing the tone of the UES and allowing expulsion of the gastric and esophageal contents via the oral cavity.

Review Questions

4. Slow wave depolarizations in the distal stomach

   A. are inhibited by elevated serum gastrin
   B. do not occur in the absence of neural stimulation
   C. are abolished by acidic chyme
   D. occur only when the stomach is distended
   E. occur in the resting (empty) stomach

5. The force required for evacuation of gastric contents during vomiting is generated by

   A. contraction of the proximal stomach
   B. reverse peristalsis in the esophagus
   C. contraction of abdominal muscles
   D. reverse peristalsis in the stomach
   E. expiring against a closed glottis

6. Gastric emptying is slowed by

   A. high fat content of chyme
   B. motilin
   C. high pH of duodenal chyme
   D. isotonic NaCl
   E. high starch content of chyme

7. The proximal stomach

   A. has a higher frequency of slow wave depolarizations than the distal stomach
   B. exhibits peristaltic contractions
   C. is not involved in emptying of liquids
   D. exhibits the property of receptive relaxation
   E. does not exhibit contractile activity

8. Contractions associated with the migrating myoelectric complex

   A. are important in preventing bacterial overgrowth in the small intestine
   B. occur during the post-prandial period
   C. are regulated by secretin
   D. are segmentation type contractions
   E. occur only in the stomach

Motility of the Small Intestine

The small intestine is the major site for digestion and absorption of most constituents of food. For efficient digestion and absorption to occur, the chyme must be mixed with various secretions and propelled down the tract. The maximum frequency and patterns of contraction in the small intestine are determined by **slow wave depolarizations**. The frequency of slow wave depolarization is highest in the proximal small intestine, about 12/min, and gradually decreases distally to become about 8/min in the terminal ileum. As in the stomach, the slow waves are myogenic and are not themselves sufficient to elicit contractions. Action potentials, superimposed upon the plateau potential of the slow wave, cause contraction of smooth muscles of the small intestine. The primary stimulus for contraction is radial stretching. The **law of the intestine** states that if the intestine is distended, it responds by contracting proximal to the distension and relaxing distal to it.

Contractile activities of the small intestine are local events affecting only a few cm at any one time. The small intestine exhibits **segmentation contractions** that divide the chyme into many segments and mix it with various secretions. Because segmentation contractions occur at a greater frequency in the proximal than in the distal small intestine, these contractions help propel the chyme distally. The small intestine also exhibits **peristalsis**. Peristaltic contractions can be initiated anywhere along the length of the small intestine but are propagated for only short distances. Peristaltic **rushes** (propagation of peristaltic contractions over long distances) are not normally observed. Peristalsis requires an intact enteric nervous system. Intestinal motility is stimulated by serotonin, gastrin, CCK, and motilin; it is reduced by sympathetic neural activity, epinephrine, secretin, and glucagon. Severe distention of one portion of the small intestine will result in inhibition of motility in the rest of the intestine. This **intestino-intestinal reflex** is mediated by extrinsic nerves.

Chyme leaves the small intestine through the **ileocecal sphincter**, a zone of elevated pressure located at the junction of the ileum and cecum. Closure of the sphincter is myogenic. The sphincter slows emptying of the ileum and prevents reflux of bacterial-laden cecal contents back into the small intestine. Distention of the ileum proximal to the ileocecal sphincter decreases sphincter tone, while distention of the cecum increases tone. The sphincter relaxes intermittently, particularly after meals, allowing ileal contents to empty into the cecum. Stimulation of ileal motility and relaxation of the sphincter after a meal has been called the **gastro-ileal reflex**. This reflex may be initiated by increased serum gastrin or via reflexes over extrinsic nerves.

Review Questions

9. The law of intestine states that

   A. slow wave depolarizations are propagated distally
   B. distension of the intestine results in contraction of the intestine proximal to the distension and relaxation distal to it
   C. segmentation contractions occur at a higher frequency in the jejunum than the ileum
   D. peristaltic contractions travel only a short distance
   E. contractile activities in the small intestine are local events

10. Cutting the extrinsic nerves to the small intestine would

   A. abolish all contracctile activity in the small intestine
   B. abolish slow wave depolarizations in the small intestine
   C. abolish the intestino-intestinal reflex
   D. result in peristaltic rushes
   E. not affect motility in the small intestine

Motility of the Colon and Rectum

The colon receives 0.5-1.0 L of chyme per day. The large intestine absorbs residual fluid and electrolytes as it forms and stores the feces. Chyme moves very slowly through the colon and is propelled by various contractile patterns. Most contractions can be classified as **segmentation type contractions** that mix colonic contents. The frequency of these contractions is higher in the middle of the large intestine than in more proximal segments, so they slow movement of chyme through the colon. Periodically segmentation contractions cease and colonic contents move distally due to peristalsis. This **mass movement** occurs one to three times a day usually during, or shortly after, ingestion of a meal. Stimulation of mass movement upon ingestion of a meal is called the **gastro-colic reflex** and may be triggered by increased serum gastrin or through extrinsic neural reflexes. In the colon the predominate function of the enteric nerve plexus is to inhibit contraction. Thus, a congenital absence of the enteric plexus in a segment of the colon **(Hirschprung's disease)** results in tonic constriction of the affected segment, constipation and dilation of the colon proximal to the constriction (megacolon).

The **rectum** normally contains little fecal material and exhibits segmentation contractions that retard entry of material from the colon. Occasionally, especially after a meal, mass movement shifts some of the colonic contents into the rectum. The resultant distention elicits the **rectosphincteric reflex** that relaxes the **internal anal sphincter**. Distention of the rectum also elicits the urge to defecate. If conditions are not appropriate for defecation, the **external anal sphincter** is voluntarily contracted, preventing expulsion of fecal material. When defecation is prevented the rectum accommodates to the distention and the internal anal sphincter regains its tone. The urge to defecate may be suppressed until the next mass movement results in additional distention of the rectum. The act of defecation is partly voluntary and partly involuntary. Involuntary movements include contraction of smooth muscles of the distal colon and relaxation of the internal anal sphincter. Voluntary movements include relaxation of the external anal sphincter and contraction of abdominal muscles to increase intra-abdominal pressure.

Review Questions

11. Mass movement refers to

    A. antral systole
    B. colonic peristalsis
    C. defecation
    D. small intestinal peristalsis
    E. oropharyngeal phase of swallowing

12. During defecation the rectosphincteric reflex causes

    A. relaxation of the external anal sphincter
    B. relaxation of the internal anal sphincter
    C. contraction of abdominal muscles
    D. relaxation of smooth muscles of the distal colon
    E. relaxation of the ileocecal sphincter

SECRETION

Salivary Secretion

Mastication stimulates secretion of saliva primarily from three pairs of **salivary glands**; the parotid, submaxillary and sublingual glands. The average adult secretes 1-2 L of saliva every day. Saliva is alkaline because of its relatively high content of **HCO_3^-**; this neutralizes acid produced by bacteria in the oral cavity and thus helps prevent dental caries. Saliva also has an elevated concentration of **K^+** and contains **ptyalin**, an **α-amylase** that starts the process of digestion of complex carbohydrates. In addition, saliva contains a number of substances (eg, lysozyme, IgA, lactoferrin) that help prevent bacterial overgrowth in the oral cavity. The acinar cells of the glands secrete the organic components of saliva while cells that line the ducts contribute most of the HCO_3^-.

Secretion of saliva is primarily under neural control. The salivary glands are innervated by both parasympathetic and sympathetic divisions of the ANS. Both **parasympathetics and sympathetics** stimulate secretion of saliva, although the parasympathetics cause a much greater volume response. Release of ACh from parasympathetic nerves increases volume and HCO_3^- concentration of saliva, as well as increasing O_2 consumption and vasodilation in the glands. In addition, parasympathetic stimulation causes myoepithelial cells to contract, facilitating movement of saliva from the ducts into the oral cavity. **Stimuli for salivation** include conditioned reflexes, taste (particularly acidic), smell, mechanical stimulation of the oral cavity, and nausea.

Review Questions

13. Stimulation of parasympathetic nerves to the parotid gland causes

    A. decreased concentration of HCO_3^- in saliva
    B. decreased O_2 consumption
    C. increased volume secretion
    D. vasoconstriction in the parotid gland
    E. nausea

14. When there is vigorous secretion of saliva, the concentration of which of the following ions is higher in saliva than in plasma?

    A. Na^+
    B. K^+
    C. Cl^-
    D. HCO_3^-
    E. K^+ and HCO_3^-

Gastric Secretion

One of the secretions of the stomach is **hydrochloric acid (HCl)**. HCl is actively secreted (Fig. 6-1) by **parietal** or oxyntic cells located in gastric glands in the fundus and body of the stomach. The luminal membranes of the parietal cells contain a **H^+/K^+-ATPase** which actively secretes H^+ into the gastric lumen in exchange for K^+. When H^+ ions are secreted, OH^- ions are left behind in the cytoplasm and must be neutralized. In parietal cells CO_2, derived from cellular metabolism and arterial blood, reacts with water to form H_2CO_3 which combines with the OH^- to form HCO_3^- and H_2O.

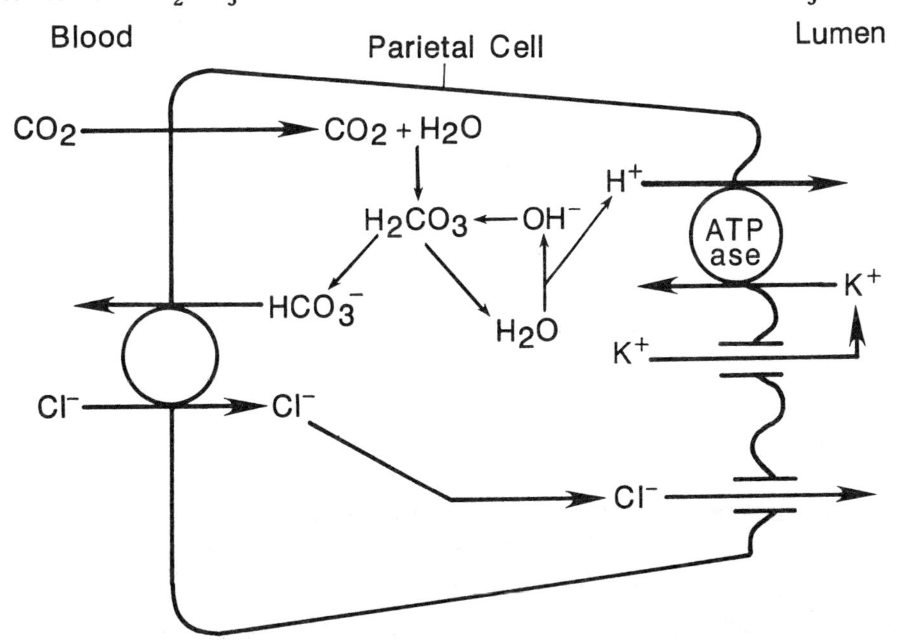

Fig. 6-1. Model for secretion of HCl by gastric parietal cells.

Formation of H_2CO_3 is catalyzed by **carbonic anhydrase**; inhibition of carbonic anhydrase with acetazolamide reduces acid secretion. The HCO_3^- formed is secreted into the blood in exchange for Cl^-. **HCO_3^-/Cl^- exchange** results in accumulation of Cl^- within the cell and Cl^- is then passively secreted into the gastric lumen. Active secretion of HCl can produce an intragastric pH < 1. Acid **functions** to break up cells, denature proteins for easier digestion, kill many ingested bacteria, and provide an acidic environment appropriate for pepsin activity. Extraction of CO_2 from the blood and secretion of HCO_3^- into it causes venous blood leaving the actively secreting stomach to be more alkaline than arterial blood (**alkaline tide**).

The glycoprotein, **intrinsic factor (IF)**, is also secreted by gastric parietal cells. Secretion of IF parallels acid secretion. IF binds **vitamin B_{12}** and is essential for absorption of vitamin B_{12} from the ileum. Lack of IF leads to **pernicious anemia**, a frequent complication of achlorhydria (absence of gastric acid secretion).

Pepsinogen is secreted by exocytosis from **chief cells** into the gastric lumen and is converted to the active form **pepsin** in the presence of HCl. Pepsin is a proteolytic enzyme with a pH optimum near 2. Mucus is secreted by mucous neck cells and surface epithelial cells and is useful for lubrication and protection of gastric epithelial cells from mechanical and chemical damage.

Review Questions

15. Which of the following would be poorly absorbed from the small intestine of an individual who is achlorhydric?

    A. Vitamin B_{12}
    B. Carbohydrate
    C. Protein
    D. Fats
    E. Bile salts

16. In the stomach, H^+ ions are secreted in exchange for

    A. Na^+
    B. K^+
    C. Ca^{2+}
    D. Cl^-
    E. HCO_3^-

17. With regard to gastric secretion of hydrochloric acid,

    A. H^+ is actively transported across the luminal membrane
    B. HCO_3^- is used to buffer excess H^+ produced within the parietal cell
    C. Cl^- is actively absorbed from the gastric lumen
    D. HCO_3^- is actively transported across the luminal membrane
    E. Cl^- is secreted into the gastric lumen in exchange for HCO_3^-

Regulation of Gastric Secretion

Both **gastrin**, secreted by **G cells** in the gastric antrum and **ACh** from vagal terminals and the enteric nerve plexus stimulate secretion of HCl from parietal cells. **Histamine**, a paracrine secretion of mast-like cells in the stomach, stimulates parietal cell secretion of HCl and potentiates the stimulatory effects of gastrin and ACh.

Control of gastric acid secretion can be separated into three phases: cephalic, gastric and intestinal. During the **cephalic phase** the taste and smell of food, mastication and swallowing all trigger **long reflexes** via the vagus, releasing ACh in the vicinity of parietal cells to stimulate HCl secretion. In addition, these long reflexes increase secretion of gastrin from G cells further stimulating HCl secretion. Cephalic phase stimulation of acid secretion is abolished by vagotomy.

During the **gastric phase** distention of the stomach triggers **long** vago-vagal reflexes and **short reflexes** that stimulate parietal cells via ACh and increase secretion of **gastrin** from G cells. In addition, certain components of food (protein digestion products and Ca^{2+}) increase secretion of gastrin. Gastrin released during the gastric phase stimulates gastric acid secretion. As the lumen becomes acidified and the pH decreases to less than 2, further secretion of gastrin is inhibited. This suppression of gastrin secretion is mediated by the paracrine secretion **somatostatin**.

During the **intestinal phase**, passage of chyme into the duodenum can either stimulate or inhibit gastric acid secretion. Distention and protein digestion products in the upper small intestine stimulate gastric acid secretion. It is proposed that **distention** causes release of an as yet unidentified hormone, **enteroxyntin**, that stimulates gastric acid secretion. Elevated **amino acid** concentrations in the blood (due to absorption of protein digestion products) also stimulate acid secretion. Furthermore, protein digestion products may stimulate gastric acid secretion by causing secretion of intestinal gastrin. **Fat, acid, and hyperosmotic solutions** in the upper small intestine inhibit gastric acid secretion. While part of this inhibition may be mediated by neural reflexes, part may be due to release of a hypothetical hormone called **enterogastrone**. Secretin, CCK and GIP have been proposed as enterogastrones, but the evidence is not compelling.

Pepsinogen secretion is stimulated by **ACh** released by long and short reflexes. Gastric acid, triggers a short cholinergic reflex and is a potent stimulus for pepsinogen secretion. Secretion of **intrinsic factor** is stimulated by those conditions which stimulate acid secretion from parietal cells. **Mucus** secretion is stimulated by cholinergic vagal reflexes and by chemical and mechanical stimulation from the lumen.

Peptic Ulcer Disease. An ulcer is a lesion or "hole" which extends through the mucosa and muscularis mucosa into the submucosa or deeper structures. Perhaps 10% of Americans suffer from peptic ulcer disease during their lifetime. Most peptic ulcers occur in the duodenum and are frequently associated with increased gastric acid secretion and rate of gastric emptying. In the stomach development of peptic ulcers is often not related to increased secretion of acid. Peptic ulcers of the esophagus result from excessive reflux of gastric contents, sometimes due to an incompetent LES. Gastric acid plays a central role in development of peptic ulcers; people who are achlorhydric do not have peptic ulcers ("no acid, no ulcer"). On the other hand, 90-95% of patients with gastrin secreting tumors (gastrinomas) have strikingly elevated gastric acid secretion and peptic ulcer disease (Zollinger-Ellison syndrome). Recent studies have provided strong evidence that infection with *H pylori* is a primary contributor to development of duodenal ulcers. Evidence is accumulating to link *H pylori* to gastric ulcers as well.

Treatment of peptic ulcer disease centers on reducing acid secretion. In the past patients were placed on bland diets. However, it was found that healing of peptic ulcers occurred as readily with normal diets. Ingestion of coffee, alcohol and large quantities of milk (which contains Ca^{2+} and protein) is discouraged since they stimulate acid secretion. Vagotomy and anticholinergic drugs can reduce acid secretion but they produce adverse side effects. Since histamine directly stimulates parietal cell secretion of acid and potentiates the effects of other secretagogues, antagonists (eg, cimetidine, ranitidine) to the histamine H_2 receptors of parietal cells are very effective. Acid secretion can be profoundly reduced by omeprazole which irreversibly inhibits the H^+, K^+-ATPase. Combined treatment with anti-secretory agents and multiple antibiotics to eradicate *H pylori* holds promise for facilitating healing and reducing recurrence of peptic ulcers.

Review Questions

18. Cutting the vagus nerves to the stomach will abolish

    A. cephalic phase stimulation of acid secretion
    B. gastric phase stimulation of acid secretion
    C. intestinal phase stimulation of acid secretion
    D. intestinal phase inhibition of acid secretion
    E. gastric phase stimulation of pepsinogen secretion

19. Which of the following stimulates secretion of pepsinogen?

    A. Enterogastrone
    B. Cholecystokinin
    C. H^+ ions in the stomach
    D. Sympathetic nerves to the stomach
    E. Secretin

20. Somatostatin

    A. stimulates parietal cell secretion of acid
    B. stimulates chief cell secretion of pepsinogen
    C. stimulates parietal cell secretion of intrinsic factor
    D. inhibits G cell secretion of gastrin
    E. stimulates G cell secretion of gastrin

21. Secretion of gastrin is **NOT**

A. stimulated by distention of the stomach
B. stimulated by protein digestion products in the stomach
C. stimulated by vagovagal reflexes
D. reduced by acidic (pH below 2.0) gastric chyme
E. stimulated by histamine

22. All of the following could contribute to development of peptic ulcer disease **EXCEPT**

A. increased gastric acid secretion
B. infection with *H pylori*
C. increased rate of gastric emptying
D. increased rate of esophageal clearance
E. decreased tone of the lower esophageal sphincter

Secretion by the Exocrine Pancreas

Two types of secretions make up pancreatic juice: 1) an isotonic solution that contains high concentrations (approx 100 mEq/L) of HCO_3^-, little Cl^- and is secreted by ductal cells, and 2) a solution rich in digestive enzymes that are secreted by exocytosis from acinar cells.

During the **cephalic** and **gastric phases** some stimulation of enzyme secretion occurs as a result of vagovagal cholinergic reflexes to the pancreas and increased serum gastrin. The **intestinal phase** accounts for 3/4 of the stimulation of pancreatic secretion. During the intestinal phase acidic chyme entering the duodenum causes secretion of **secretin** by S cells in the upper small intestine. Secretin stimulates ductal cells to increase volume and HCO_3^- secretion. The HCO_3^- functions to neutralize acid, thus removing the stimulus for further secretion of secretin. Fat and protein digestion products entering the duodenum stimulate secretion of **CCK**, which in turn stimulates enzyme secretion. These enzymes hydrolyze proteins, polysaccharides and fats. Secretin and CCK potentiate the stimulatory effects of one another. ACh, from parasympathetic innervation to the pancreas, potentiates the effects of CCK and secretin. Thus vagotomy may decrease the pancreatic secretory response to a meal by more than 50%.

Review Questions

23. Vagal stimulation of the exocrine pancreas is physiologically important, because it

A. causes a large increase in enzyme secretion
B. causes a large increase in HCO_3^- secretion
C. counteracts the inhibitory effects of gastrin
D. potentiates the effects of secretin and cholecystokinin (CCK)
E. increases the ratio of trypsin to chymotrypsin in pancreatic secretions

24. The presence of which of the following substances in the duodenum will give the greatest volume of pancreatic secretion?

    A. Fat
    B. HCO_3^- ions
    C. H^+ ions
    D. Amino acids
    E. H^+ ions plus fat

Biliary Secretion

Bile is secreted continuously by the liver. The rate of secretion depends on the state of alimentation. Bile contains bile salts (also called bile acids), lecithin (a phospholipid), cholesterol, bile pigments (eg, bilirubin), and electrolytes. The major **bile salts** in man are glycine and taurine conjugates of cholate and chenodeoxy-cholate, which are **primary bile salts** synthesized by the liver. Bile also contains conjugated deoxycholate and lithocholate, which are **secondary bile salts** produced by bacterial alteration of primary bile salts. The glycine conjugates predominate in adults. Bile salts and Na^+ are actively secreted by hepatocytes into bile canaliculi. This produces an osmotic gradient, resulting in secretion of water accompanied by passive secretion of electrolytes. As the bile passes down the ductules and ducts, its electrolyte content is altered primarily by addition of HCO_3^-. **Secretin** stimulates secretion of HCO_3^- into bile. The concentration of bile salts in bile is greater than their critical micellar concentration. Consequently, bile salts, as well as phospholipid and cholesterol, form macromolecular structures called **micelles**. In the micelles the hydrophilic portions of bile salts and phospholipids are exposed at the outer surface of the micelle, while the hydrophobic portions of those molecules and cholesterol are sequestered in the hydrophobic interior. Formation of micelles is important in solubilizing the cholesterol in bile and preventing formation of gallstones.

When fasting, the sphincter of Oddi is closed and bile passes into the gallbladder where it is stored and concentrated. Na^+ is actively reabsorbed, Cl^- and HCO_3^- follow passively, while water is reabsorbed by osmosis. Entrance of a lipid-rich meal into the upper small intestine increases secretion of **CCK** which causes the sphincter of Oddi to relax and the gallbladder to contract. Gallbladder bile is gradually expelled into the small intestine. In the small intestine the bile salts aid in digestion and absorption of lipids and lipid soluble vitamins, while the HCO_3^- in bile helps neutralize acidic chyme. Most primary and secondary bile salts are reabsorbed from the small intestine and only small quantities are excreted in the feces. While some primary and secondary bile salts are passively reabsorbed in the jejunum, most are reabsorbed in the **ileum** by **Na^+-dependent secondary active transport** (see Fig. 6-2, p. 187). The liver efficiently extracts bile salts from portal plasma, and most bile salts are removed from plasma in one passage through the liver. Return of bile salts to the liver is the major stimulus for hepatic secretion of bile. Reabsorbed bile salts (primary and secondary) are again secreted into bile canaliculi. This recycling of bile salts is called the **enterohepatic circulation** of bile salts.

The bile pigment, **bilirubin**, is a breakdown product of hemoglobin that causes both jaundice and CNS damage if not excreted. Bilirubin is hydrophobic and is present in plasma bound to albumin. Hepatocytes extract bilirubin from plasma and conjugate it with glucuronic acid to make it water soluble.

Bilirubin-diglucuronide is then secreted into bile. Unlike bile salts, most bile pigment is not reabsorbed but is excreted in the feces. The GI tract is the major route for elimination of bilirubin.

Bile is the only route for excretion of **cholesterol** from the body. Cholesterol is relatively insoluble in water and is solubilized in bile by formation of bile salt/lecithin/cholesterol micelles. When the amount of cholesterol in bile exceeds the ability of bile salts and lecithin to solubilize it, cholesterol may precipitate out of solution to form cholesterol **gallstones**. Thus, **lithogenic** (stone forming) bile can result from either excessive secretion of cholesterol or inadequate secretion of bile salts and/or lecithin. While most gallstones are primarily composed of cholesterol, some contain mainly calcium salts of bilirubin. Pigmented stones form when water soluble bilirubin-diglucuronide is made insoluble due to deconjugation by naturally occurring ß-glucuronidase or by bacteria infecting the biliary tract.

Review Questions

25. All of the following are present in micelles **EXCEPT** which one?

    A. Primary bile salts
    B. Phospholipid
    C. Cholesterol
    D. Bilirubin-diglucuronide
    E. Secondary bile salts

26. The most important factor determining the rate of bile secretion by the liver is

    A. serum cholecystokinin
    B. the rate of return of bile salts to the liver via the portal circulation
    C. the rate of synthesis of bile salts
    D. serum secretin
    E. serum cholesterol

27. Resection of the ileum would cause or contribute to an **increase** in all of the following **EXCEPT**

    A. fecal excretion of bile salts
    B. serum cholesterol
    C. hepatic synthesis of bile salts
    D. risk of formation of gallstones
    E. risk of fat soluble vitamin deficiency

DIGESTION AND ABSORPTION

Digestion and Absorption of Carbohydrates

Carbohydrates typically contribute more than 50% of caloric intake. Starches make up about 60% of carbohydrates, sucrose about 30%, and the remainder are disaccharides, such as lactose, maltose, and trehalose. Salivary amylase begins the process of digestion of polysaccharides like starch. However, since salivary amylase is inactivated by acid in the stomach, its effect is limited. Most digestion of starch takes place in the small intestine by the action of **pancreatic amylase** that attacks α 1-4 bonds. Pancreatic amylase hydrolyses amylose to produce maltose and the trisaccharide maltotriose. When branched polysaccharides (eg, amylopectin and glycogen) are hydrolyzed, α-limit dextrins are also produced. These products of amylase digestion are further digested by saccharidases located in the brush border of intestinal epithelial cells. **Brush border saccharidases** include α-dextrinase and maltase (which hydrolyze α-limit dextrins, maltotriose and maltose) as well as sucrase, lactase, and trehalase. The major products of digestion are the monosaccharides glucose, galactose, and fructose. Fructose is absorbed by facilitated diffusion.

D-Glucose and D-galactose are absorbed by **Na^+-dependent secondary active transport** (Fig. 6-2). The intracellular concentration of Na^+ in enterocytes is maintained at a low level due to the activity of a Na^+-K^+ ATPase, which transports sodium from the cytoplasm across the basolateral membrane into interstitial fluid. In addition, there is an electrical potential difference across the brush border with the cell interior negative with respect to the lumen. Thus the electrochemical gradient favors movement of Na^+ from the lumen into the cell. In Na^+-dependent secondary active transport, the energy in the luminal to cytoplasmic electrochemical gradient for Na^+ is used to drive an organic solute (eg, glucose) from the lumen into the cell against its electrochemical gradient.

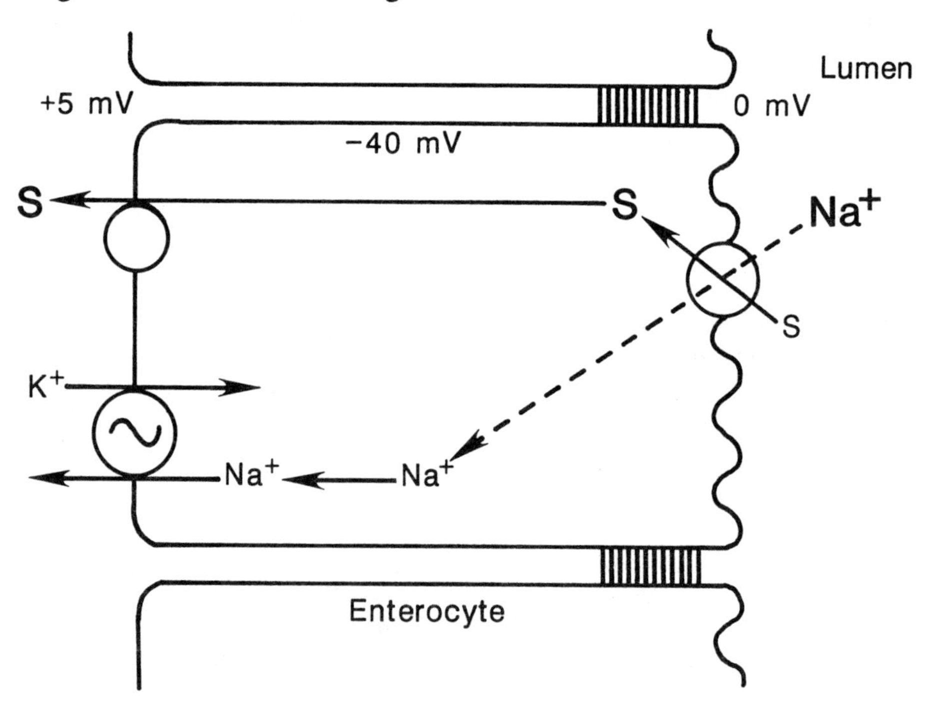

Figure 6-2. Model for Na^+-dependent secondary active transport.

Glucose and Na^+ both bind to a **carrier** at the luminal surface of the brush border membrane. Inward movement of Na^+ down an electrochemical gradient into the cell is coupled with, and provides the energy for, translocation of glucose in the same direction, resulting in intracellular accumulation of glucose. Once intracellular glucose reaches a concentration greater than that in interstitial fluid, it crosses the basolateral membrane by facilitated diffusion. The result is secondary active transport of glucose from the lumen to the interstitial fluid and portal circulation. Na^+-dependent secondary active transport of sugars is found throughout the length of the small intestine. The rate of absorption per cm of intestine is greatest in the proximal intestine and decreases distally. Under normal conditions all carbohydrates have been absorbed by the time that chyme reaches mid-jejunum.

Review Questions

28. Which of the following would be an effective source of calories in the absence of secretions from the exocrine pancreas?

    A. Sucrose
    B. Glycogen
    C. Cellulose
    D. Starch
    E. Amylose

29. In Na^+-dependent secondary active transport of glucose, metabolic energy is **directly** used to

    A. transport Na^+ across the basolateral membrane
    B. transport Na^+ across the brush border membrane
    C. transport glucose across the brush border membrane
    D. transport glucose across the basolateral membrane
    E. phosphorylate glucose

Digestion and Absorption of Proteins

Gastric HCl and pepsin digest some proteins in the stomach. Peptic digestion ceases when pepsin is mixed with alkaline pancreatic secretions in the small intestine. Most digestion of **protein** takes place in the **small intestine**. Proteolytic enzymes are secreted by the pancreas into the small intestine in inactive form. Initially, **trypsinogen** is converted to **trypsin** by the action of **enterokinase**, an intestinal protease. Trypsin then activates additional trypsinogen as well as the other proenzymes: chymotrypsinogen, procarboxypeptidase and proelastase. **Trypsin**, **chymotrypsin** and **elastase** are endopeptidases, while **carboxypeptidase** is an exopeptidase. Digestion produces some free amino acids (30%) but mostly peptides (70%). The final step in digestion takes place at the **brush border membrane** which contains a number of peptidases. Digestion at the brush border produces free amino acids, di- and tripeptides.

Free amino acids are absorbed by **Na^+-dependent secondary active transport**. There are several different Na^+-dependent carrier systems for different classes of amino acids, all of which are distinct from the carrier system for monosaccharides. In addition, some **di-** and **tripeptides** are absorbed intact by an unknown mechanism. Once inside enterocytes, di and tripeptides are hydrolyzed and pass into the portal circulation as free amino acids. Absorption of peptides is of physiological importance, since amino acids are more readily absorbed when supplied as peptides than as free amino acids. Most amino acids and peptides are absorbed in the jejunum, but some may be absorbed in the ileum.

Review Questions

30. Chymotrypsinogen is activated by

    A. enterokinase
    B. trypsin
    C. pepsin
    D. carboxypeptidase
    E. amylase

31. Which of the following statements regarding digestion and absorption of protein and its digestion products is correct?

    A. Digestion of protein by pancreatic enzymes produces mainly free amino acids
    B. There is only one transport system for amino acids
    C. Some dipeptides are absorbed intact into enterocytes
    D. All amino acids are more effectively absorbed from a mixture of amino acids than from a mixture of peptides
    E. Amino acids and sugars compete for the same transport system

32. All of the following are absorbed by Na^+-dependent secondary active transport **EXCEPT** which one?

    A. D-glucose
    B. bile salts
    C. dipeptides
    D. L-alanine
    E. L-phenylalanine

Digestion and Absorption of Lipids

The most abundant lipids in the diet are **triglycerides**, but significant amounts of **phospholipid** and **cholesterol** are often present. Since fats and their digestion products are not very soluble in water, special mechanisms are necessary to digest and absorb them in the aqueous environment of the gastrointestinal tract.

Lipids enter the stomach as large drops. Powerful contractions of the stomach convert these drops into an **emulsion**. Perhaps 5-10% of the triglycerides are digested in the stomach by **lipases** secreted by glands in the base of the tongue and the stomach. Only small quantities of lipids are absorbed from the stomach, primarily fatty acids with short hydrocarbon chains.

Most digestion and absorption of **lipids** occurs in the **small intestine.** In the intestine the lipid emulsion is stabilized by **bile salts** that coat the emulsion droplets and prevent them from coalescing. This coating of bile salts also reduces the adherence of **pancreatic lipase** to the emulsion droplets and reduces its ability to digest the triglycerides. Adherence of pancreatic lipase to the emulsion is facilitated by colipase. **Colipase**, a polypeptide secreted by the pancreas in an inactive form, is activated by trypsin. Once colipase adheres to the surface of an emulsion droplet, pancreatic lipase binds to a site on the colipase molecule and hydrolyses triglycerides to form **2-monoglycerides** and **free fatty acids (FFA)**. Phospholipids are digested by **phospholipase A_2**, which is secreted by the pancreas in inactive form and is activated by trypsin. Esterified cholesterol and other esters are hydrolyzed by **cholesterol esterase** (also secreted by the pancreas). The products of lipid digestion are solubilized by incorporation into **mixed micelles** composed of bile salts, monoglycerides, FFA, lysophospholipids, cholesterol, and fat soluble vitamins.

Mixed micelles aid movement of fat soluble substances through the unstirred water layer adjacent to the luminal surface of the brush border membrane thereby facilitating absorption of lipids and lipid digestion products. Lipids in the micelles are at equilibrium with lipids in the unstirred layer. Cholesterol, lysophospholipids, monoglycerides, FFA, and fat soluble vitamins in the unstirred layer are absorbed into enterocytes by simple diffusion. Removal of lipids from the unstirred layer shifts the equilibrium so that additional lipids are released from mixed micelles and are absorbed. The bile salts are later reabsorbed from the ileum by Na^+-dependent secondary active transport.

Once inside enterocytes, absorbed monoglycerides and FFA are resynthesized into triglycerides, and most cholesterol is esterified. The newly synthesized triglycerides and cholesterol esters aggregate to form lipid droplets in the cytoplasm of enterocytes. A layer of phospholipid and protein is added to form **chylomicrons**, which pass through the basolateral membrane and enter lymphatics.

Review Questions

33. Pancreatic lipase

    A. hydrolyses cholesterol esters
    B. is secreted in inactive form
    C. hydrolyses phospholipids
    D. hydrolyses triglycerides
    E. hydrolyses fat soluble vitamins

34. Colipase

    A. is secreted in an inactive form
    B. is activated by enterokinase
    C. blocks adherence of pancreatic lipase to lipid emulsions
    D. hydrolyses monoglycerides
    E. hydrolyses triglycerides

35. All of the following are properties of bile salts **EXCEPT** which one?

A. Aid in solubilization of lipids
B. Stabilize lipid emulsions
C. Aid movement of lipids across the unstirred water layer
D. Inhibit adherence of pancreatic lipase to lipid emulsion droplets
E Digest lipids

36. Which of the following is absorbed by either active or secondary active transport?

A. Monoglycerides
B. Fat soluble vitamins
C. Free fatty acids
D. Phospholipids
E. Bile salts

Absorption of Vitamins

Fat-soluble vitamins are absorbed by simple diffusion. Efficient absorption of these vitamins requires bile salts and mixed micelles. Once absorbed, fat soluble vitamins are incorporated into chylomicrons and pass into lymphatic lacteals. All **water-soluble vitamins** are absorbed, at least in part, by carrier-mediated transport. The exact nature of the transport systems for some of the water-soluble vitamins is not clear, but many appear to be absorbed by Na^+-dependent secondary active transport. **Vitamin B_{12}** is absorbed by an active mechanism. Absorption of vitamin B_{12} requires complexing with **intrinsic factor** (IF), a glycoprotein secreted by gastric parietal cells. IF-Vitamin B_{12} complex remains in the lumen of the GI tract until it reaches the ileum where it binds to specific receptors on ileal enterocytes. After binding, vitamin B_{12} is actively absorbed and eventually appears in the blood complexed to another protein, **transcobalamin.** Inability to efficiently absorb vitamin B_{12} results in **pernicious anemia.**

Absorption of Fluid and Electrolytes

The ability of the small intestine to absorb fluid and electrolytes is, in the short term, more important than its ability to absorb organic nutrients. The actual load of fluid and electrolytes that must be absorbed is much larger than the amount ingested. An adult ingests in 1-2 liters of water a day, but the fluid load to the small intestine is 9-10 liters, 8-9 liters being added by secretions of the GI system. Large quantities of electrolytes also enter the GI tract with these secretions. Thus, absorption and reabsorption of fluids and electrolytes is important not only for replacing what was lost in other processes such as urination, perspiration, and respiration, but also to recover the large quantities of fluid and electrolytes secreted into the GI tract each day.

Most of the fluid and electrolytes are absorbed from the **small intestine.** Movement of water occurs by osmosis in response to osmotic gradients. When chyme first enters the duodenum, it is adjusted to **isotonicity** by net secretion of water across the mucosa (for a hypertonic meal) or by net absorption (for a hypotonic meal). After adjustment to isotonicity, the volume is reduced by absorption of an **isotonic**

fluid. This absorption of fluid is secondary to and dependent upon absorption of solutes, primarily Na^+. Active transport of Na^+ from the lumen into the lateral intercellular space raises the osmotic pressure in this restricted region. An osmotic gradient is thus produced and water flows from the lumen into the lateral intercellular space. This flow of water increases hydrostatic pressure in the lateral intercellular space causing isotonic fluid to move from the lateral intercellular space into the interstitial fluid and blood. Glucose facilitates movement of Na^+ from the lumen into enterocytes and thereby stimulates Na^+ and water absorption. This is the rationale for including sugar in per oral rehydration fluids.

The major electrolytes absorbed from the small intestine are Na^+, K^+, and Cl^-. **Na^+** is absorbed (Fig. 6-3) from the lumen into the enterocytes by 1) passive diffusion, 2) co-transport with organic solutes (Na^+-dependent secondary active transport), 3) co-transport with Cl^- and 4) exchange with H^+. After entering the enterocyte by any of these routes, Na^+ is **actively transported** across the basolateral membrane by the **Na^+-K^+ ATPase.** Since some Na^+ absorption is electrogenic, it generates an small electrical gradient across the epithelium favoring passive absorption of Cl^-. Cl^- can cross the brush border membrane (Fig. 6-3) via co-transport with Na^+ or in exchange for HCO_3^-. After entering the enterocytes, Cl^- passively diffuses across the basolateral membrane into the interstitial fluid and eventually the blood. **K^+** is also absorbed passively. As Na^+ and water are absorbed, the volume of the luminal contents decreases, resulting in an increase in K^+ concentration and passive absorption of this ion.

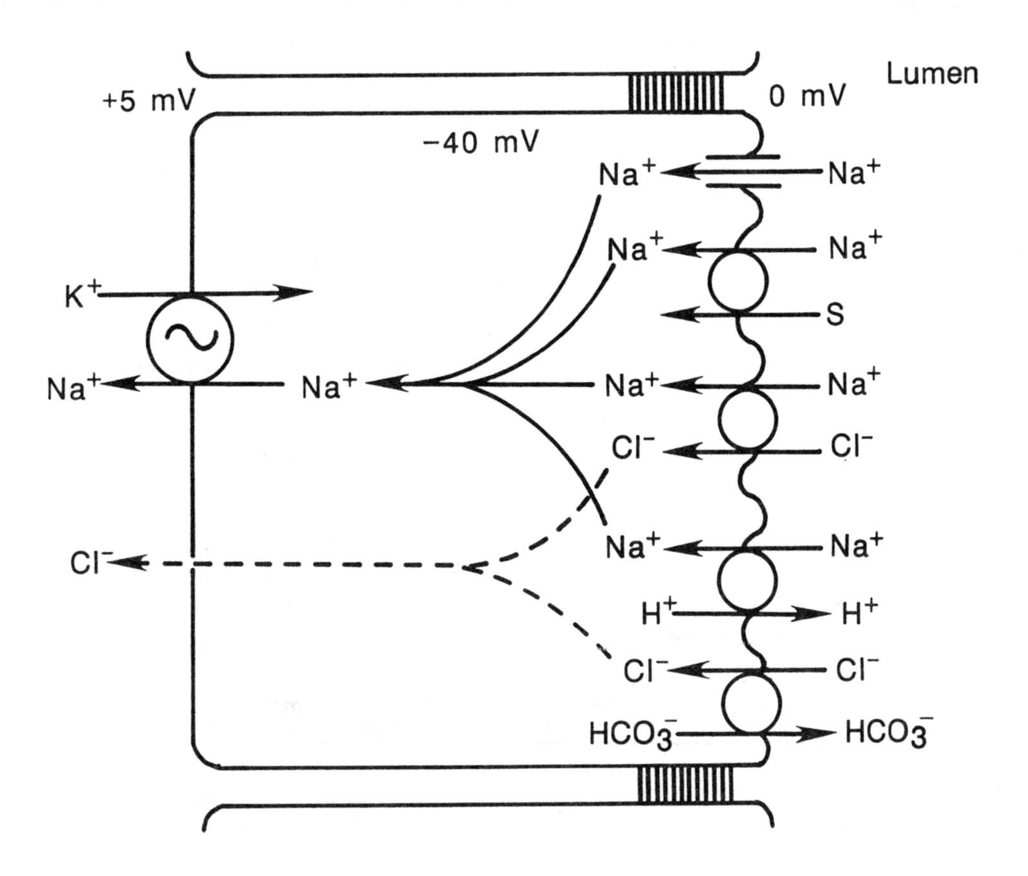

Figure 6-3. Model for NaCl absorption in the small intestine.

Calcium is absorbed by active transport in the small intestine. Ca^{2+} absorption is localized in the proximal small intestine and is enhanced by **1,25-dihydroxycholecalciferol**, a derivative of vitamin D that stimulates synthesis of a **calcium binding protein** in enterocytes. **Calcium binding protein** facilitates passive uptake of Ca^{2+} into the enterocyte. Ca^{2+} is then actively transported across the basolateral membrane by a **Ca^{2+}-ATPase**, which is also regulated by 1,25-dihydroxycholecalciferol.

Most **iron** absorption also occurs by active transport in the proximal small intestine. In the lumen of the stomach and upper small intestine organic acids (eg, ascorbic and citric acids) reduce ferric ions (Fe^{3+}) to ferrous ions (Fe^{2+}) which are absorbed more efficiently. Iron then binds to a specific receptor on the brush border membrane and is transported into the cell. Heme iron is absorbed intact into enterocytes where it is hydrolyzed to release free iron. Some of the iron which enters enterocytes is rapidly transported into the blood to complex with a protein called **transferrin**, while some becomes bound to **apoferritin** in the cell to form **ferritin**. Over time some of the iron in ferritin is converted to free iron and transferred to the blood. The amount of iron transferred from enterocytes to the blood, versus that lost when enterocytes are exfoliated, depends upon the body's need for iron.

While the chyme is in the **colon**, residual ions and water are absorbed. **Na^+** is passively transported across the brush border membrane and actively transported across the basolateral membrane (Fig. 6-4). **Cl^-** is passively absorbed in exchange for **HCO_3^-**. Because of the large potential difference which develops across the epithelium, **K^+** is passively secreted into the lumen. Unlike electrolyte absorption in most of the small intestine, absorption of electrolytes in the colon responds to changes in serum aldosterone; **aldosterone** stimulates both Na^+ absorption and K^+ secretion.

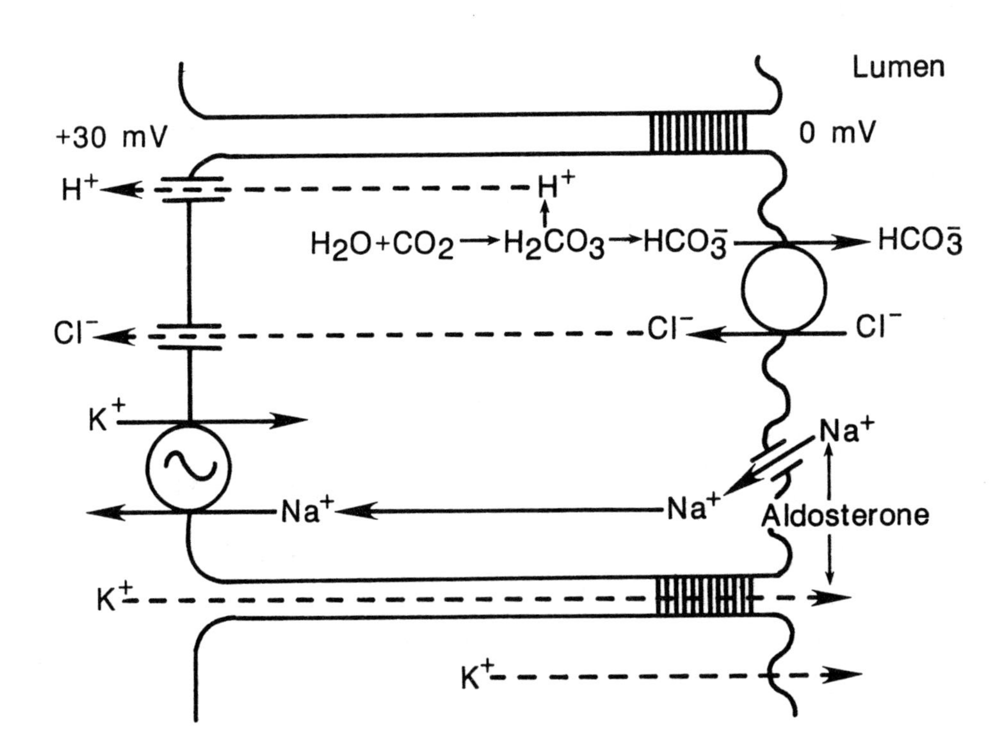

Figure 6-4. Model for absorption of Na^+ and Cl^- and secretion of K^+ and HCO_3^- in the colon.

Review Questions

37. The site for active absorption of vitamin B_{12} is the

    A. stomach
    B. duodenum
    C. jejunum
    D. ileum
    E. colon

38. When chyme first enters the duodenum, the direction of net flux of water is primarily determined by the

    A. osmolarity of the chyme
    B. rate of Na^+ absorption
    C. rate of organic solute absorption
    D. rate of Cl^- absorption
    E. serum secretin concentration

39. Each of the following is a mechanism for transporting Na^+ across brush border membranes **EXCEPT** which one?

    A. Co-transport with organic solutes
    B. Co-transport with Cl^-
    C. Transport by the Na^+-K^+ ATPase
    D. Passive diffusion
    E. In exchange for H^+

40. The colon normally secretes

    A. Na^+
    B. Cl^-
    C. H^+
    D. K^+
    E. aldosterone

41. Most iron stored in the enterocyte is

    A. present as heme iron
    B. bound to transferrin
    C. free in the cytoplasm
    D. bound to apoferritin
    E. eventually transported across the brush border membrane into the lumen

DIARRHEA

Diarrhea is the frequent passage of soft, unformed stools and is sometimes defined as passage of stool volume greater than 500 ml per day. Diarrhea occurs when the volume of fluid delivered to the colon exceeds its normal absorptive capacity or when the absorptive capacity of the colon has been reduced. Diarrheas are often divided into **three categories**: 1) osmotic, 2) secretory and 3) exudative.

Osmotic diarrhea occurs when osmotically active solutes which have been ingested are retained in the lumen of the GI tract. Retention of solutes in the lumen of the small intestine requires that water also be retained to maintain isotonicity. Thus, the volume of fluid delivered to the colon may exceed its absorptive capacity. Osmotic diarrhea may result from 1) ingestion of substances which are inherently difficult to absorb (eg, $MgSO_4$ which is commonly found in laxatives), 2) primary malabsorption due to congenital absence of a specific transport system (eg, glucose-galactose malabsorption) and 3) malabsorption secondary to maldigestion (eg, lactose intolerance).

Secretory diarrhea refers to disorders in which there is active intestinal and/or colonic secretion of fluids and electrolytes. Diarrhea occurs when the absorptive capacity of the gut distal to the site of secretion is exceeded. Bacterial infections of the GI tract can cause secretory diarrheas. Infection with *V. cholerae* produces a particularly virulent secretory diarrhea. A subunit of the enterotoxin produced by this microorganism enters enterocytes and activates adenylate cyclase resulting in increased cAMP. cAMP in turn inhibits NaCl absorption and stimulates secretion of electrolytes and fluid. Cholera patients may lose as much as a liter of fluid per hour leading to dehydration, volume depletion and death within hours of the onset of symptoms. Fluid replacement, often by oral administration of electrolyte solutions containing glucose, is critical. Elevated cAMP and secretory diarrhea are also seen with tumors which secrete excessive amounts of vasoactive intestinal peptide (eg, pancreatic cholera syndrome). Malabsorption of bile salts and fats can result in secretory diarrhea. Bacteria in the colon metabolize bile salts and fatty acids to produce derivatives (eg, dehydroxy bile salts and hydroxylated fatty acids) which inhibit colonic absorption and stimulate colonic secretion of fluids and electrolytes.

Exudative diarrheas are a consequence of gross structural damage to the intestinal or colonic mucosa, resulting in impaired absorption and exudation of serum proteins, white cells, mucus and blood into the lumen. These diarrheas are characterized by frequent passage of small volume stools. Examples of causes of exudative diarrheas are inflammatory bowel diseases (eg, ulcerative colitis and Crohn's disease), mucosal invasion by protozoa (eg, *Entamoeba histolytica*) and certain bacterial infections (eg, shigella and salmonella).

Review Question

42. A patient has a diarrhea which is characterized by passage of a large volume of fluid. The diarrhea is not alleviated by fasting. Which of the following is the most likely cause of this diarrhea?

    A. Lactose intolerance
    B. Glucose-galactose malabsorption
    C. Cholera enterotoxin
    D. Ulcerative colitis
    E. Crohn's disease

For Questions 43-48. Match the correct hormone or chemical from the list below with each question.

A. Aldosterone
B. Bilirubin
C. Cholecystokinin
D. Gastrin
E. GIP
F. Histamine
G. Motilin
H. Secretin
I. Somatostatin

43. ___ Hormone which is secreted in response to vagal reflexes.

44. ___ Paracrine which stimulates gastric secretion of HCl.

45. ___ Hormone which stimulates colonic absorption of sodium and secretion of potassium.

46. ___ Hormone which stimulates contraction of the gallbladder.

47. ___ Paracrine which inhibits secretion of gastrin.

48. ___ Hormone which stimulates pancreatic secretion of bicarbonate.

ANSWERS TO GASTROINTESTINAL PHYSIOLOGY QUESTIONS

1. C. Most extrinsic nerves to the GI tract act through the enteric nerve plexus.

2. A. The upper esophagus and UES are composed of skeletal muscle and require extrinsic innervation to contract (A, C). The LES and distal esophagus are composed of smooth muscle and can contract in the absence of extrinsic innervation (B, D).

3. B. Since the esophagus is in the thoracic cavity and subject to intrapleural pressures, air would enter the esophagus with each inspiration in the absence of the UES (B). The UES is closed between swallows (E) and restricts movement of material from the esophagus into the pharynx (D). The LES prevents regurgitation of acid from the stomach (C).

4. E. Slow wave depolarizations are myogenic and are generated in the absence of external stimuli.

5. C. Contraction of abdominal muscles increases intra-abdominal and intragastric pressures producing a pressure gradient favoring evacuation.

6. A. Fat (A) and acid (C) inhibit gastric emptying. Isotonic NaCl empties more readily (D) than solutions that are hyper- or hypotonic. Motilin (B) stimulates gastric emptying during the interdigestive period.

7. D. During deglutition the proximal stomach exhibits receptive relaxation (D) allowing the swallowed bolus to enter the stomach. Once distended, the proximal stomach has sustained contractions (E) which occur in the absence of slow wave depolarizations (A) and which are important in emptying of liquids (C).

8. A. Contractions associated with the MMC remove mucus, cells and bacteria from the small intestine and thus help prevent bacterial overgrowth (A). They are peristaltic contractions (D) occurring during fasting, not after eating (B). Phase 3 is controlled by motilin (C).

9. B. Radial stretching or distension is the major stimulus for contractions in the intestine. The other answers, A and C-E, are true but are not the law of the intestine.

10. C. The intestino-intestinal reflex is mediated by extrinsic nerves (C). While activity in extrinsic nerves modifies intestinal motility (E), normal electrical (B) and contractile activity (A,D) can occur in the absence of extrinsic innervation.

11. B. Mass movement is a type of colonic peristalsis that moves colonic contents distally.

12. B. Contraction of the abdominal muscles (C) and relaxation of the external anal sphincter (A) are voluntary. The reflex stimulates contraction of smooth muscles of the colon (D).

13. C. Parasympathetic efferents stimulate salivation, resulting in increased volume secretion (C), O_2 consumption (B) and vasodilation (D). In addition, the concentration of HCO_3^- in saliva increases as salivary secretion is increased (A).

14. E. The concentrations of K^+ and HCO_3^- in saliva during vigorous secretion are greater than in plasma (E), whereas those of Na^+ and Cl^-, are less than plasma concentrations (A, C).

15. A. Parietal cells secrete intrinsic factor that is required for efficient absorption of vitamin B_{12}. The other substances are well absorbed in the absence of gastric secretions.

16. B. H^+ is secreted in exchange for K^+.

17. A. After active secretion of H^+ (A) into the lumen, OH^- is neutralized by reaction with H_2CO_3 (B). The HCO_3^- that is formed is transported into the interstitial fluid and blood (D & E).

18. A. All cephalic phase effects on gastric acid secretion are mediated by the vagus nerves (A). While vagal reflexes may be involved in gastric and intestinal regulation of secretion, other mechanisms are also operative (B-E).

19. C. Acid in the stomach stimulates pepsinogen secretion via a cholinergic reflex (C).

20. D. Somatostatin, which is secreted by paracrine cells in response to acid chyme, inhibits G cell secretion of gastrin (D).

21. E. Histamine (E) stimulates parietal cell secretion of acid but does not affect G cell secretion of gastrin. Distention of the stomach (A) triggers vagovagal (C) and enteric reflexes that stimulate gastrin secretion. Peptides also stimulate G cells to secrete gastrin (B), while gastric acid inhibits gastrin release (D).

22. D. Increased acid entering the duodenum (A & C) and esophagus (E) would promote developmnet of duodenal and esophageal ulcers while infection with *H pylori* would promote development of gastric and duodenal ulcers (B).

23. D. While vagal reflexes to the pancreas cause a small enzyme secretory response (A), the major effect of these reflexes is to potentiate the stimulatory effects of secretin and CCK (D).

24. E. H^+ and fat in the duodenum stimulate secretion of secretin and CCK, respectively. CCK potentiates the stimulatory effects of secretin on the volume of secretion.

25. D. Amphiphilic and hydrophobic solutes (eg, phospholipids, cholesterol and bile salts, A-C) are present in micelles, but hydrophilic solutes (eg, bilirubin-diglucuronide) are not (D).

26. B. The return of reabsorbed bile salts to the liver in the portal circulation is the most potent stimulant of hepatic secretion of bile (B).

27. B. Removal of the ileum would interupt the enterohepatic circulation increasing fecal excretion of bile salts (A). The resulting reduction of the bile salt pool would cause increased hepatic synthesis of bile salts (C) and increase the risk of gallstone formation (D) and fat soluable vitamin deficiency (E). Serum cholesterol (B) might decrease as it is utilized for hepatic synthesis of bile salts.

28. A. Sucrose can be digested by saccharidases in the brush border of enterocytes and the products absorbed (A). Other carbohydrates listed are either not digested (C) or require pancreatic amylase for complete digestion (B, D and E).

29. A. ATP is hydrolyzed by the basolateral Na^+, K^+-ATPase for active transport of Na^+ across the basolateral membrane into the interstitial fluid. Other transport steps (B-D) are passive or secondary active and do not utilize metabolic energy (ATP).

30. B. Trypsinogen is activated by enterokinase (A), while chymotrypsinogen is activated by trypsin (B).

31. C. Some di- and tri-peptides can be absorbed intact (C). There are several systems for transport of amino acids (B) and these are distinct from that for sugars (E). Some amino acids are more readily absorbed when they are supplied to the intestine in the form of peptides (D).

32. C. Dipeptides are absorbed by a mechanism that it not Na^+-dependent.

33. D. Pancreatic lipase is specific for triglycerides (D) and is secreted in active form (A).

34. A. Colipase is secreted in inactive form (A), is activated by trypsin (B), facilitates adherence of lipase to lipid droplets (C) but does not hydrolyze lipids (D and E).

35. E. Bile salts facilitate digestion and absorption of lipids (A-D) but do not themselves hydrolyse lipids (E).

36. E. Bile salts are absorbed by secondary active transport (E). All the listed lipids and lipid digestion products (A-D) are absorbed by simple diffusion.

37. D. IF-B_{12} complex binds to specific receptors in the ileum (D) where vitamin B_{12} is actively absorbed.

38. A. Chyme is **initially** adjusted to isotonicity by secretion (hypertonic chyme) or absorption (hypotonic chyme) of water (A). After adjustment to isotonicity, absorption of water is dependent upon solute absorption (B-D).

39. C. The Na^+, K^+-ATPase transports Na^+ across the basolateral membrane, not the brush border membrane (C).

40. D. The colon normally secretes K^+ (D) and bicarbonate. It normally absorbs Na^+ (A) and Cl^- (B).

41. D. In the enterocyte, iron binds to apoferritin to form ferritin.

42. C. Cholera enterotoxin (C) causes a voluminous diarrhea which continues during fasting. Diarrheas caused by lactose intolerance (A) and glucose-galactose malabsorption (B) are alleviated by fasting, while those caused by ulcerative colitis (D) and Crohn's disease (E) are of small volume.

43. D; 44. F; 45. A; 46. C; 47. I; 48. H.

ENDOCRINOLOGY

Introduction

Synthesis. Hormones are amino acid derivatives such as catecholamines and thyroid hormones, steroids derived from cholesterol, and peptide and protein hormones composed of three to hundreds of amino acids.

The biosynthesis of catecholamines in the adrenal medulla, which begins with tyrosine and ends with epinephrine, is controlled by neural and hormonal regulation of the enzymes in the pathway. Tetraiodothyronine and triiodothyronine are also synthesized from tyrosine with the addition of iodide radicals in the thyroid gland.

Steroids are synthesized from cholesterol. Even though all steroid molecules have the 4-ring "steroid nucleus" in common, enzyme-catalyzed reactions made at key points (eg, addition of hydroxyl, keto, methyl, or aldehyde groups or sites of unsaturation) confer estrogenic, androgenic, progestin, or corticoid activity.

Peptide and protein hormone synthesis begins with the transcription of the genetic code for the hormone sequence into messenger RNA, which is subsequently translated into protein on the endoplasmic reticulum. Most of these hormones are synthesized with a polypeptide signal sequence at the beginning of the chain. In this **pre**hormone form the signal sequence guides the newly synthesized chain into the endoplasmic reticulum, where the signal sequence is cleaved. At this point many peptide and protein hormones are still inactive and are termed **pro**hormones. Subsequent enzymatic cleavage into smaller segments and/or chemical modification (eg, glycosylation, phosphorylation, etc.) convert the molecule into an active hormone. The active form of the hormone is packaged into secretory vesicles.

The term "prohormone" also applies to hormones acted upon by target cells to convert them to a more active form. For example, testosterone is converted by prostatic cells into dihydrotestosterone, and thyroxine (tetraiodothyronine) is converted to triiodothyronine.

Secretion. Hormones that are packaged into secretory vesicles at the end of their synthesis, eg, catecholamines, peptides and proteins, have a similar mechanism of secretion. This mechanism is called coupled secretion or **stimulus-secretion coupling**; it involves a complex sequence of events including Ca^{2+} influx and ATP-activation of myotubules. Other hormones, such as thyroid and steroids, are secreted by diffusion of the end product from the gland. Steroid hormones are not stored at their site of synthesis, while thyroid hormones are stored as colloid or thyroglobulin in the lumen of the follicle.

Transport. The steroids, thyroid hormones, and some peptide and protein hormones are bound to plasma proteins when secreted. Plasma proteins are synthesized in the liver and transport specific hormones. The purpose of plasma transport of hormones is not to increase hormone solubility. Rather plasma proteins retard metabolism of hormones and generally act to form a plasma "pool" of hormone which buffers sudden changes in hormone concentration.

Action. A hormone first interacts with a specific protein of target cells, called the receptor, which triggers a response. A receptor consists of two parts: 1) a binding site which recognizes and binds to a specific hormone and 2) an effector site which initiates cellular responses to the hormone. Catecholamine, peptide and protein hormones bind to receptors in the membrane of their target cells. These can be further classified into membrane receptors which have enzymatic activity, are coupled to an enzyme via a G protein, or are connected to a channel or pore. Thyroid and steroid hormones bind to cytoplasmic or nuclear receptors which interact with chromatin to alter transcription of specific genes.

The signal or message conveyed by hormones must be interpreted or transduced by the target cell into specific responses, eg, protein synthesis, secretion of another cell product, change in ionic potential, etc. The binding of thyroid and steroid hormones to their specific intracellular receptor directs the receptor to bind to a specific region of the genome, resulting in the increased or decreased transcription of specific genes. The signal or message that resides in hormones which bind to membrane receptors must be transduced or interpreted in other ways. This involves G protein-dependent mechanisms. Basically, when the hormone binds to its membrane receptor, a G protein (also found in the cell membrane) will stimulate or inhibit cAMP, stimulate cGMP, activate diacylglycerol and inositol triphosphate, or alter K^+ or Ca^{2+} channels. In many cases G protein activation causes protein phosphorylation. There are several known protein kinases regulated by hormones - protein kinase A, protein kinase C, etc. Some phosphatases which dephosphorylate proteins are also regulated by hormones. When enzymes are phosphorylated or dephosphorylated, their activity changes. Some become more catalytically active and some less, depending on the enzyme. So this is another place where hormones control cell function. Lastly, some genes have regulatory sites for some of the intracellular second messengers, eg, cAMP. Thus, hormones that bind to membrane receptors activate or inactivate enzymes by second messenger signal transducers and alter transcription rates of specific genes.

Endocrinopathies. Dysfunctions of the endocrine system are due to 1) abnormal concentrations of hormone, 2) altered metabolism, or 3) receptor errors. First, hyposecretion and hypersecretion occur by either non-tumoral glandular activity or neoplastic production of a hormone. Enzymatic defects can result in the overproduction of precursors rather than the normal end product. For example, congenital adrenal hyperplasia involves hypersecretion of androgens due to deficient production of glucocorticoids and a secondary excess of ACTH. Second, altered hormone metabolism can result in conditions which symptomatically resemble hypo- or hyper- secretion of a hormone. For example, target cells that metabolize a hormone to a more active form may no longer function normally. Third, alterations in receptor function give rise to pathological conditions involving hormones. Type II diabetes mellitus is an example of such a clinical problem, as is testicular feminization (due to absence of functional androgen receptor). In such cases a relative lack of receptors causes unresponsiveness to the hormone, so the symptoms are of hormone lack. Other endocrinopathies may be due to postreceptor defects (altered signal transduction pathways), improper fetal differentiation, growth and development, neuroendocrinopathies involving altered secretion of neurohormone(s) or altered responsiveness to hormones, autoimmune diseases and certain cancers.

Review Questions

1. **Match** the action or actions in column B that fit with each hormone in column A.

	Column A		Column B
____	1. Acetylcholine	A.	Binding to receptor activates a second messenger
____	2. Glucagon	B.	Binds to an intracellular receptor
____	3. Testosterone	C.	Affects the rate of transcription
____	4. Triiodothyronine	D.	Not stored in the cells of synthesis

2. Chronically elevated plasma levels of a hormone may ultimately result in symptoms of hormone insufficiency or deficiency. Explain why this occurs.

HYPOTHALAMUS - ANTERIOR PITUITARY RELATIONSHIPS

The hypothalamus and the anterior lobe of the pituitary gland communicate via a **portal** capillary bed. Long and short vessels carry blood from the median eminence of the hypothalamus to the adenohypophysis. Neurons of specific nuclei in the hypothalamus synthesize specific **releasing hormones** and **release-inhibiting hormones** that affect trophic cells of the pituitary to increase or to decrease their secretory activity. These releasing and release-inhibiting hormones are secreted from nerve endings in the **median eminence** near the superior aspect of the portal system. They are transported to the anterior pituitary where they exert their effects on specific **pituicytes** (pituitary cells).

Releasing Hormones

Each of the six major trophic hormones of pituitary origin is regulated by hormones from the hypothalamus. Four releasing factors and two release-inhibiting factors have been identified (top of Fig. 7-1). Many chemical factors stimulate prolactin release.

Corticotropin-releasing hormone (CRH) was the first hypothalamic hypophysiotrophic hormone to be named. It is a large polypeptide (MW about 4500, 41 amino acids) and a potent stimulator of cAMP synthesis and ACTH secretion in corticotrophs. Although CRH is present within and outside the central nervous system, its greatest concentration is in the median eminence and pituitary stalk. Synthesized in the paraventricular nuclei, its secretion is influenced by a number of hormonal and neural inputs.

Thyrotrophin-releasing hormone (TRH) from paraventricular nuclei is a potent tripeptide, being effective in nanogram quantities. TRH controls secretion of thyroid stimulating hormone (TSH); it is stimulated by NE and inhibited by somatostatin. TRH is also a putative neurotransmitter. Although TRH also stimulates secretion of prolactin and growth hormone, it is not their physiological stimulus.

Gonadotrophin-releasing hormone (LHRH, FSHRH, or GnRH) is a decapeptide synthesized by neurosecretory cells in the preoptic-arcuate region. It is secreted episodically over a 10-fold concentration range under noradrenergic and endorphin control from higher CNS centers. It is responsible for the cyclic fluctuations in gonadotrophin output seen during the menstrual cycle. GnRH stimulates the physiological release of both LH and FSH. Mechanisms involved in their differential release are not known, but probably involve the actions of estrogens and progestins on the pituitary.

Somatostatin (GHIH), growth hormone release inhibitory hormone from the paraventricular nuclei, is a 14 amino acid peptide found in many parts of the body; the spinal cord, pancreas, GI tract, and thyroid gland. **GHIH** suppresses secretion of insulin, glucagon, growth hormone, TSH, and many gastrointestinal secretions; it also inhibits GI motility and has behavioral effects. Its action is largely **paracrine**, ie, it influences adjacent cells by local diffusion. While its mechanism of action is unknown, its inhibition of cAMP and its effects on calcium ion fluxes may be common to its inhibition of neurons, smooth muscle and GH secretion.

Growth hormone-releasing hormone (GHRH; somatocrinin) is a 44 amino acid peptide secreted by the hypothalamic arcuate nucleus that possesses high potency and specificity in the stimulation of growth hormone (GH) secretion.

Prolactin (PRL) secretion is regulated by a release-inhibiting hormone (PIH; dopamine), which is produced in arcuate nuclei and travels via fibers in tuberoinfundibular pathways to the median eminence and via hypophyseal portal vessels to lactotrophic cells in the anterior pituitary.

Prolactin secretion is also controlled by prolactin-releasing factors (PRFs). Oxytocin is a physiologically significant PRF, as is vasoactive intestinal polypeptide (VIP) and peptide histidine isoleucine (PHI). Several other peptides such as angiotensin II, substance P, and neurotensin also stimulate prolactin release. TRH stimulates not only TSH but also prolactin release.

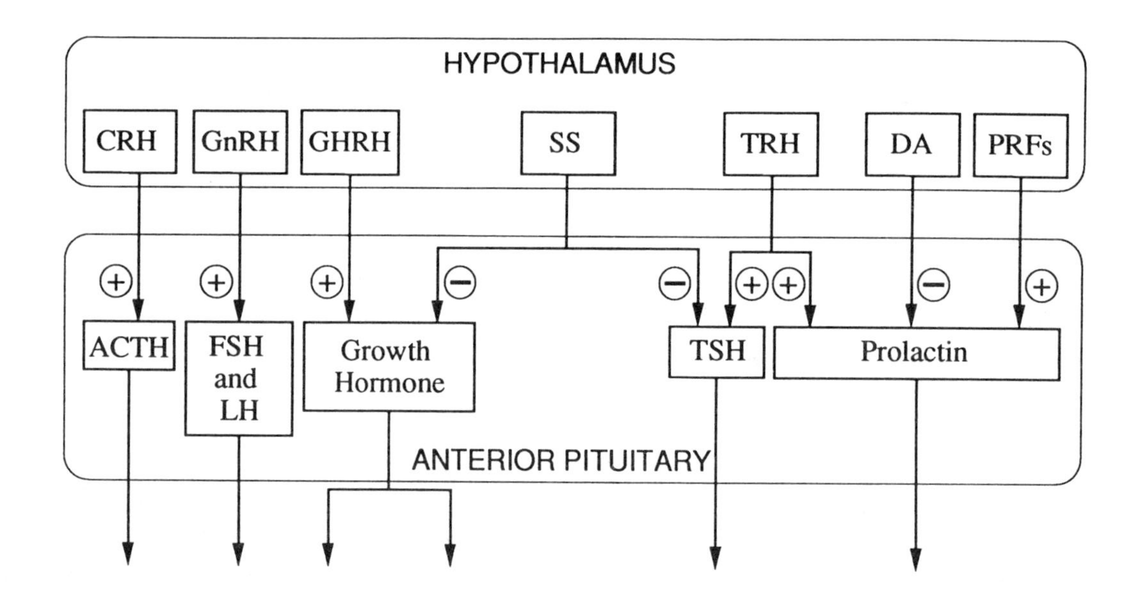

Figure 7-1. Schematic drawing for the effects of hypothalamic releasing hormones on release of anterior pituitary hormones. Key: + for stimulation; - for inhibition.

Secretory activity of hypophysiotrophic cells is controlled by the balance between signals from many extrahypothalamic areas of the CNS and feedback from the periphery. **Peptidergic neurons** of the hypothalamus are affected by a complex set of excitatory and inhibitory neuronal inputs. The monoamine neurotransmitters (dopamine, norepinephrine, and serotonin) are involved in the control of certain hypothalamic peptidergic neurons. Gamma aminobutyric acid (GABA) and ACh have also been implicated in the secretion of some releasing factors. For example, a tract that carries sleep/activity-related information may arise in a different place and use another transmitter than one activated during extreme fright. Yet activity in both tracts may result in a stimulatory message to one peptidergic neuron (eg, CRH) and an inhibitory signal to another neuron (eg, GnRH).

Anterior Pituitary Hormones

The six major anterior pituitary hormones (bottom of Fig. 7-1) can be classified into three groups: corticotropin-related peptides, the somatomammotropins, and the glycoproteins. Firstly, the **corticotropin-related hormones** of major importance are single chains of polypeptides of about 10,000 m.w. or less and derived from a common precursor. **ACTH** is the most important hormone in this group, although the group also includes β-lipotropin, α- and β-MSH, and β-endorphin. Secondly, the **somatomammotropins** are growth hormone and prolactin (and human placental lactogen); these may have evolved from a common hormone. They are single chain polypeptides of similar size (about 22,000 m.w.), possess two or three disulfide bridges, and share lactogenic and growth-promoting activities. Thirdly, the **glycoprotein hormones** are the largest hypophysial hormones (about 29,000 m.w.). This class of hormones includes LH, FSH, TSH (and human chorionic gonadotropin). Their molecule consists of a similar alpha chain in each of the hormones and a beta chain which confers specificity to each molecule.

Hypothalamic and pituitary hormones are not released in a steady discharge but are secreted **episodically**, with the pulses occurring at regular intervals, such as a few times per hour. Since the number of receptors on target cells varies inversely with plasma levels of their respective hormones, this episodic secretion may stimulate target organs with less suppression of receptors, ie, it avoids **down regulation.**

Negative Feedback Regulation

Both the hypothalamus and the hypophysis can participate in negative feedback regulation. Each hormone that is secreted by peripheral endocrine glands has its own pattern of feedback regulation (long loop). These are as follows:

1) TSH secretion is diminished by **thyroxine** and **triiodothyronine** which render the thyrotrophe refractory to the effects of TRH.

2) LH secretion is under complex control in the human female. At one stage of the menstrual cycle, **estrogen** stimulates GnRH and sensitizes gonadotrophes to the action of GnRH. At other times estrogen and progesterone inhibit GnRH secretion. Both the hypothalamus and pituitary are involved. In the male testosterone acts at both hypothalamic and hypophyseal levels to **back regulate** LH secretion.

3) The female pattern of FSH secretion is also complex. Since GnRH stimulates the secretion of both LH and FSH, a differential control mechanism may exist. The pattern of circulating steroid hormones

may alter the relative sensitivity of the pituitary, so that either FSH or LH is released or FSH secretion may be back regulated by **inhibin**, a non-steroidal product of ovary and testis.

4) **Growth hormone** (GH) secretion is governed by peptidergic neurons in the hypothalamus whose activity is regulated by various inputs. GH itself may regulate its own secretion. The peptidergic neurons are regulated primarily by the blood glucose level acting upon a **glucostat** in the hypothalamus and by plasma amino acid levels. Other factors that affect GH secretion are sleep, exercise, emotional states, and plasma somatomedin levels.

5) **Prolactin** (PRL) is also thought to be regulated by two hypothalamic factors whose secretion is influenced by peripheral signals. Prolactin secretion may be under the tonic inhibitory control of PIH (dopamine). To increase PRL secretion would involve either interferring with PIH release and/or increasing the secretion of PRF.

6) **ACTH** secretion is inhibited by cortisol, acting primarily at the hypothalamus but also at the pituitary.

In addition, the pituitary trophic hormones themselves may feed back (short loop) to inhibit the secretory activity of the neuron releasing the hormone. This is possible because of **retrograde blood flow** in certain loops of the portal system which transports pituitary secretions to the area of the median eminence. Furthermore, the secretory activity of the pituitary trophic cells varies with the circadian rhythm. GH, PRL, ACTH, and gonadotrophin secretions increase during sleep, whereas thyroid hormones diminish.

In summary, anterior pituitary activity depends upon factors from the hypophysiotropic area and upon the feedback effects of pituitary and systemic hormones. In addition, anterior pituicytes are affected by 1) posterior pituitary hormones which gain access via capillary loops in the median eminence and 2) direct capillary communication between the posterior and anterior lobes. For example, vasopressin is a powerful facilitator of ACTH secretion. Therefore, the responsiveness of anterior pituitary cells to hypophysiotrophic hormones is "fine-tuned" by hormonal input from the posterior pituitary, the periphery (negative feedback), and from other regions of the brain. The hypophysiotrophic cells themselves are influenced from various brain regions by neurotransmitters, brain peptides (VIP, enkephalins and endorphins, gastrointestinal peptides, etc.), anterior pituitary and systemic hormones.

Endorphins and Enkephalins

The discovery of specific binding sites in the brain for morphine and related opiates suggested that the body had its own "endogenous opiates"; the endorphins and enkephalins were soon found. The enkephalins are pentapeptides first extracted from brain tissue. In addition, pituitary extracts contain opioid activity in a larger molecule. An enkephalin sequence occurs at amino acids 61-65 of β-lipotrophin (β-LPH) from the pituitary. Certain pituitary basophils synthesize a large prohormone molecule that undergoes post-translational modification (proteolytic cleavages) to yield ACTH and β-LPH. β-LPH is further cleaved to yield β-endorphin, which corresponds to amino acids 61-91 of β-LPH. ACTH, β-LPH and endorphins are also found in the brain, but enkephalins are not secreted from the pituitary, even though they correspond to the first five amino acids of β-endorphin. Post-translation modification is a phenomenon common to all protein hormones (eg, insulin, parathyroid hormone, etc.). In the same manner a molecule of the precursor, pro-enkephalin, gives rise to met-enkephalin, leu-enkephalin, heptapeptide, and octapeptide. Likewise, a molecule of the precursor, prodynorphin, gives rise to dynorphin A and B and α- and β-neoendorphin.

Enkephalins are widespread in the brain, particularly in the hypothalamus, globus pallidus, caudate, thalamus, and spinal cord. Besides the pituitary, endorphins are localized in the hypothalamus, amygdala, periaqueductal grey, and locus ceruleus. Opioids interact with specific, high affinity receptors on their target cells. Morphine, its antagonist naloxone, and beta-endorphin bind preferentially to mu opioid receptors, enkephalins to delta, and dynorphins to kappa opioid receptors. Opioids decrease motor activity and responsiveness to noxious stimuli. They are synthesized in neurons with other neurotransmitters and neuropeptides. Opioids are abundant in the hypothalamus, where they decrease the frequency of LH pulses during the luteal phase and modify the secretion of several anterior pituitary hormones. More functions of opioids are being discovered all the time, which fits with the wide distribution of their receptors, their multiple chemical structures and sites of synthesis, and their inability to pass the blood-brain barrier.

Growth Hormone (Somatotrophin)

Growth hormone (GH) content of the anterior pituitary is high, making up 5-10% of pituitary weight, and it remains constant with age. Humans require human or primate GH and do not respond to porcine or bovine GH. GH is secreted episodically at 20-30 minute cycles superimposed on a diurnal rhythm, with peaks during Stage 4 sleep.

The major stimuli for GH secretion besides sleep are **hypoglycemia**, stress, amino acids, free fatty acids, and GHRH, which is sensed by glucose-sensitive cells in the CNS. Other stimuli that evoke GH secretion are exercise, pyrogens, vasopressin, and opioids. Plasma levels fluctuate from a basal level of less than 3 ng/ml to peaks of 100 ng/ml. Women demonstrate greater GH responses to appropriate stimuli, which is likely due to sensitization of pituitary cells by estrogens.

Secretory activity of the GH-secreting pituicytes is controlled by somatostatin and GHRH. For example, hypoglycemia may "turn on" α-noradrenergic receptors in cells of the ventromedial nucleus of the hypothalamus which then stimulate GHRH-producing cells, thus increasing GHRH secretion. **Hyperglycemia** may act via α-noradrenergic receptors on cells of the ventromedial nucleus of the hypothalamus to diminish GHRH secretion. Exercise, arginine, and vasopressin enhance serotonergic pathways to increase GHRH secretion. Growth hormone and GHRH inhibit the secretion of GH from the pituitary by desensitizing the pituitary to somatocrinin and by enhancing the secretion of hypothalamic somatostatin.

In muscle and skeletal tissue the short-term (minutes to hours) effect of growth hormone is **insulin-like**, ie, facilitating glucose uptake, enhancing amino acid transport and protein synthesis. Its long-term effects (hours to days), in association with glucocorticoids, are to 1) increase blood glucose (by inhibiting peripheral utilization and by supporting gluconeogenesis and glucose-6-phosphatase activity), 2) stimulate lipolysis, and 3) increase amino acid transport and subsequent protein synthesis. These long-term effects give GH an **anti-insulin** or "diabetogenic" action. The effects of GH on growth of skeletal and muscle tissues are probably mediated by insulin-like growth factor-I (IGF-I), whereas the effects on carbohydrate and fat metabolism are direct effects of GH itself. The functions of growth hormone are summarized in Figure 7-2.

The primary site of IGF (somatomedin) production is the liver. Administration of GH stimulates the appearance in the blood of a family of insulin-like peptides. IGF-I, the most concentrated and most studied of the family, circulates bound to a specific plasma protein with a half-life of 2-4 hours. The biologic effects of somatomedins are mediated by their interaction with somatomedin receptors. Binding to such receptors in cartilage results in the following effects: 1) SO_4^{2-} incorporation into proteoglycans, 2) thymidine incorporation onto DNA, 3) RNA synthesis, and 4) protein synthesis. Thus, they are mitogenic. Binding of somatomedins to receptors in skeletal muscle causes their metabolic effects, eg, stimulation of meiosis and protein synthesis (hypertrophy and hyperplasia).

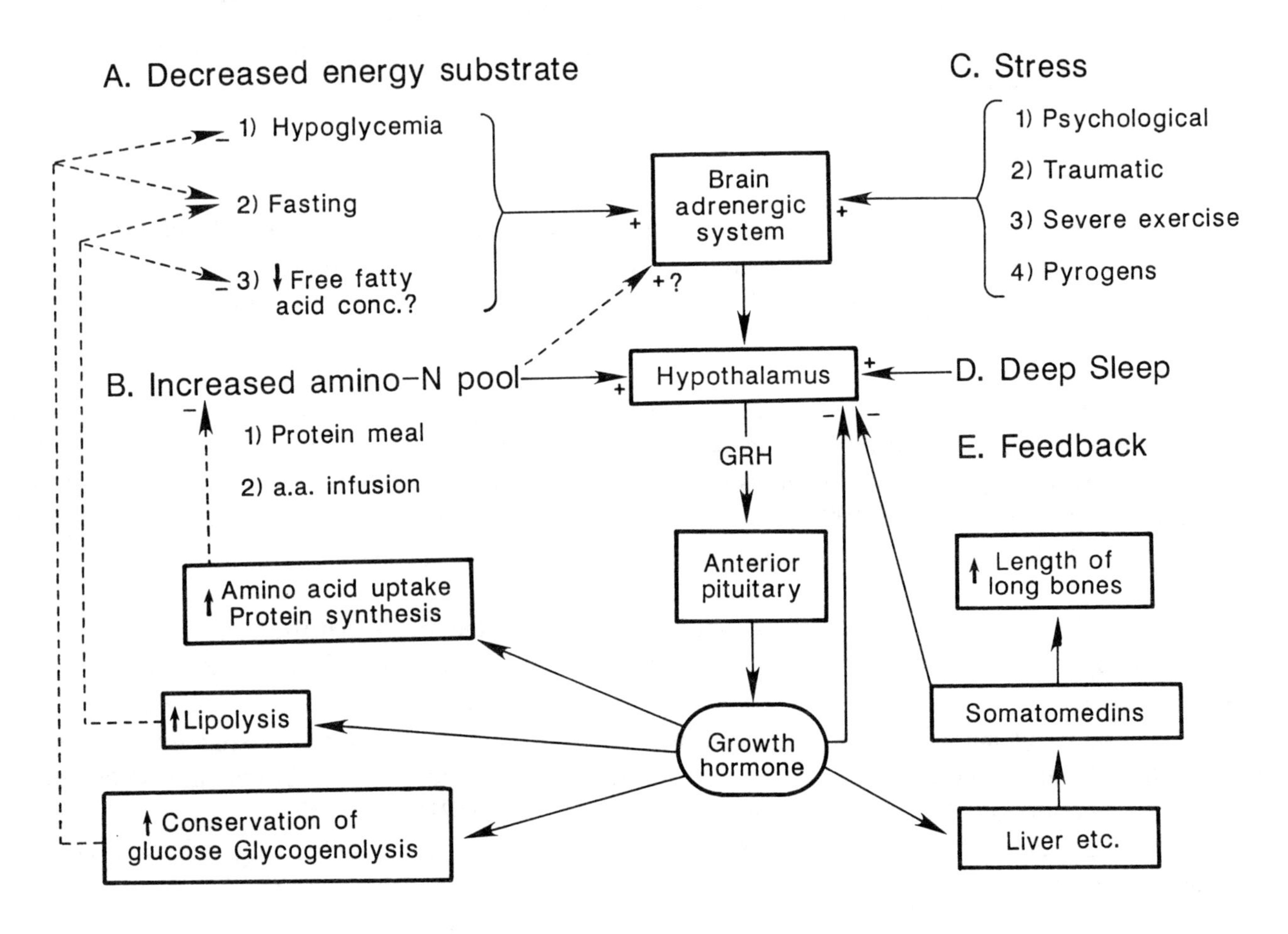

Figure 7-2. Functions of growth hormone.

Feedback Control. IGF-1, secreted by the liver in response to GH, inhibits GH production via effects on the hypothalamus and pituitary. GH itself affects the level of its own secretion by negative feedback on the hypothalamus. Stress, exercise, and emotional states markedly affect circulating levels of GH, depending on other factors such as food intake and sleep. Nutritional factors have important effects on GH secretion. Hyperglycemia and increased free fatty acids tend to suppress GH secretion, which may explain why GH secretion is reduced in response to provocative stimuli in obese patients.

Pathophysiology

Tests are used to establish the ability of the pituitary to respond to provocative stimuli and to distinguish between hypothalamic and pituitary disorders. Administration of insulin produces a powerful decrease in blood glucose, followed by a peak of GH. Additional tests include the infusion of arginine, L-DOPA, clonidine (an alpha-noradrenergic agonist) and glucagon. Pretreatment with propranolol or estrogen tends to potentiate the response, which usually occurs 60-180 min after introduction of the stimulus. The administration of GHRH may help distinguish GH deficiency due to a pituitary defect from a hypothalamic disorder. The measurement of IGF-I concentrations by radioimmunoassay may also aid diagnosis of growth-related problems.

A deficency of GH production results in **dwarfism**, of which there are three types; 1) those due to pituitary disease, 2) those due to a hypothalamic disorder, and 3) those with normal hypothalamic-pituitary function which fail to respond to GH. Primary pituitary dysfunction may be due to genetic syndromes with deficient GH secretion, to tumor destruction of the pituitary, or to physical trauma. Some patients secrete a GH molecule which behaves normal immunologically, but which is not biologically active. A hyposecretion of GH secondary to hypothalamic damage is generally the result of a perinatal insult, but it may also be due to infections and tumors. African pygmies display resistance to GH despite elevated GH levels. Plasma IGF-I levels are low and do not respond to exogenous GH. The defect is most likely in the GH receptor or in the ability of the liver to produce IGF-I.

Growth hormone excess results in **gigantism** in children and **acromegaly** in adults. These conditions usually result from a pituitary adenoma, a GHRH-producing pancreatic tumor or an ectopic GH-producing tumor. Gigantism is characterized by tall stature, in some cases approaching 8 feet. Acromegaly involves coarsened features of the face, hands, and feet. Exaggerated growth of the mandible and frontal and nasal bones also contribute to coarse facies. Connective tissue of the heart, liver and kidney hypertrophies, and glucose tolerance is reduced. Most often plasma GH and IGF-I levels are elevated several-fold.

Review Questions

3. Which hypothalamic hormone(s) or factor(s) control the following pituitary hormones?

   A. Corticotropin ____________

   B. Growth hormone ____________

   C. Thyrotropin ____________

   D. Prolactin ____________

4. **Match** the releasing hormones in column B with the hormones that inhibit them in column A.

	Column A	Column B
____	1. Cortisol	A. CRH
____	2. Estradiol	B. GHRH
____	3. IGF-1	C. GnRH
____	4. Triiodothyronine	D. TRH

5. The hypothalamic-pituitary capillary portal system plays a major role in the regulation of all of the following pituitary hormones **EXCEPT**

   A. GH
   B. FSH
   C. prolactin
   D. vasopressin
   E. ACTH

6. Which of the following are biologically active peptides related to corticotropin that originate in the anterior pituitary?

   A. α-MSH
   B. β-lipotropin
   C. ACTH
   D. Vasopressin
   E. Endorphin

7. Nine year old Tawny is small for her age and has had difficulty staying awake and paying attention in school. A pituitary reserve test revealed normal growth hormone secretion with subsequent plasma IGF-I levels within the normal range. Her CNS development is normal, and she has no apparent illness. Her physical exam is unremarkable except for several obvious bruises in various stages of healing, which Tawny's mother dismisses as due to clumsiness. The most probable diagnosis is

   A. hypersecretion of somatostatin
   B. hypothyroidism
   C. defective GHRH receptors in anterior pituicytes
   D. Inadequate sleep due to child abuse
   E. Insufficient GHRH secretion

8. Which of the following results from destruction of the anterior pituitary gland?

   A. Elevated BMR
   B. Hypogonadism
   C. Cushing's Syndrome
   D. Normal plasma IGF-I levels
   E. Goiter

9. All of the following are effects of growth hormone **EXCEPT** that it

   A. inhibits insulin action at peripheral tissues
   B. decreases urinary urea
   C. enhances amino acid uptake
   D. promotes ketogenesis
   E. enhances lipogenesis

10. Which anterior pituitary cell type secretes more than one hormone?

   A. Lactotrophs
   B. Gonadotrophs
   C. Corticotrophs
   D. Thyrotrophs
   E. Somatotrophs

11. The glycoprotein alpha subunit is secreted by all of the following cells **EXCEPT**

   A. gonadotrophs
   B. thyrotrophs
   C. trophoblasts
   D. lactotrophs

12. The "insulin-like" actions of growth hormone

   A. are mediated by somatomedins
   B. include protein synthesis
   C. are due to the actions of growth hormone on the insulin receptor
   D. can curtail ketogenesis in diabetic patients
   E. include gluconeogenesis

13. Which hormone would have a higher blood concentration at 7 am after a night of sleep?

   A. Insulin
   B. Cortisol
   C. Progesterone
   D. Thyroid hormone
   E. Vasopressin

14. Somatomedin stimulates ____________________ in the epiphyseal plate of long bones; secretion is ________________________ (enhanced/diminished) by androgen and feeds back to negatively inhibit ____________________________ at the hypothalamic level.

HYPOTHALAMUS - POSTERIOR PITUITARY RELATIONSHIPS

The neurophypophysis is an extension of the nervous system rather than a discrete gland. The axons of neurons in the supraoptic and paraventricular nuclei of the anterior hypothalamus extend down the pituitary stalk to the posterior lobe where their terminals discharge secretory products into the blood. The two hormones synthesized in the hypothalamus and secreted from the posterior pituitary are arginine **vasopressin** (also called **antidiuretic hormone** or ADH) and **oxytocin.**

Both hormones are nonapeptides (9 amino acids) with a disulfide bridge between amino acids 1 and 6. Arginine vasopressin (AVP) differs from oxytocin in two ways; a free amine group on the lst cysteine amino acid and arginine in place of leucine as the 8th amino acid.

Both of these hormones are synthesized in neurons of the supraoptic and paraventricular nuclei and packaged in secretory granules with a carrier protein, **neurophysin.** Neurophysin I specifically binds vasopressin, and neurophysin II binds oxytocin. A neurophysin and its hormone are synthesized simultaneously as a polypeptide precursor (analogous to opiomelanocortin) and packaged within secretory granules. Subsequent proteolytic activity (post-translational enzymatic cleavage) in the granule releases hormone and neurophysin. The neurophysins have overlapping affinities for the two hormones. Different types of signals evoke a neurohypophyseal secretion specific for one hormone without significant secretion of the other. The granules migrate down the nerve fibers at a rate of about 3 mm/hr and accumulate at the nerve endings in the neurohypophysis. Upon stimulation, the granules migrate to the plasma membrane of the terminal and release their contents via exocytosis into the blood stream. The peptide hormones are split from their carrier neurophysin, and both enter the capillary circulation. The neurophysins have no physiological action.

Vasopressin-containing fibers from the paraventricular nucleus also terminate in the median eminence, in contact with capillaries of the hypophyseal portal system. The hormone released here is found first in the anterior pituitary, then in the general circulation. Vasopressin and oxytocin fibers project into numerous areas of the brain and spinal cord. Depolarization induces the release of these products from the axonal endings.

ADH

The brain has osmoreceptors that respond to changes of plasma osmotic pressure by stimulating or inhibiting the release of ADH. The osmotic threshold is the plasma osmolality where ADH is secreted, a fairly constant 287 mOsm/Kg. Hemorrhage (hypovolemia) is also a powerful stimulus of ADH secretion. It appears that volume loss acting through volume receptors (cardiopulmonary) is the adequate stimulus rather than hypotension acting via baroreceptors.

Osmoregulation is a more sensitive stimulus than hypovolemia, since only a 1% change in plasma osmolality is allowed before corrective ADH secretion occurs, whereas a 7-15% fall in blood volume is required to trigger an ADH response. However, when conflicting signals are received (blood loss with hyponatremia), volume regulation overrides the osmoregulatory mechanism. Furthermore, hemorrhage can result in ADH levels 10-100 times that achieved by an osmotic stimulus. At these elevated concentrations ADH exerts a vasopressor effect.

Two types of receptors, V_1 and V_2, have been described. V_2 receptors have been identified in renal tubules and are coupled to adenylate cyclase. Binding of vasopressin by these receptors and the subsequent increase in cAMP leads to phosphorylation of luminal membrane components. This results in increased permeability to water, and thus facilitates water movement across the distal nephron and collecting duct. In the absence of ADH, urine flow can be as high as 15-20 ml/min, and urine osmolality as low as 30 mOsm/kg. In the presence of ADH, urine flow may be reduced to 0.5 ml/min, and urine may be concentrated to 1200 mOsm/kg. V_1 receptors are not coupled to adenylate cyclase. Their occupation by vasopressin results in an increase in the cytosol concentration of calcium ions. All extrarenal effects of vasopressin appear to be mediated by vasopressin receptors of the V_1 type.

Stress and other nociceptive stimuli increase vasopressin secretion. Hemorrhage and ether anesthesia are powerful stimuli for vasopressin release by the neurohypophysis. Vasopressin strongly stimulates ACTH secretion by pituitary cells and potentiates the action of CRF.

Oxytocin

Milk let-down (ejection) is the primary physiologic effect of oxytocin. It is released in response to neural signals arising in the nipple upon suckling the breast. It can also be released by psychological inputs such as the anticipation of nursing. Oxytocin stimulates myoepithelial cells to contract, expressing milk into the duct system. Estrogen sensitizes the myometrium to stimulation by oxytocin, and progesterone makes it more resistant. A physiological role for oxytocin in labor has not been demonstrated. However, it is used clinically to stimulate uterine contractions during and after labor.

Oxytocin binds to membrane receptors in the uterus and mammary gland. Estrogens increase the population of oxytocin receptors in the myometrium. The posterior pituitary exhibits cyclical changes in oxytocin secretion in response to changes of circulating estrogen. Calcium ions are probably the second messenger for oxytocin.

Pathophysiology of ADH Secretion

The Schwartz-Bartter syndrome, the syndrome of exaggerated or inappropriate vasopressin (ADH) secretion, consists of hyponatremia accompanied by normal or elevated natriuresis. Such secretion could come from the neurohypophysis or from malignant tumors which synthesize vasopressin and its neurophysin. The principal causes of this syndrome, besides tumors, are disorders of the central nervous system (eg, meningitis), pulmonary disorders (eg, pneumonia), and iatrogenic hyponatremia (eg, diuretics). The amount of vasopressin varies enormously from individual to individual, but it is always inappropriate for their plasma osmolality. Excess vasopressin accounts for the hypo-osmolality, but what accounts for the natriuresis? Hypotonic expansion of the extracellular space increases GFR and inhibits proximal tubular reabsorption of Na^+. It is a kind of "escape" similar to that associated with mineralocorticoid and may involve natriuretic factors. Corrective therapy for this condition may include 1) water restriction, 2) perfusion of hypertonic salt solution, 3) administration of lithium, which blocks vasopressin action in the kidney, or 4) use of a vasopressin antagonist.

Diabetes insipidus is a permanant state of polyuria, as opposed to transient polyuria due to excessive fluid ingestion. The term, diabetes, means "to void a large volume of urine" and the term, insipidus, refers to the tasteless nature of the urine voided. This condition is in sharp contrast to diabetes mellitus, in which large volumes of urine are produced with sweet taste (mellitus = sweet). It can be due to a deficit of vasopressin secretion (neurogenic diabetes insipidus) or to insensitivity of the renal tubule to vasopressin (nephrogenic diabetes insipidus). In general, this disease is benign.

Review Questions

15. All of the following statements regarding oxytocin and vasopressin are true **EXCEPT** which one?

    A. They are synthesized separately from neurophysins.
    B. They are synthesized in the choroid plexuses.
    C. They are synthesized in the supraoptic and paraventricular nuclei.
    D. They depend on the hypothalamic-pituitary portal capillaries for transport to the posterior pituitary.
    E. Oxytocin, but not vasopressin, has modulating actions on anterior pituitary secretion.

16. Oxytocin

    A. stimulates mammary myoepithelial contraction
    B. stimulates myometrial contraction when plasma progesterone levels are high
    C. stimulates the production of neurophysin by the liver
    D. is a prolactin release-inhibiting factor
    E. initiates labor

17. A water-deprived patient with diabetes insipidus of hypothalamic origin would exhibit all of the following **EXCEPT**

    A. increased thirst drive
    B. increased plasma osmolality
    C. increased plasma angiotensin II levels
    D. increased hematocrit
    E. reduced number of renal vasopressin receptors

18. Rank the following factors in order of their importance in controlling vasopressin secretion [most potent (1) to least potent (5)].

    _____ A. Blood pressure
    _____ B. Plasma osmolality
    _____ C. Blood volume
    _____ D. Angiotensin II
    _____ E. Stress

19. The effects upon vasopressin and its actions include all of the following **EXCEPT**

   A. is stimulated by blood loss, pain, and anxiety
   B. is inhibited by alcohol consumption
   C. increases sodium transport in the ascending loop of Henle
   D. increases water permeability of the collecting duct
   E. increases vascular smooth muscle tone

20. The paraventricular nucleus synthesizes all of the following **EXCEPT**

   A. ADH
   B. TRH
   C. oxytocin
   D. CRH
   E. vaspressin

THYROID HORMONES

The thyroid follicular cells synthesize and secrete **thyroxine** (T_4) and **triiodothyronine** (T_3). These hormones are derivatives of tyrosine and contain elemental iodine, which is required for hormone synthesis and action. The thyroid gland has large amounts of hormone stored in the follicular lumen as thyroglobulin.

The average person takes in about 500 μg of iodide (I^-) per day. Iodine in the diet is reduced to iodide in the GI tract and rapidly absorbed. About 120 μg I^- per day is taken up by the thyroid gland, which contains about 90% of the body stores of iodide. Whatever is not taken up and stored in hormonal form is excreted by the kidneys. In the absence of adequate dietary iodine less T_4 and T_3 are formed, and the thyroid hypertrophies (develops goiter) from stimulation by TSH. The hypertrophy is a compensatory mechanism, an effort to normalize plasma T_4 and T_3 levels. If successful, the individual will have a **euthyroid goiter**. If not, the individual will have a **hypothyroid goiter**.

Synthesis

The protein **thyroglobulin** (TG) is synthesized by the thyroid and serves as the storage form of thyroid hormones. It is a large glycoprotein (about 10% carbohydrate) molecule of four peptide chains. Post-translational modification involves the addition of carbohydrate moieties to the nascent TG molecule and iodination of tyrosyl residues. **Thyroid peroxidase** catalyzes the iodination of tyrosyl groups in the apical border of the cell, before the TG is secreted unto the lumen (Fig. 7-3 on the next page).

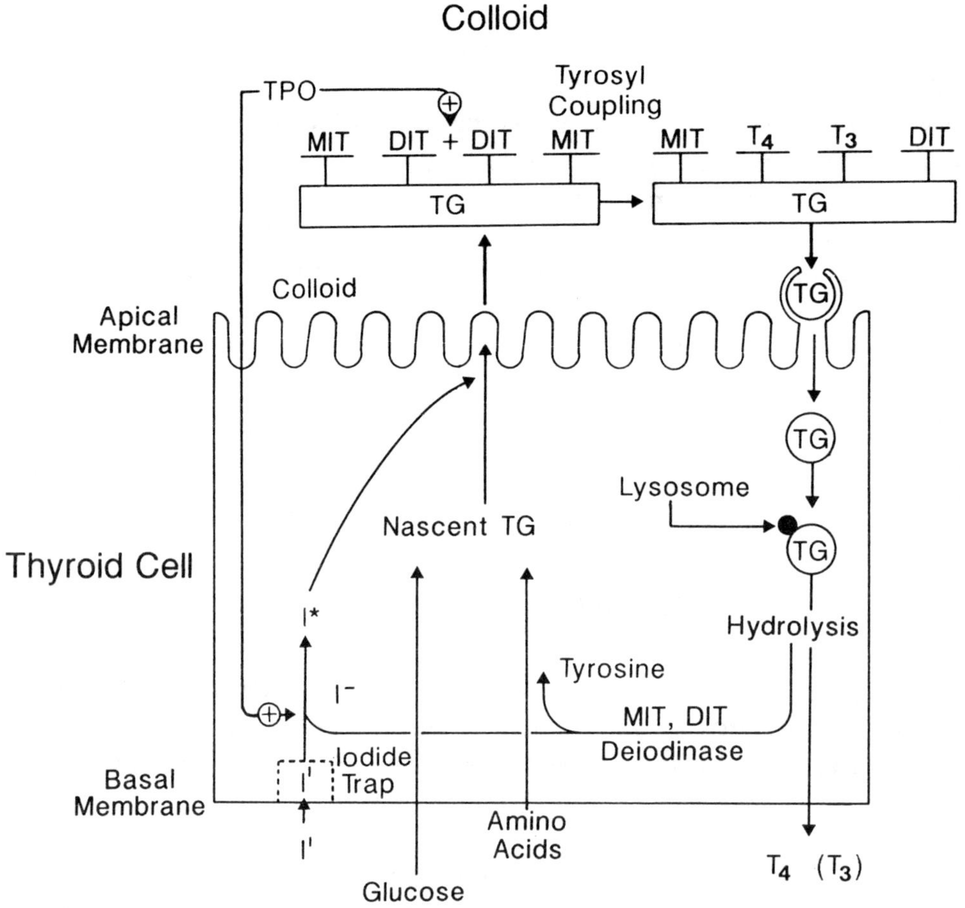

Figure 7-3. The intrathyroidal iodide cycle. MIT = monoiodotyrosine; DIT = diiodotyrosine; T_4 = thyroxine; T_3 = triiodothyronine; I* = "active" I^-; TPO = thyroid peroxidase; TG = thyroglobulin (Modified with permission from Tepperman, J: *Metabolic and Endocrine Physiology*, 5th ed. Copyright 1987 by Year Book Medical Publs., Chicago).

Iodide is actively transported into the thyroid cell at basal membranes and is concentrated to 30-40 times the serum level. The trap or iodide pump is coupled with a Na^+-K^+ ATPase and is inhibited by ouabain and by large excesses of iodide. Thiocyanates and perchlorates competitively inhibit the pump mechanism.

Iodide is rapidly oxidized by thyroid peroxidase to an active form (I^+ radical) and combines with tyrosine in the TG molecule, also catalyzed by thyroid peroxidase, to form monoiodotyrosines (MIT) and diiodotyrosines (DIT). The enzyme, which is stimulated by TSH, also catalyzes the coupling of molecules of MIT and DIT in random order. TG secreted from the cell into the follicle lumen is a glycoprotein containing iodinated tyrosines, with many coupled together to form triiodothyronine (T_3) or tetraiodothyronine (T_4).

Within a few minutes after TSH binds to a thyroid cell membrane receptor, colloid droplets appear in the cytoplasm by pinocytosis. Lysosomes migrate from the basal end of the cell to the apical end. Fusion of lysosome and droplet occur, with proteolysis of TG within the lysosome. T_4 and T_3 diffuse out of the cell in a ratio of about 15:1, depending on the iodide content of the cell and the prior degree of iodination of TG. MIT and DIT are substrates for an intracellular deiodinase which cleaves the I-tyrosine bond. Iodide and tyrosine residues are then recycled by incorporation into new TG (Fig. 7-3).

Control

The level of thyroid hormone in plasma is regulated by the anterior pituitary and hypothalamus in classic negative feedback manner. **Thyroid stimulating hormone** (TSH) is secreted by the anterior pituitary and stimulates almost all phases of thyroid activity. Its release is stimulated by a hypothalamic releasing hormone and is inhibited by thyroid hormones. Thyroid hormones diminish secretion of **thyrotrophin releasing hormone** (TRH), the number of TRH receptors on the anterior pituitary cell membrane, and the secretion of TSH. TSH is a glycoprotein with a plasma half-life of about 1 hr. Its secretion declines during sleep and non-specific stress and increases upon exposure to cold. Estrogens increase the sensitivity of the thyrotrophe to the action of TRH, whereas cortisol and growth hormone decrease its sensitivity. TRH binds to the adenohypophyseal membrane receptor, resulting in increased cAMP levels.

The thyroid gland has an ill-defined autoregulatory mechanism. Iodide transport is inversely proportional to dietary iodide intake. Furthermore, the hypophysectomized animal or hypopituitary patient still show some low thyroid I^- trapping ability. The human thyroid is innervated by sympathetic nerves, but their physiological role is unclear beyond their positive influence on hormone secretion.

Thyroxine (T_4) is the major circulating form of hormone with 99.97% bound to protein, and the remainder freely diffusible. Sixty per cent of circulating T_4 is bound to **thyroid binding globulin** (TBG), 30% to thyroxine binding prealbumin, and the remainder is bound non-specifically to albumin. **Triiodothyronine (T_3)** is also largely bound in the circulation, but to a slightly lesser degree than T_4, 0.3% being freely diffusable. T_3 is mainly (70%) bound to TBG, but with a lower affinity than the TBG-T_4 binding. T_3 is not bound to prealbumin but is bound to albumin (30%). The binding proteins are influenced by drugs and hormones. Androgens and glucocorticoids decrease TBG concentration, whereas estrogens increase it. Salicylates and diphenylhydantoin displace T_4 from binding proteins and thereby give the appearance of hypothyroidism (low total plasma T_4 and T_3) in a person with a normally functioning gland.

Thyroid hormone is inactivated via deiodination in the liver and kidneys (80%). About 20% is lost in the feces after hepatic conjugation with glucuronate and sulfate and biliary secretion. Target cells also deiodinate T_4 to **T_3**, which is the more active form and accounts for most of the biological activity. Seventy to 90% of plasma T_3 arises from peripheral deiodination. The other product of deiodination is reverse T_3 (rT_3). rT_3 is a biologically inactive molecule and is increased in starvation, severe illness, trauma, liver and kidney diseases, and the presence of glucocorticoids. The balance between formation of T_3 and rT_3 from T_4 is an important point for regulation of thyroid hormone activity (Fig. 7-4).

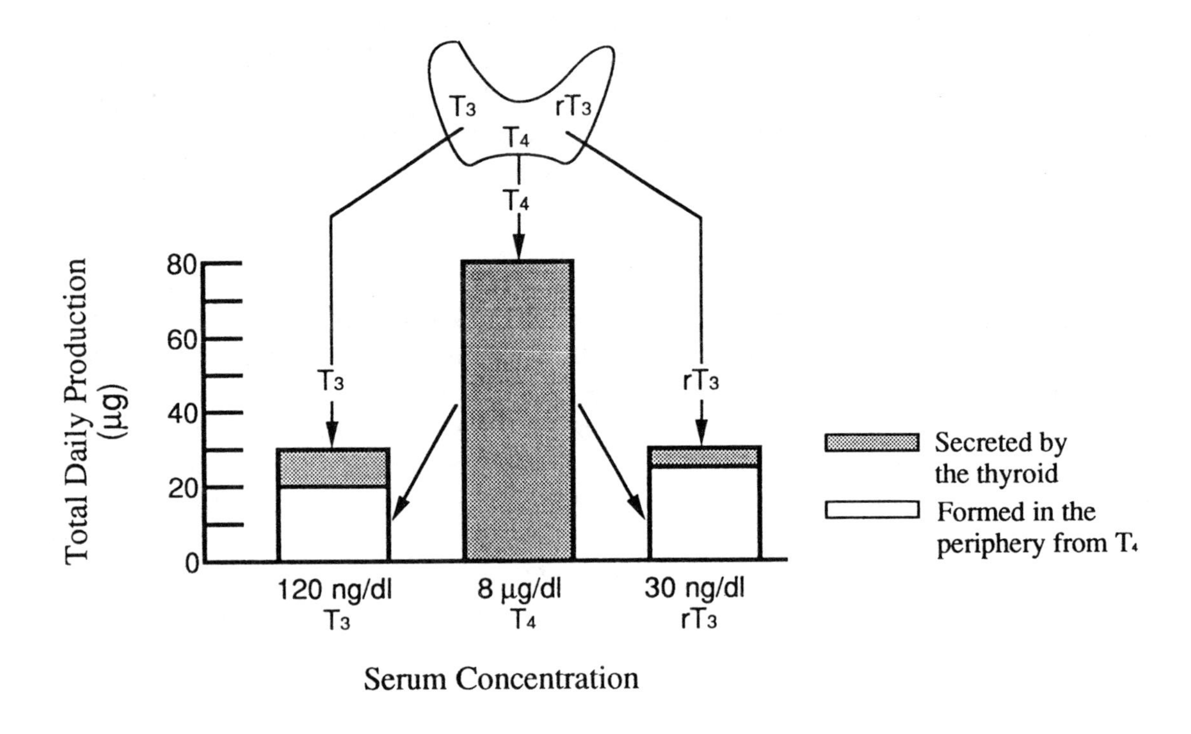

Figure 7-4. Dynamics of thyroid hormone production and mean serum concentrations. (Modified with permission from Schimmel, M and Utiger, RD, *Ann Intern Med* **87**:760, 1977).

Actions

Thyroid hormones have a similar mechanism of action to steroid hormones, since the hormone binds to a nuclear receptor with subsequent stimulation of transcription of specific messenger RNAs. Most of the major effects of thyroid hormones are mediated by the nuclear receptor. The production of all major classes of mRNA increases in response to thyroid hormones. The nuclear thyroid hormone receptor has a 10-fold higher affinity for T_3 than for T_4. There is a direct relationship between receptor number and responsiveness to thyroid hormones. Animal studies have shown that about 85% of the total hormone bound to liver and kidney nuclei is T_3 and the remaining 15% is T_4. Although most thyroid hormone effects are mediated by nuclear events, these hormones also have direct effects on mitochondria to enhance oxidative phosphorylation (ATP production).

The actions of thyroid hormones can be divided into two major categories; 1) effects on growth and differentiation and 2) metabolic effects. Fetal or neonatal thyroid deficiency leads to **cretinism**, a condition of abnormal facial appearance and structure, reduced neuron size and content, and delayed skeletal maturation. Many of the CNS effects of thyroid hormone deficiency are developmental failures. Deficiency in the adult produces mental and physical sluggishness, and somnolence as in a cretinous child. Mental and psychiatric symptoms in an adult with thyrotoxicosis subside with hormone replacement. Thyroid hormones are not required for prenatal growth, but they are necessary for differentiation. Postnatally, the thyroid is required for both processes. Thyroxine is a

growth-promoting hormone in replacement doses. Growth failure occurs in hypothyroid children, and tissue wasting occurs in hyperthyroidism due to excessive protein catabolism. Growth and differentiation of bone marrow, mammary glands, and teeth are also stimulated by thyroid hormones.

Thyroid hormones increase oxygen consumption and heat production. The **calorigenic effect** is most marked on heart, skeletal muscle, liver, and kidney, but is absent in brain, retina, spleen, and gonads. A large fraction of their calorigenic action results from stimulating Na^+-K^+ ATPase activity and utilization of ATP in hormone-sensitive cells. Many of the effects on intermediary metabolism can be interpreted as effects secondary to an **increased basal metabolic rate** (BMR). Thyroid hormones promote glucose absorption from the gut (probably a direct effect) and gluconeogenesis and glycogenolysis (probably an indirect effect of the actions of epinephrine). Lipolysis is increased, as is cholesterol turnover and plasma clearance of cholesterol. Normal thyroid hormone levels promote protein synthesis. T_4 and T_3 promote normal contraction of cardiac and skeletal muscles.

Pathophysiology of Thyroid Function

Hypothyroidism occurs in about 2% of women and less frequently in men. It may be due to disorders of the thyroid (primary), of the anterior pituitary or hypothalamus (secondary), or of peripheral target tissues (hormone resistance). Primary hypothyroidism accounts for over 95% of patients with the disease and may be associated with thyroid tissue atrophy or enlargement. Both forms are commonly the result of an autoimmune thyroiditis leading to destruction of the parenchyma. Primary hypothyroidism is commonly due to destruction of the gland by surgery or radioactive iodine treatment. Tissue resistance to the actions of the hormone is due to a receptor defect or post-receptor defect. This leads to increased thyroid hormone productions in the absence of hyperthyroidism. In such cases plasma TSH is normal and responsive to TRH.

Severe thyroid hormone deficiency in infancy is termed **cretinism** and is characterized by mental and growth retardation. When hypothyroidism occurs later in childhood, mental retardation is less prominent and impairment of linear growth is the major feature. So bone age is less than chronological age.

Hypothyroidism causes hypotonia, sluggishness and muscular weakness. Neuromuscular excitability is decreased with consequent increased reflex time. There is decreased glucose absorption with decreased BMR.

Onset of hypothyroidism is very slow in the adult. Early symptoms including lethargy, tiredness and constipation may develop. Eventually mental function and motor activity slow, weight gain occurs, and cold intolerance is seen. Other symptoms include hair loss, hoarse voice, myxedema, muscle stiffness, delayed tendon reflexcs and abnormal menstrual function in women.

Hyperthyroidism also occurs in about 2% of women and much less frequently in men. The major cause is diffuse toxic goiter or Graves' disease. In this case an immunoglobulin interacts with the TSH receptor and activates it. A goiter is accompanied by ophthalmopathy (exophthalmia). Early manifestations of hyperthyroidism are nervousness, irritability, tachycardia, and heat intolerance. Weight loss despite normal or increased food intake also is common. Abnormal menstrual periods, increased GI motility, skeletal muscle weakness and cardiovascular irregularities may develop. Hyperthyroidism stimulates BMR and protein breakdown, leading to increased pulse pressure by increased systolic pressure, negative nitrogen balance, muscle wasting, weakness, tremors, and twitches. Neuromuscular excitability is increased in hyperthyroidism with consequent twitches and tremors. There is increased glucose absorption with increased BMR and diarrhea.

Some of the symptoms of hyperthyroidism suggest hyperactivity of the sympathetic nervous system. Hyperthyroid patients are more sensitive to catecholamines, and many manifestations of thyroid excess are controlled by administration of β-adrenergic blockers. Thyroid hormones increase the number of β-adrenergic receptors in heart muscle cells, while hepatocyte receptors are not affected. Not all actions of thyroid hormone, eg, increased O_2 consumption, are prevented by β-receptor blockade.

Review Questions

For Questions 21-25. A male in his early forties comes to your endocrine practice presenting with lethargy; evidence of intellectual impairment; dry, scaly skin; sparse, dry hair; dull expression with droopy eyelids, puffiness of the face and periorbital swelling; bradycardia (60 b/min); blood pressure of 90/70; hematocrit of 27; enlarged heart, constipation and hypothermia.

Assays of plasma concentrations of T_4 and T_3 reveal the following:

	T_4	T_3
Total	2.5 ug/dL	0.25 ng/dL
Free	0.6 ng/dL	0.10 ng/dL

Radioimmunoassay (RIA) of peripheral blood reveals elevated TSH levels. A TSH challenge test does not increase the secretion of thyroid hormones from the thyroid gland.

Refer to p. 280 for normal values to understand this patient.

21. Is this a primary or secondary disorder?

    Is more information needed to make this diagnosis? If so, what is needed?

22. Describe the feedback loop involved.

23. List several defects that could cause the problem.

24. Would you expect to find a palpable goiter? Explain.

25. Describe a suitable treatment for this individual.

26. Which of the following are characteristics of thyroxine (T_4)?

   A. About 90% of the hormone in the circulation is bound to plasma protein.
   B. Its half-life in blood is less than one hour.
   C. Large amounts are stored intracellularly.
   D. Conversion to triiodothyronine (T_3) takes place within target tissue cells.

27. An inhibitor of thyroid peroxidase may

   A. inhibit the intracellular binding of T_4 or T_3
   B. result in goiter
   C. increase iodide uptake into the thyroid gland
   D. diminish the sensitivity of pituitary cells to TRH

CALCIUM METABOLISM

Neural excitability, muscle contraction, blood coagulation, the function of specific enzymes and hormones, and other cellular processes depend upon Ca^{2+}. Of the 2-5 mmoles of intracellular Ca^{2+}/kg, more than 99.9% is bound, sequestered to membranes and organelles. All cells have pump mechanisms to extrude excess Ca^{2+}, so free intracellular calcium amounts to less than 1 μM. The low ECF Ca^{2+} allows cells to better regulate their intracellular Ca^{2+}. Plasma calcium is closely regulated at 9-11 mg%. Forty-five per cent of plasma calcium is bound to protein, 10% is complexed to phosphates and citrates, and 45% is free. The free fraction in plasma (5 mg% or 2.3 mM) is subject to hormonal influences. **Bone calcium** is 99% of total body calcium. But only 0.05% of Ca^{2+} is exchangeable with ECF. Normally, there is a balanced steady state between GI intake and renal loss. Regulation of ECF (plasma) Ca^{2+} depends on hormonal influences on bone, GI tract and kidneys.

Three hormones are the primary regulators of calcium balance; parathyroid hormone, **1,25-dihydroxycholecalciferol** (vitamin D), and calcitonin. Other hormones also influence calcium metabolism. An excess of glucocorticoids (eg, Cushing's disease) leads to osteopenia by 1) promoting breakdown of bone matrix, 2) decreasing the number of osteoblasts, 3) antagonizing the GI absorption of Ca^{2+}, and 4) stimulating parathyroid hormone secretion. Estrogens inhibit bone resorption and are effective in treatment of osteoporosis of postmenopausal women. GH (somatomedin), androgens, thyroxine, and insulin also promote bone growth by different mechanisms. However, day-to-day regulation of calcium balance is controlled by the three major hormones.

Parathyroid hormone (PTH) is synthesized in parathyroid cells. Upon stimulation by low ECF Ca^{2+}, secretory vesicles release the 84 amino acid residue PTH into blood. The primary regulator of PTH secretion is Ca^{2+}. When ECF Ca^{2+} is high, the degree of intracellular turnover of PTH is high, 80% being degraded. When ECF Ca^{2+} is low, the turnover rate declines, with 60% being degraded and the extra 20% secreted.

Calcitonin is synthesized in "C cells" of the thyroid that come from the neural crest. This hormone is secreted as a 32 amino acid polypeptide chain. Its secretion is stimulated by high Ca^{2+} and inhibited by low ECF Ca^{2+}. It is also stimulated by gastrin, whose secretion, in turn, is inhibited by calcitonin.

The synthesis of **vitamin D** is more complex (Fig. 7-5 below). Ultraviolet light on the skin converts 7-dehydrocholesterol to pre-vitamin D_3. Pre-vitamin D_3 is confined to the skin because of its high lipid solubility and its very low affinity for vitamin D-binding protein. However, at normal body temperature pre-vitamin D_3 is slowly converted to D_3, which is readily bound by plasma binding protein and carried by the circulation to the liver. The liver has a high level of 25-hydroxylase enzyme activity which converts vitamin D_3 to 25-hydroxy D_3, which is then released from the liver. In the kidney it is converted to 1,25-dihydroxy D_3 or 24,25-dihydroxy D_3. The former **(1,25)** is the **most active** vitamin D molecule in the body, whereas the latter (24,25) is much less active. Vitamin D or D_3 is more accurately called cholecalciferol, ie, 1,25-dihydroxy D_3 = 1,25-dihydroxy cholecalciferol. The 1-hydroxylase enzyme in the kidney is an important point of regulation of the pathway. Its activity, and thus the formation of 1,25-dihydroxy D_3, is stimulated by low plasma phosphate (inorganic phosphorus) and by low Ca^{2+}. The mechanism of increasing enzyme activity by phosphate is unknown. Low Ca^{2+} stimulates the secretion of PTH, and then PTH directly stimulates the 1-hydroxylase. In normocalcemia and normophosphatemia a 24-hydroxylase enzyme is more active, and 25-hydroxy D_3 is converted to 24,25-dihydroxy D_3. Diet and exposure to sunlight are also important in the biosynthesis of vitamin D.

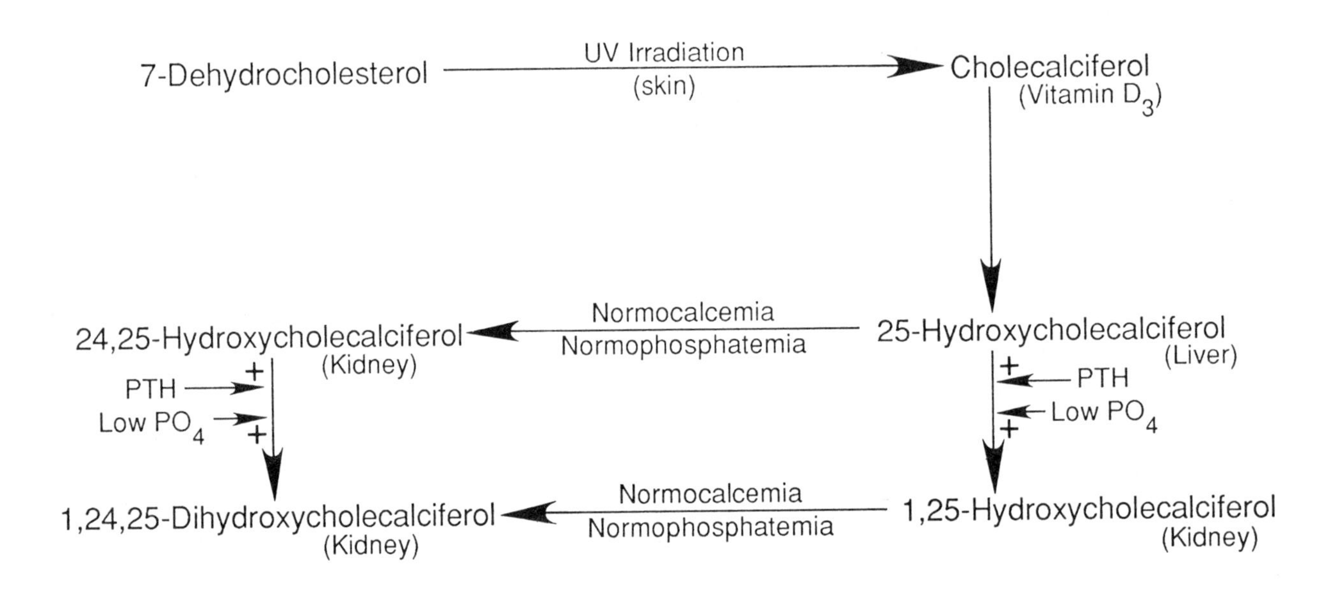

Figure 7-5. Sites of vitamin D formation and its metabolites.

In the intestine 1,25-dihydroxy D_3 stimulates the absorption of Ca^{2+} and PO_4^{3-}. Ca^{2+} absorption occurs primarily in the duodenum by active transport against an electrochemical gradient. 1,25-dihydroxy D_3 enters the intestinal mucosal cell where it is bound to a specific receptor, forming a complex that subsequently stimulates transcription of specific mRNA. Translation of this mRNA into a specific protein, **calcium binding protein**, facilitates transcellular calcium transport. Vitamin D_3 also increases absorption of phosphate by a separate mechanism. Glucocorticoids antagonize the effect of D_3, thus inhibiting Ca^{2+} absorption. PTH has no direct effect on gut absorption of Ca^{2+}, but it has an indirect effect by stimulating the formation of 1,25-dihydroxy D_3 in the kidney. PTH increases gastric acid and pepsin secretion, which may be the basis for the high incidence of peptic ulcer in patients with hyperparathyroidism. Food intake may prepare the body for a Ca^{2+} load by the action of gastrin in stimulating calcitonin secretion. Calcitonin, in turn, has a minor negative effect on gastrin and HCl secretion.

Parathyroid hormone is the most active hormone in the kidney that affects Ca^{2+} and PO_4^{3-}. PTH stimulates **Ca^{2+} reabsorption** by the distal tubule and blocks PO_4^{3-} reabsorption in the proximal tubule. These actions may be mediated by cAMP subsequent to PTH binding to a specific receptor on the tubule cell membrane. Calcitonin may or may not decrease Ca^{2+} and PO_4^{3-} reabsorption. Vitamin (1,25) D increases phosphate reabsorption, which might make it antagonistic to the action of PTH. However, in normocalcemia and hypophosphatemia when PTH levels would be low to normal, hypophosphatemia increases the formation of 1,25-dihydroxy D_3. Vitamin D then increases renal (and intestinal) absorption of PO_4^{3-}, thus raising blood levels.

PTH and vitamin D act synergistically in bone to stimulate Ca^{2+} resorption; either one alone will have a similar action in high doses. The plasma level of PTH required for an effect on bone is higher than that for its renal effect; so, the first level of action of PTH is renal. PTH and Vitamin D affect bone by increasing ECF PO_4^{3-} and Ca^{2+} and decreasing bone mineralization. Bone mineralization (largely an osteo**blastic** function) and demineralization (largely an osteo**clastic** function) is a dynamic process. PTH and vitamin D cause a net increase in osteoclastic activity and a net increase in bone resorption. Calcitonin has the opposite effect, inhibiting osteoclastic activity and diminishing the efflux of labile Ca^{2+}.

Pathophysiology of Calcium Metabolism

Parathyroid Hormone. Parathyroid insufficiency, regardless of the age of onset, leads to hypocalcemia, hypocalciuria, hyperphosphatemia with increased tubular reabsorption of phosphorus, and a low to normal alkaline phosphatase. It is accompanied by a fall in plasma vitamin D_3. The principal manifestations are convulsions, either generalized or localized, tetany, laryngospasm, intracranial hypertension, changes in teeth and nails, cataract, and depressive or psychotic states. In newborns and infants the most severe problems associated with parathyroid insufficiency are tremors, convulsions, and cardiomegaly.

Primary hyperparathyroidism is common. Patients present with hypercalcemia, thirst, polyuria, mental confusion, headache, calcified cornea, radiological signs of demineralization, renal disorders, gastric ulcers, and pancreatitis. Diagnosis is based on hypercalcemia, hypophosphatemia with reduced tubular reabsorption of phosphorus, occasional elevated alkaline phosphatases and increased urinary hydroxyproline.

Calcitonin. A large number of thyroid cancers produce calcitonin. Patients with either acute or chronic renal insufficiency often have elevated levels of calcitonin. Marked elevations of immunoreactive calcitonin in serum have been observed in pancreatitis, in acute hepatitis, and in infantile meningococcemia. Reduced or absent calcitonin production is sometimes observed in congenital athyreosis and in patients whose thyroid has been destroyed.

The hypocalcemic effect of calcitonin is more significant when calcium levels are elevated. Bone renewal is more rapid, and the exchangeable calcium pool is smaller. Calcitonin is used in the treatment of hypercalcemia. It is useful in the treatment of hyperparathyroidism, bone metastases, vitamin D intoxication, and idiopathic hypercalcemia in children. Calcitonin has a protective effect on bone which has undergone an exaggerated osteolysis.

Vitamin D_3. An insufficiency of vitamin D provokes hypocalcemia, hypophosphatemia, elevation of serum alkaline phosphatases, reduction in calciuria and hyperaminoaciduria. The first step in clinical evaluation is to recognize the existence of rickets or osteomalacia. Signs associated with reduced vitamin D activity include a reduction in growth and development of osteoid tissue in bone.

Several syndromes, including hypercalcemia, diffuse calcinosis (corneal, renal, pulmonary, myocardial and vascular, digestive) and skeletal changes, seem to be either directly or indirectly related to excessive vitamin D. Extremely high levels of vitamin D lead to hypercalcemia, with polydipsia, excessive diuresis, absence of appetite, vomiting, and constipation. Fever, pallor, somnolence, abdominal pains, growth arrest and weightless may be observed. Arterial hypertension is frequent.

Actions of Other Hormones on Calcium Metabolism. Thyroid hormones increase urinary and plasma calcium. They augment phosphatemia and reduce phosphaturia. In addition, T_3 and T_4 inhibit intestinal absorption of calcium, stimulating the renewal of bone, but with a greater effect on resorption. A negative calcium balance accompanies hyperthyroidism. Thyroid insufficiency in the infant is characterized by a localized or generalized densification of the skeleton, increased digestive absorption and a reduced urinary excretion of calcium.

Hypercorticism leads to osteoporosis, reduction in osteoblastic function, and slowing of growth. It is accompanied by normal calcemia with increased plasma PTH. Glucocorticoids reduce intestinal absorption of calcium, opposing the action of vitamin D. Osteoporosis is associated with poor intestinal absorption of calcium, reduced renal retention of calcium, reduced bone formation and especially of collagen synthesis, and increased resorption of bone. Glucocorticoid deficiency may lead to hypercalcemia.

Estrogens reduce the response of bone tissue to PTH. Short-term treatment with estrogens reduces calcemia, phosphatemia, and calciuria, increases retention of calcium in bone, reduces urinary hydroxyproline, and reduces the resorptive surface of bone. Estrogens may augment the activity of 1-hydroxylase, thus increasing the synthesis of vitamin D. Growth hormone is a potent factor in bone growth via IGF and stimulates 1-hydroxylation of 25-OH-D_3.

Review Questions

For Questions 28-30: A 55-year-old female had a total thyroidectomy followed by thyroid hormone replacement. Forty-eight hours later she developed laryngeal spasms, tetany, and cramps in the muscles of the hand and arms. Laboratory tests revealed the following:

Plasma calcium	6.5 mg/dL
Plasma phosphorus	5.0 mg/dL
Urine calcium	30 mg/dL
Urine phosphorus	0.1 g/day

Daily oral calcium gluconate and vitamin D alleviated the tetany and laryngeal spasms.
See p. 280 for normal values.

28. What is the endocrinopathy?

29. What purpose is served by simultaneous administration of vitamin D with calcium?

30. What caused the tetany and spasms?

31. The following is a JEOPARDY-type problem. That is, the answer is given, you are asked to supply the question.
    Answer: Renal 1-hydroxylase enzyme.

    Question: ______________________________

32. The plasma parathyroid hormone level in pseudohypoparathyroidism is normal to slightly elevated. Yet the patient demonstrates symptoms of low parathyroidism, such as decreased plasma levels of Ca^{2+} and $1,24(OH)_2D$ and increased plasma PO_4^{3-} levels. What might account for this condition?

33. All the following statements are true **EXCEPT** which one?

    A. Phosphate is an essential part of the thyroid hormone molecule.
    B. Plasma phosphate levels are important for bone mineralization.
    C. Low levels of plasma phosphate stimulate conversion of 25 (OH) cholecalciferol to $1,25(OH)_2$ cholecalciferol.
    D. Hyperphosphatemia is not a common medical problem.
    E. Phosphate excretion by the kidney and its absorption from the GI tract are hormonally regulated.

34. **Match** each hormone in column B with its target tissue(s) in column A.

	Column A	Column B
___	1. Bone	A. Calcitonin
___	2. Gut	B. Parathyroid hormone
___	3. Kidney	C. Vitamin D

If the above hormones have more than one action, what is the primary or most important action in each case?

PANCREATIC HORMONES

The pancreatic islets, representing about 3% of the volume of the organ, are composed of four cell types. A (alpha) cells secrete glucagon; B (beta) cells secrete insulin; D (delta) cells secrete somatostatin, and F cells secrete pancreatic polypeptide.

Insulin is synthesized in **B cells** as a large pre-pro-insulin molecule. Proinsulin is confined within membranous granules containing zinc and a proteolytic enzyme. This enzyme acts at two specific sites in the chain to cleave it into a connecting peptide and the two-chained insulin molecule. Apparently the connecting peptide is necessary to permit proper alignment of the molecule to form the A and B polypeptide chains connected by disulfide bonds. Insulin, connecting peptide, and a small quantity of proinsulin are secreted together. Connecting peptide has no biological activity.

Insulin secretion is stimulated primarily by glucose. Glucose interacts with B cell membranes and intracellular metabolic machinery to stimulate the active secretion and synthesis of insulin. The combination of glucose and insulin inhibits glucagon secretion. Glucose stimulation of beta cells evokes increases in cytosolic Ca^{2+} and cAMP, resulting in extrusion of secretory vesicle contents into ECF. Peripheral insulin concentration rises from its basal level (0.4 ng/ml) about 10 minutes after ingesting a meal and reaches a peak (usually not more than 4 ng/ml) in blood about a half hour later. Insulin secretion is a biphasic response; an immediate increase peaking within a few minutes, followed by a second, slower increase which reaches a higher level of secretion. The first phase represents the release of stored insulin, and the second phase represents the secretion of stored insulin plus newly synthesized insulin. Intravenous infusion of glucose gives a more rapid response, with elevated peripheral insulin within 2 min, peaking by 4 min. The insulin response to glucose is sigmoidal, with no measureable insulin release until plasma glucose concentration exceeds about 90 mg/100 ml. Amino acids and ACh stimulate insulin secretion, while catecholamines inhibit release via α-receptors. The stimulation of α-adrenergic receptors inhibits insulin secretion, whereas the stimulation of β-adrenergic receptors stimulates insulin secretion. The net effect of catecholamines is to inhibit insulin secretion. Activation of the ventromedial area of the hypothalamus has an inhibitory effect on insulin secretion, while the ventrolateral region stimulates insulin secretion and increases vagal tone. Oral ingestion of glucose produces a higher serum insulin concentration and lower serum glucose than does a similar dose of intravenous glucose. Of the gastrointestinal peptides capable of increasing insulin secretion (gastrin, secretin, CCK-PZ, and gastric inhibitory peptide (GIP)), only GIP is capable of potentiating the insulin secretory effect of glucose and amino acids at physiological concentrations.

Insulin binds to the alpha subunit of its receptor, which results in phosphorylation of the beta subunit and activation of the beta subunit tyrosine-dependent protein kinase activity. Phosphorylation of the receptor subunit and activation of its kinase activity are initial steps in transmembrane signalling and subsequent biological effects of insulin. Following binding and initiation of post-receptor events, insulin is taken up (translocated) into the cell. This internalization of the hormone appears to degrade it.

The major action of insulin is to **promote storage** of **energy substrates**. In the liver insulin stimulates glycogen formation by stimulating glucokinase and glycogen synthetase and inhibiting phosphorylase. It also increases triglyceride synthesis and very low density lipoprotein formation. In muscle insulin increases amino acid transport and ribosomal protein synthesis as well as glycogen synthesis. In adipose tissue it promotes glucose transport, esterification of glycerol with fatty acids, and transport and triglyceride synthesis. It also stimulates potassium uptake in muscle and adipose tissues.

Glucagon is synthesized as a precursor molecule that undergoes progressive post-translational cleavage to yield a 29 amino acid protein. Glucagon and three gut hormones, **secretin**, vasoactive inhibitory peptide **(VIP)**, and gastric inhibitory peptide **(GIP)**, share certain homologous amino acid sequences and the ability to potentiate the stimulatory effect of glucose on B cell insulin secretion. Glucagon is also synthesized in alpha cells of the gut as well as the pancreas.

Glucagon release is inhibited by glucose and by fatty acids. Glucagon release is stimulated by amino acids, hypoglycemia, β-catecholamine release induced by exercise and stress, and ACh. Beta-adrenergic agonists stimulate the secretion of glucagon, as does stimulation of the ventromedial hypothalamus. Cholecystokinin (pancreozymin), whose secretion is stimulated by amino acids, powerfully potentiates glucagon secretion.

Glucagon is a "counter-regulatory" hormone for **making energy substrates available** for metabolism. In liver cells glucagon stimulates glycogenolysis, gluconeogenesis, and ketogenesis. In muscle and adipose tissue it promotes lipolysis of triglycerides and lysosomal breakdown of proteins. Whereas insulin acts alone to lower blood sugar and to promote energy substrate storage, glucagon acts with glucocorticoids, growth hormones, and catecholamines to 1) raise blood sugar, 2) promote energy substrate mobilization, and 3) shift muscle metabolism to fatty acids and ketones while maintaining glucose availability for the central nervous system. Glucagon acts primarily via activation of adenylate cyclase and subsequent intracellular increases of cAMP.

Somatostatin (GHIH in Fig. 7-1) is synthesized in D cells of the pancreas and in the brain. Stimuli that release insulin also promote somatostatin release, eg, glucose and amino acids. Somatostatin **inhibits** the **secretion** of insulin and glucagon as well as glucose absorption from the GI tract. Vagal stimulation and enteric hormones regulate secretion of **pancreatic polypeptide** after a meal. Somatostatin inhibits the secretion of pancreatic polypeptide, a 36 amino acid protein.

Each islet cell type does not respond autonomously, but influences and is influenced by the activity of adjacent cell types. The three major hormones (not pancreatic polypeptide) are secreted locally and diffuse to neighboring cells and into the blood supply. Such local effects due to diffusion within a finite space are characterized as **paracrine** actions of a hormone. In this manner insulin inhibits glucagon secretion, glucagon stimulates insulin and somatostatin secretion, and somatostatin inhibits insulin and glucagon secretion.

Food Deprivation

The fasting state reproduces the hormonal profile of **diabetes**; reduced insulin and elevated secretion of glucogen and other counter-regulatory hormones. Acute (minutes to hours) adjustments to starvation involve redistribution of substrates via hormone actions. Chronic (hours to days) adjustments include elaborate resetting of cellular enzyme levels via hormone actions. While acute elevation of the blood glucose level is not life-threatening, a rapidly falling blood glucose level may be. In a fed state hepatic glucose is utilized by conversion to glycogen or by metabolism via glycolysis with insulin regulation. However, in a fasted or diabetic state the four counter-regulatory hormones are dominant and increase glucose production via glycogenolysis (epinephrine and growth hormone) and gluconeogenesis (glucagon and cortisol). Both processes result in increased levels of glucose-6-phosphate in liver but not in muscle, and release of free glucose into the circulation. Glucagon and the glucocorticoids play an important role in mobilizing amino acids from peripheral tissue proteins (principally muscle) that are used for long-term maintenance of blood glucose via hepatic gluconeogenesis. Cortisol and growth hormone also block peripheral carbohydrate oxidation and shift the tissues to fatty acid utilization.

In early starvation muscle yields a mixture of **amino acids** but especially alanine and glutamine. Later on nitrogen excretion declines as brain β-hydroxybutyrate dehydrogenase activity increases, since brain uses ketone bodies instead of amino acids. Alanine is the principal gluconeogenic substrate and is a prime stimulator of glucagon release which then stimulates gluconeogenesis (together with cortisol).

Epinephrine, cortisol, and growth hormone also stimulate the mobilization of **fatty acids** from adipose tissue. Increased rates of fatty oxidation and correspondingly decreased rates of glucose oxidation have been described during starvation and diabetes. As free fatty acid levels increase, they inhibit carbohydrate metabolism. A decrease in insulin availability and an excess of insulin antagonists, which occur in diabetes mellitus and in starvation, result in free fatty acid mobilization and ketone production. Elevated free fatty acids and the hormone-stimulated dominance of lipolysis over lipogenesis in liver cells lead to 1) a fall in malonyl CoA concentration, 2) enhanced transport of fatty acids into mitochondria, and 3) generation of acetyl CoA leading to formation of acetone, β-hydroxybutyrate and acetoacetate.

To summarize, a decline in the insulin:glucagon ratio, whether brought about by drastic changes in diet or by availability of insulin, causes shifts in substrate availability. Oxidation of glucose by peripheral tissues declines, protein catabolism furnishes substrates for gluconeogenesis, and mobilization of stored fat and subsequent oxidation of fatty acids increases. Mobilization of stored fat provides reducing equivalents to sustain gluconeogenesis, in addition to furnishing fatty acids and ketone bodies for use by peripheral tissues. The action of hormones and substrates together influence the necessary alterations in enzyme activities to channel substrates into new pathways (eg, alanine to gluconeogenesis).

What happens in the insulin-deprived **diabetic** is a violent caricature of the adaptation to starvation. The primary event is relative insulin withdrawal, either by a decrease in the amount of insulin available or by an increase in insulin requirement. The sequence of metabolic and pathophysiologic events that follows insulin deficiency is shown in Figure 7-6. Restoration of the required insulin (and thus the proper insulin:glucagon ratio) results in adjustments similar to re-feeding after starvation. Anabolic processes prevail, and the proportion of carbohydrate to fat in the fuel mixture increases in muscle and adipose tissue.

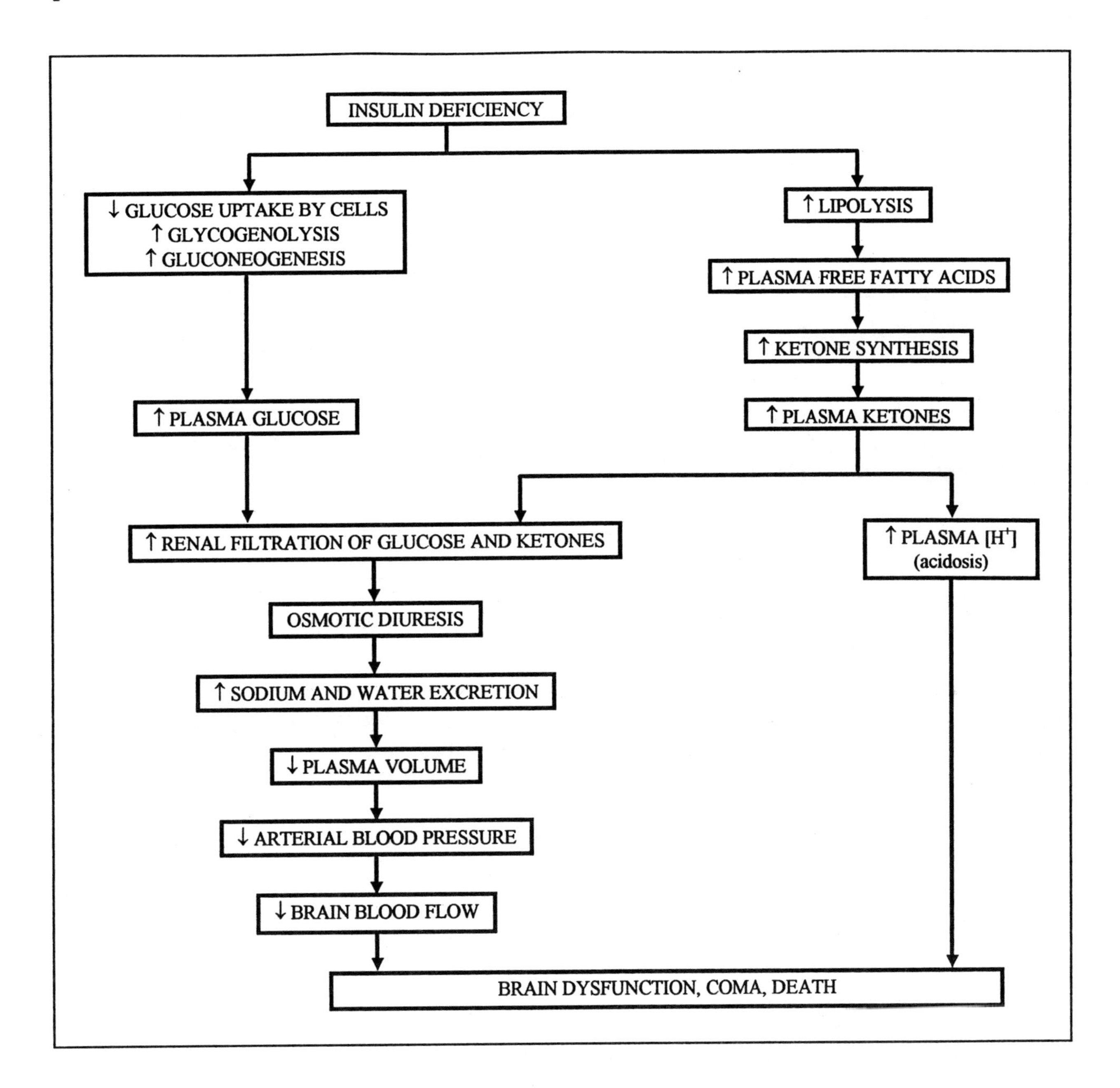

Figure 7-6. Summary of pathophysiology of diabetic acidosis.

Ingested glucose is distributed to three types of tissues. 1) About 55% of ingested glucose is taken up by the **liver** and much of that is stored as glycogen. Secretion of glucose by the liver is suppressed. The liver also uses fatty acids and glucose to synthesize very low density lipoproteins, which are stored in adipose tissues. 2) Insulin simultaneously inhibits lipolysis in adipocytes. Only about 15% of ingested glucose is taken up by insulin-dependent tissues (adipose and skeletal muscle). 3) In skeletal muscle there is glycogen deposition and a shift from protein catabolism to protein synthesis. About 25% of ingested glucose is extracted by insulin-independent tissues; brain, nerves, blood cells, renal medulla, and germinal epithelium of testis.

Pathophysiology of Diabetes Mellitus

15-20% of all diabetics require exogenous insulin, ie, they are **insulin-dependent**. This form of diabetes most frequently affects subjects younger than 40 years of age. It results from a combination of genetic predisposition and environmental factors. The acquired and environmental factors include autoantibodies against the islets of Langerhans and viral infections resulting in B cell alteration.

Non-insulin dependent diabetes mellitus represents the most common form of the disease, comprising 80-85% of all diabetics. It is characterized by the association of a relative insulin deficiency and insulin resistance. A family history of diabetes is more frequent in non-insulin dependent than in insulin-dependent diabetes. Environmental factors are related to nutrition and life style, with obesity and physical inactivity being predominant risk factors. Malnutrition-related diabetes mellitus occurs mostly in developing tropical countries, where both protein malnutrition and the predominance of certain foods (eg, cassava) play a major role.

Other types of diabetes mellitus include diabetes secondary to pancreatic disease or other endocrinopathy (eg, Cushing's syndrome) and iatrogenic diabetes (eg, glucocorticoid therapy). Gestational diabetes is an alteration of glucose tolerance that can appear during pregnancy. Besides the risks of fetal morbidity, gestational diabetes also carries the risk of subsequent insulin-dependent diabetes in the mother.

Glucagon secretion is often inappropriately high in diabetic states, despite hyperglycemia. Therefore, insensitivity of A cells to glucose is probably involved in diabetes.

Hypercholesterolemia

Elevation of plasma cholesterol can cause premature atherosclerosis with heart attacks at an early age. In homozygous familial hypercholesterolemia plasma cholesterol levels are in the range of 600-1200 mg/dl (normal = <200 mg/dl), while heterozygous individuals have levels of 300-550 mg/dl. Cholesterol deposits, called xanthomas, occur in skin, tendons, and over bony prominences in these individuals and are strong evidence of hypercholesterolemia.

The genetic defect is an abnormal LDL receptor. A spectrum of defects have been seen, from no receptors, degradation of receptors before they can be inserted into the plasma membrane, receptors that are unable to bind LDL, to defective receptor arrangements on the membrane. Giving bile-salt-binding resins to increases LDL receptors in heterozygous patients. Blocking cholesterol synthesis by nicotinic acid or HMG-CoA reductase (lovastatin) substantially drops plasma cholesterol levels. These drugs do not work in homozygous individuals, since no normal gene is available to stimulate.

Review Questions

For Questions 35-38: A type I diabetic was found in a coma. Blood and urine glucose and blood and urine ketones were all high; serum bicarbonate was <12 mEq/L. Respirations were quick and deep with an acetone breath. Blood pressure was 95/61, and the pulse at 119 was weak and rapid.

35. Is this person in ketoacidosis or insulin shock? Explain.

36. Why is the serum bicarbonate low? What is the acid-base status of this patient?

37. What is the cause of the dyspnea, hypotension, and tachycardia?

38. What is the indicated treatment?

For Questions 39-40. A beta-blocker, followed by a large dose of insulin, is administered to an individual.

39. What would you expect the plasma glucose response to be over the next two hours?

    A. Plasma glucose concentration would decrease.
    B. Plasma glucose concentration would increase.
    C. Plasma glucose concentration would decrease and then increase.
    D. Plasma glucose concentration would increase and then decrease.

40. If somatostatin were given with the beta-blocker, what would you expect the plasma glucose response to be over the next two hours?

    A. Plasma glucose concentration would decrease.
    B. Plasma glucose concentration would increase.
    C. Plasma glucose concentration would decrease then increase.
    D. Plasma glucose concentration would increase then decrease.

41. A patient with severe hypoglycemia and recurring fainting is suspected of having either an insulin-secreting tumor or inadvertent insulin administration. To distinguish between the two, you would measure the serum

    A. glucagon concentration
    B. C-peptide concentration
    C. potassium concentration
    D. insulin concentration

42. On the accompanying graph of the glucose tolerance test in a normal individual, draw the line that would represent the effect of hypercortisolemia.

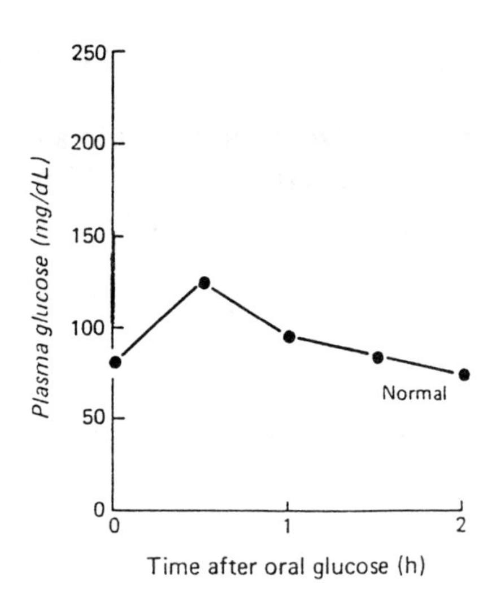

43. All of the following statements are true **EXCEPT** which one?

    A. The principal signal for switching from anabolism to catabolism is decreased I/G.
    B. In prolonged starvation the body's demands for calories is decreased.
    C. In prolonged starvation lipolysis supplies fatty acids to the liver.
    D. In prolonged starvation the utilization of ketones by the central nervous system increases.
    E. The rate of gluconeogenesis remains high throughout prolonged starvation.

44. Untreated type I diabetics have severe ketosis. Why don't type II diabetics also have it?

ADRENAL MEDULLARY HORMONES

The adrenal medulla is a modified sympathetic ganglion whose cells are innervated by preganglionic sympathetic axons (See Fig. 1-6, p. 22). These cells secrete two biologically active catecholamines, **epinephrine** and **norepinephrine**, into the general circulation. Biosynthesis of catecholamines is initiated when tyrosine is taken up by cells and converted in the cytoplasm to DOPA. This is the rate-limiting step in the biosynthetic chain and is well-regulated (see below). DOPA is converted to dopamine, which is converted to norepinephrine **(NE)** by the action of dopamine-β-hydroxylase **(DA-OHase)**. Norepinephrine is methylated by phenylethanolamine-N-methyl transferase **(PNMT)** to form epinephrine **(EPI)**.

Acute nerve stimulation results in hormone secretion, increased rate of catecholamine synthesis, and the release of tyrosine hydroxylase from tonic inhibition by high levels of NE. Chronic stimulation (24-48 hours) causes increases in enzyme activities due to new protein synthesis. Stimuli such as hypoglycemia, immobilization, cold and stress lead to enzyme induction, via neural and ACTH action, of both tyrosine hydroxylase and DA-OHase. Cortisol influences DA-OHase and is a powerful inducer of PNMT. Blood that originates in the cortical sinusoids traverses the adrenal medulla before entering the general circulation; this helps maintain the biochemical capability of the medulla.

Mature granules (called **chromaffin granules** because of their histological properties) contain catecholamines, ATP, DA-OHase, and an acidic protein, chromogranin A. The very high concentrations of basic catecholamines within granules are probably stabilized and bound by the acidic nucleotides and chromogranin A. Acetylcholine interacts with receptors on the chromaffin cell to increase membrane permeability to Na^+ and Ca^{2+}, thus depolarizing the membrane potential. Increased cytosolic Ca^{2+} is responsible for movement of the granules to the membrane and exocytosis.

The ratio of EPI to NE secreted by the resting adult medulla is about 4:1. This ratio is much lower in the fetus and begins to rise soon after birth with the influence of cortisol on PNMT. All of the EPI in the general circulation of the adult comes from the adrenal medulla; however, most NE comes from "spillover" from sympathetic axon terminals. After bilateral adrenalectomy plasma EPI falls to zero, but plasma NE concentrations are unchanged.

The disappearance of catecholamines from the circulation is rapid, with a biological half-life of 1-2 minutes. Circulating NE and EPI are primarily **metabolized** by catechol-0-methyl-transferase (COMT) to the inactive 0-methylated compounds normetanephrine and metanephrine. These compounds are then deaminated by monoamine oxidase to form VMA (vanillylmandelic acid). Intracellular EPI and NE are exposed to monoamine oxidase and broken down to inactive, deaminated metabolites. These metabolites leave the cell and are then 0-methylated under the influence of COMT to form VMA.

The binding of an agonist, eg, epinephrine, to the β-adrenergic receptor leads to an increase in adenylate cyclase activity and to an increase in cAMP. Further transduction of adrenergic signals are typical of adenylate cyclase systems described earlier. For example, protein kinase A is involved in the activation of muscle and hepatic phosphorylases and adipose tissue lipases, as well as in the inactivation of glycogen synthase and other enzymes.

The action of catecholamines at $alpha_1$ receptors appears to be independent of cAMP, but it is mediated by an elevation of cytosol Ca^{2+}. Current hypotheses propose that the second messenger of α_1-adrenergic effects is inositol triphosphate. How the presynaptic α_2 receptor exerts its effect on nerve terminals is not known. In blood platelets it inhibits adenylate cyclase.

Overall, the effects of EPI are generally sympathetic and might be considered as appropriate preparation for a sustained effort. Cardiovascular, pulmonary, and GI effects have been discussed. Metabolic effects include glycogenolysis (liver and muscle), gluconeogenesis (liver), lipolysis (fat cells), thermogenesis, inhibition of insulin secretion, and facilitation of glucagon secretion. The result is increased availability of glucose and free fatty acids to brain, heart, and muscle. Figure 7-7 summarizes the physiological effects of EPI.

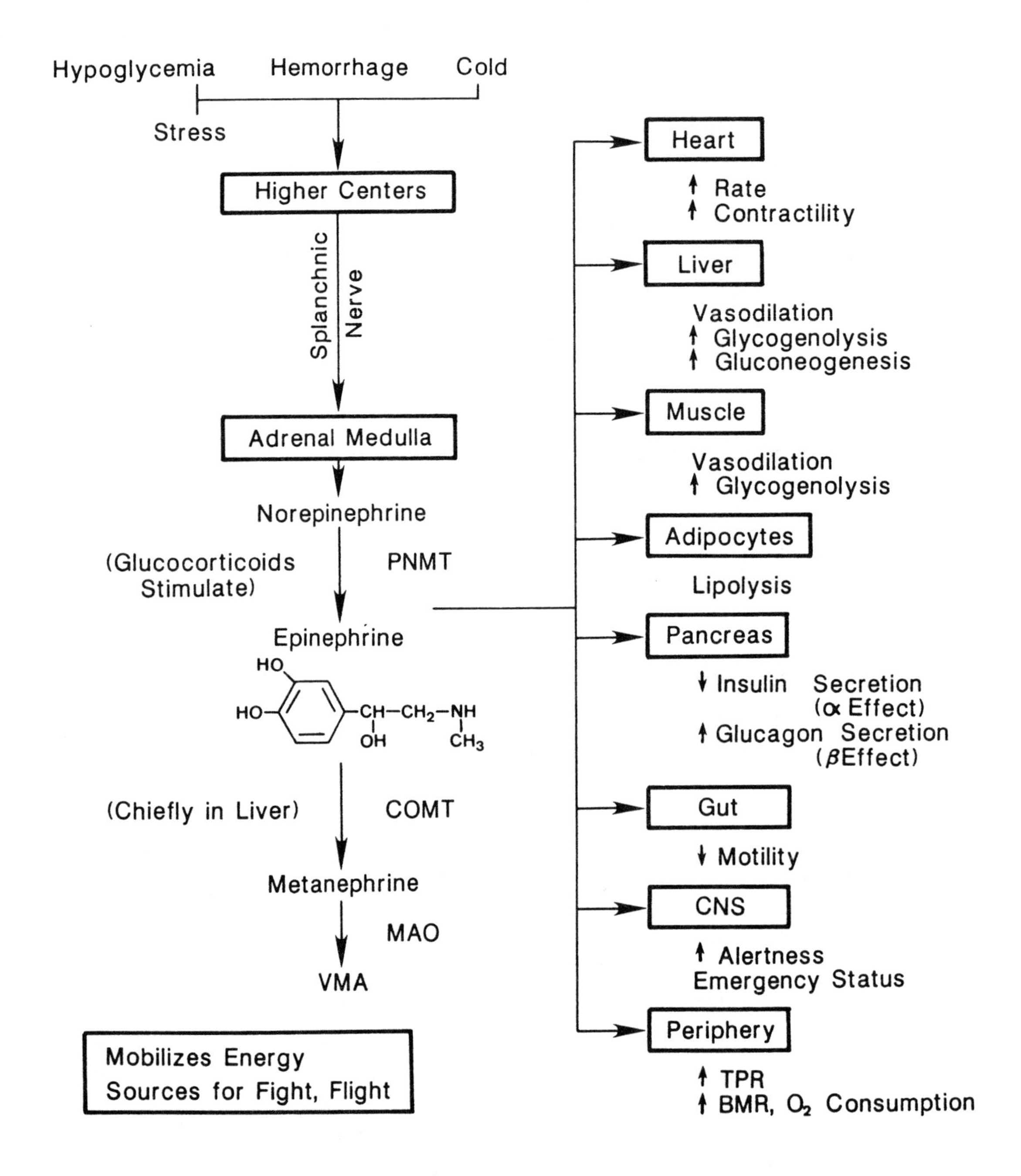

Figure 7-7. Physiological effects of epinephrine.

Pathophysiology of Adrenal Medulla

Pathophysiology of the adrenal medulla is restricted to one rare disease, pheochromocytoma. This solitary tumor develops from adrenal chromaffin tissue. Symptoms include sweating attacks, tachycardia, and headaches associated with arterial hypertension. Adrenal medullary insufficiency is unknown.

Review Questions

45. The removal of epinephrine from the synapse principally involves

    A. uptake into post-ganglionic neurons
    B. storage in preganglionic sympathetic neurons
    C. enzymatic degradation within the synapse
    D. uptake by extraneuronal tissues
    E. diffusion into adjacent spaces

46. The synthesis of epinephrine depends on

    A. angiotensin-II secretion
    B. high local cortisol concentration
    C. dehydroepiandrosterone
    D. an intact zona glomerulosa

47. The metabolic rate per kg of body weight of a resting individual decreases during

    A. intravenous injection of epinephrine
    B. ingestion of food
    C. exposure to low ambient temperature
    D. prolonged fasting
    E. intravenous injection of glucagon

48. Secretion of catecholamines by cells of the adrenal medulla is dependent upon

    A. the increased concentration of Ca^{2+} in the cytoplasm
    B. exocytosis of calmodulin
    C. inactivation of the release of somatostatin
    D. hyperpolarization of the cell membrane
    E. the release of norepinephrine from preganglionic neurons

49. In target cells stimulation of cyclic AMP by epinephrine

    A. requires translocation of epinephrine into the nucleus.
    B. is mediated by a protein that binds GTP
    C. is amplified by prior long-term exposure of the cells to norepinephrine
    D. is inversely related to the number of adrenergic receptors present on the cells.
    E. Depends upon the internalization of epinephrine-receptor complexes

ADRENAL CORTICAL HORMONES

Synthesis and Transport

The cortical layers of the adrenal gland are the source of steroid hormones with potent actions on glucose, fat, and protein metabolism **(glucocorticoids)** and on sodium and water balance **(mineralocorticoids)**. This gland also secretes a weak androgen, dehydroepiandrosterone, in both sexes.

Mineralo- and glucocorticoids are 21 carbon steroids synthesized from cholesterol. The rate-limiting step is the conversion of cholesterol to pregnenolone which is then converted into the major biologically active corticosteroids. The relative activities of various enzymes determine which steroids are produced. The **zona fasciculata** is the major site for production of cortisol, the major glucocorticoid. The **zona glomerulosa** synthesizes steroids, such as aldosterone, with mineralocorticoid activity, since it lacks the enzyme 17 β-hydroxylase to produce cortisol.

ACTH and cortisol are secreted in sporadic bursts, producing a **diurnal** pattern of plasma cortisol concentration; most of the ACTH-cortisol bursts occur between 4 and 8 AM. Plasma aldosterone concentration shows less circadian variations. Plasma renin and aldosterone concentrations change with posture. An upright position shifts fluid to the lower extremities and away from central volume receptors of the kidney; this elevates renin and aldosterone. Basal levels return in the horizontal position.

Cortisol is reversibly bound in plasma with high affinity to an α_2-globulin (cortisol binding globulin) and with low affinity to albumin. Only about 6% is free or loosely bound to albumin. As free steroid is used by peripheral tissues, it is replaced from the bound fraction, which is replenished by adrenal secretion. Globulin-bound steroid is not biologically active, because it is confined to the circulation and cannot diffuse into tissues. Pregnancy and estrogen administration increase cortisol binding globulin concentrations, raising the total plasma cortisol, but not the free biologically active fraction. Decreased binding may occur with administration of androgen and in certain diseases. Aldosterone is bound to albumin and has a lower affinity for cortisol binding globulin.

The disappearance rates of cortisol and aldosterone from plasma are closely related to their plasma binding. The **biological half-life** of cortisol is 80 min and for aldosterone is 30 min. The effect of raising cortisol binding globulin is much greater on cortisol than on aldosterone and increases the half-life of cortisol more than that of aldosterone. Metabolism of corticosteroids occurs mainly in the liver. The two main steps are side chain removal and conjugation to glucuronic acid or to sulphates. The conjugates are water soluble and thus appear in the urine.

ACTH stimulation of cortisol production involves ACTH binding to fasciculata cell membrane receptors, stimulation of adenylate cyclase and cAMP, and a resultant increase in activity of the enzymes that convert cholesterol to pregnenolone.

ACTH has two effects on the adrenals. Its rapid effect immediate releases adrenocortical hormones into the circulation. Its slower action leads to hypertrophy of the adrenal cortex. ACTH binds to a receptor at the level of the plasma membrane, leading to activation of adenylate cyclase.

Regulation of Corticosteroids

CRH controls ACTH production. Secretion of ACTH and cortisol follows a circadian pattern, largely dependent upon the sleep/wake cycle. CRH is secreted by cells in the paraventricular nuclei, which receive various inputs from neurons secreting ACh, NE, and 5-HT. If CRH-producing cells are destroyed, the ability of adrenal cortex to respond to many stressors is blocked. The amygdala sends signals to the hypothalamus regarding emotional stress, fear, anxiety, and apprehension, causing increased CRH-ACTH secretion. The **biologic clock** responsible for the diurnal changes in ACTH secretion is located in the **suprachiasmatic** nuclei of the hypothalamus. Responses to injury are mediated by nociceptive pathways and the reticular formation. Baroreceptors send inhibitory signals via nucleus tractus solitarius to the CRH-secreting cells. Therefore, activity of the **CRH-ACTH-cortisol axis** varies with the individual and the strength and type of stimulus, such as trauma, shock, infection, hypoglycemia, surgery, and emotional stress. ACTH production is feedback-inhibited by glucocorticoids acting at both the pituitary and the hypothalamus. Thus, the rate of ACTH secretion is determined by two inputs; 1) stimuli converging through the median eminence to increase ACTH secretion, and 2) the braking action on ACTH secretion by glucocorticoids, which is proportional to their circulating levels.

Most of the actions of **cortisol** are initiated by its binding to a specific cytoplasmic receptor in target cells. This interaction causes accumulation of hormone-receptor complexes in the nucleus where it stimulates synthesis of specific mRNA molecules, resulting in synthesis of new protein. Secondary actions of glucocorticoids, eg, their "permissive" effects, appear to involve other mechanisms. For example, these steroids enhance binding of catecholamines to their receptors when given in large doses.

Effects of Glucocorticoids

The following tissues are affected by glucocorticoids:

- **Liver** - increased protein synthesis (gluconeogenic enzymes); enhancement of the actions of glucagon and epinephrine
- **Skeletal muscle** - protein catabolism
- **Adipose tissue** - lipolysis; promote action of epinephrine
- **Brain** - increased membrane excitability, emotional lability, arousal, and enhanced learning
- **Pituitary** - inhibition of ACTH and prolactin; increased synthesis of growth hormone
- **Lymphocytes and fibroblasts** - decreased metabolism, cell death
- **Bone** - decreased protein synthesis
- **Lung** - stimulation of surfactant production in the fetus

Thus, the metabolic consequences of glucocorticoids are **mobilization** of energy sources from a number of tissues during prolonged inaccessibility to glucose. Even the stimulation of glycogen accumulation in the liver can be considered as a preparation for intervention by glycogenolytic hormones. Certain tissues other than liver, especially brain, are spared the inhibitory effects of glucocorticoids on protein and carbohydrate metabolism. Glucocorticoids have other actions on the CNS in addition to their negative feedback action on ACTH. Some patients with **Cushing's syndrome** may be euphoric and have an increased appetite; others may become depressed or psychotic. **Addison's disease** is associated with depression and a general feeling of ill-health.

Glucocorticoids have several influences on the cardiovascular system, and they are crucial for the maintenance of normal blood pressure in the presence of stress. Glucocorticoids enhance cardiac contractility, increase vascular tone and reactivity, and increase the level of renin substrate (angiotensinogen). Glucocorticoids can suppress virtually every phase of the immunological and inflammatory response. For example, they 1) decrease capillary membrane permeability, 2) stabilize lysosomal membranes, 3) inhibit fibroblastic activity, and 4) reduce secretion of lymphokine and monokine immune mediators. Glucocorticoids also affect cellular responses to other hormones. Glucocorticoid excess is associated with secondary hyperinsulinism. Whereas glucocorticoids promote lipolysis, the accumulation of fat in such areas as the trunk may be the result of hyperinsulinemia. Other effects include increased sensitivity of the stomach to acid-producing stimuli, increased intraocular pressure, and redistribution of body calcium.

Pathophysiology of Glucocorticoids

Patients with **Addison's disease** (adrenal insufficiency, Table 7-1) present with excessive fatigue and pigmentation (resulting from hypersecretion of ACTH and β-MSH due to reduced cortisol). Episodes of hypoglycemia are accompanied by hypotension (especially orthostatic), weight loss, anorexia, vomiting, and diarrhea. Adrenal insufficiency of hypothalamic-hypophyseal origin, which involves a deficit of ACTH, is characterized by the same symptoms as Addison's disease, except hyperpigmentation is absent. Iatrogenic adrenal insufficiency is seen in patients treated with large doses of glucocorticoids for prolonged periods. When the treatment is discontinued, the hypothalamic-hypophyseal-adrenal axis remains altered.

Table 7-1. Distinctions between primary and secondary adrenal insufficiency.

	Primary	**Secondary**
Site	Adrenal	Hypothalamic-pituitary
ACTH secretion	Increased	Decreased
Pigmentation	Increased	Decreased
Headaches, visual loss	Rare	Frequent
Body weight	Decreased	Variable
Other pituitary hormones		
Growth hormone	No change	Decreased
Gonadotropin	No change	Decreased
Adrenal hormones		
Aldosterone	Deficient	Normal
Androgens	Deficient	Variable
Cortisol	Markedly decreased	Moderately reduced
Responses to exogenous ACTH	None	Sluggish

Hypercortisolism, or **Cushing's syndrome**, is characterized by the redistribution of body fat toward the face and trunk, whereas the limbs are thin. Hypertension, thin skin, muscle catabolism, poor wound healing, and osteoporosis are also frequent findings. It may be primary, ie, an autonomous adrenal tumor, or secondary, ie, in response to hypersecretion of ACTH by the pituitary or an ectopic tumor. Cushing patients have some degree of virilization, but this symptom is especially evident in virilizing adrenal tumors and congenital enzyme blocks. Virilizing tumors secrete large amounts of dehydroepiandrosterone sulfate (DHEAS) and androstenedione. These precursors are metabolized to testosterone which masculinizes the patient. The most frequent enzymatic alteration is congenital adrenal hyperplasia characterized by deficient 21- or 11β-hydroxylase enzymes. Typically, the enzymatic block results in low cortisol levels and concomitantly high ACTH secretion. The ACTH stimulation of the adrenal cortex results in hypersecretion of precursors with androgenic activity. Dexamethasone treatment returns androgen levels to normal.

Effects of Mineralocorticoids

The **volume of extracellular fluid** is regulated mainly via mechanisms that control renal excretion of **sodium.** Primarily excretion and secondarily intake of Na^+ and water are regulated by 1) the renin-angiotensin system, 2) aldosterone and 3) antidiuretic hormone.

Renin-angiotensin system (Figure 7-8). Following its release from juxtaglomerular cells into the blood, the renal enzyme, renin, acts on angiotensinogen, a glycoprotein that is synthesized in the liver. This reaction produces the biologically inactive decapeptide, angiotensin I. Angiotensin I is rapidly converted to the biologically active octapeptide, **angiotensin II**, in the pulmonary circulation. Angiotensin II elevates arterial pressure and stimulates aldosterone secretion. Angiotensin II is degraded by peptidases collectively referred to as angiotensinases. One of the metabolites of angiotensin II, (des-Asp^1) angiotensin II, sometimes referred to as angiotensin III, is biologically active, especially in aldosterone secretion. Each component of the renin-angiotensin system is discussed in greater detail below.

Renin is a carboxyl protease (40,000 MW) which splits the Leu^{10}-Leu^{11} bond in angiotensinogen to release angiotensin I. Renal renin is synthesized in the juxtaglomerular apparatus composed of juxtaglomerular cells and the macula densa. The **juxtaglomerular cells** are modified smooth muscle cells in the media of afferent, and to a lesser extent, efferent arterioles of nephrons. These cells synthesize, store and release renin. The juxtaglomerular cells and smooth muscle cells of afferent arterioles are innervated by renal nerves. The **macula densa** is an area of modified renal tubular epithelium at the junction between the ascending loop of Henle and the distal convoluted tubule, in close contact with afferent and efferent arterioles. Renin secretion is controlled by a renal vascular receptor, the macula densa, sympathetic input and a variety of humoral agents including angiotensin II, vasopressin and potassium (Fig. 7-8).

There is a reciprocal relationship between renal perfusion pressure and the rate of renin secretion. The afferent arteriole contains a receptor that responds to changes in the wall tension of the arteriole; decreased tension is associated with increased renin secretion. The **renalvascular receptor** is responsible for elevations in renin secretion seen in several physiological and pathological situations, eg,

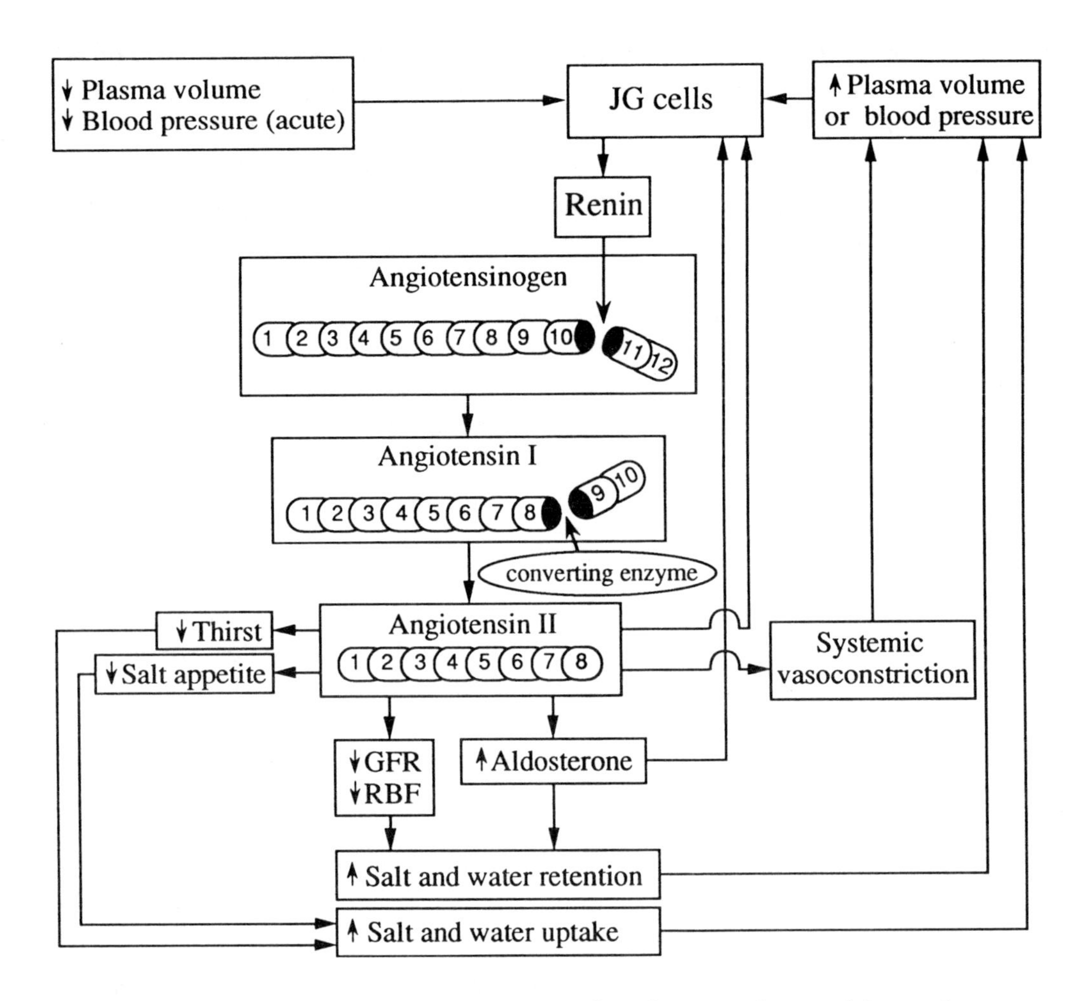

Figure 7-8. Summary of the control of renin secretion and its actions.

renal artery constriction or obstruction and hypovolumic orthostatic changes. The amount of Na^+ delivered to distal tubules is probably detected by the macula densa, and there is a reciprocal relationship between Na^+ load and renin release. The macula densa likely mediates changes in renin secretion associated with alterations in filtered load of Na^+ or in proximal tubular reabsorption of Na^+ ions (see also p. 164).

Sympathetic innervation also modulates renin secretion. Renin secretion is increased by increased sympathetic activity, from electrical stimulation of certain areas of the brain, carotid sinus hypotension, non-hypotensive hemorrhage, exercise and standing. In addition, renin secretion increases with electrical stimulation of renal nerves or with administration of catecholamines; it decreases when renal nerves are sectioned. Catecholamines, either circulating or released locally from renal sympathetic terminals, stimulate renin secretion by acting directly on juxtaglomerular cells. This action is mediated by β-adrenoceptors and may involve activation of adenylate cyclase and formation of cyclic AMP. Alpha adrenoceptors may also play a role. Stimulation of α-adrenoceptors may increase renin secretion by constricting afferent arterioles with resultant activation of renal vascular receptors. Alpha adrenoceptor stimulation would also decrease the delivery of Na^+ and Cl^- to the macula densa by decreasing glomerular filtration rate and by increasing proximal tubular Na^+ and Cl^- reabsorption.

Angiotensin II, vasopressin and K^+ all **inhibit renin secretion**, apparently by intrarenal mechanisms. Angiotensin II is a negative feedback inhibitor of renin secretion, since administration of specific angiotensin II antagonists increases the rate of renin secretion. Vasopressin also modulates renin secretion, since elevations of plasma vasopressin concentration decrease the rate of renin secretion. The suppression of renin secretion by K^+ results from diminished proximal tubular Na^+ reabsorption with a consequent increase in the delivery of Na^+ to the macula densa.

Angiotensinogen is the protein substrate for enzymatic cleavage by renin to form angiotensin I. Most angiotensinogen is produced by the liver. It is also present in the renal cortex, cerebrospinal fluid, lymph and amniotic fluid; the origin and function of this angiotensinogen is unknown. Angiotensinogen is a glycoprotein (60,000 MW); its production is regulated by adrenocortical steroids, estrogens and angiotensin II. The **plasma angiotensinogen** concentration is about 1 μM/L. The concentration of angiotensinogen in plasma decreases following hypophysectomy or adrenalectomy and increases following administration of ACTH or adrenocortical steroids. The ability of adrenocortical steroids to increase angiotensinogen production is related to their glucocorticoid rather than their mineralocorticoid activity. Therefore, plasma angiotensinogen levels are increased in patients with Cushing's syndrome but not in patients with primary aldosteronism. Estrogens act on the liver to increase the production of many proteins including angiotensinogen. Consequently, the concentration of angiotensinogen in plasma is increased during pregnancy and in women taking estrogen-containing oral contraceptives.

Administration of **angiotensin II** increases the concentration of angiotensinogen in plasma. This increase can also be produced by administration of renin, so it does not result from decreased metabolism of angiotensinogen because of suppression of renin secretion by angiotensin. The increase appears to result from stimulation of the production of angiotensinogen by the liver. The stimulatory effect of angiotensin II on angiotensinogen production may be a positive feedback mechanism, which prevents plasma angiotensinogen concentration from falling when utilization of angiotensinogen is increased during periods of increased renin secretion. Angiotensin II is removed rapidly from the circulation; its half-life is less than 1 min. The peptide is degraded to its component amino acids in the capillary beds of most tissues by many enzymes: aminopeptidases, endopeptidases and carboxypeptidases. These enzymes are collectively referred to as **angiotensinases**, although there is no evidence that degradation of angiotensin is a specific function of these enzymes.

Angiotensin II is the biologically active component of the renin-angiotensin system. Most of its physiological actions are concerned with control of blood pressure and regulation of the volume and composition of extracellular fluid. On a weight basis it is 10-20 times more potent as a pressor substance than norepinephrine. The increase in blood pressure produced by angiotensin results primarily from direct constriction of vascular smooth muscle. However, angiotensin II also increases blood pressure indirectly by acting on the central nervous system, probably in the **area postrema**, a region of the medulla oblongata without a blood brain barrier. The centrally-mediated pressor effect of angiotensin results from a combination of increased sympathetic tone and decreased parasympathetic tone to the cardiovascular system. Angiotensin also stimulates the release of catecholamines from the adrenal medulla and increases the release of NE from sympathetic nerve endings. The renin-angiotensin system is also a major regulator of aldosterone secretion, as discussed below.

Aldosterone is a potent **mineralocorticoid** that is synthesized and secreted by the zona glomerulosa of the adrenal cortex. Aldosterone (and other steroids with mineralocorticoid activity) increases reabsorption of Na^+ from urine, sweat, saliva and gastric juice. In the kidney aldosterone acts primarily on the epithelium of the distal tubule and collecting duct to increase reabsorption of Na^+ and promote secretion of K^+ and H^+. Excessive secretion of aldosterone causes 1) retention of Na^+, 2) expansion of extracellular fluid volume, 3) depletion of K^+ and 4) metabolic alkalosis. The increase of extracellular fluid volume by aldosterone is limited by an "escape phenomenon." This phenomenon describes increased Na^+ excretion despite the continued action of aldosterone after volume expansion passes a certain point. This increase is mostly due to decreased reabsorption of Na^+ in the proximal tubule, possibly by the action of **atrial natriuretic peptide** from the atria of the heart. Deficiency of aldosterone results in loss of Na^+, elevated plasma K^+ and acidosis.

The major regulator of aldosterone secretion is the renin-angiotensin system. Angiotensin II is a potent stimulator of aldosterone secretion; plasma aldosterone levels increase promptly when the peptide is infused intravenously or directly into the arterial supply of the adrenal. In low doses angiotensin II produces a selective increase in aldosterone secretion, but large doses increase the secretion of cortisol as well as aldosterone. Angiotensin II acts early in the aldosterone biosynthetic pathway, increasing the conversion of cholesterol to pregnenolone. However, the conversion of corticosterone to aldosterone may also be facilitated during prolonged exposure to angiotensin II.

ACTH acts directly on the zona glomerulosa to increase the rate of aldosterone secretion. However, the zona glomerulosa appears to be less sensitive to ACTH than the inner two zones, since larger doses of ACTH are needed to increase aldosterone output than to produce maximum glucocorticoid output. Nevertheless, stressed animals readily secrete enough ACTH to produce an increase in aldosterone secretion, and hypophysectomy blocks the increase in aldosterone secretion produced by various stressful stimuli. ACTH increases the conversion of cholesterol to pregnenolone in the zona glomerulosa; there is evidence that this effect is mediated via the activation of adenylate cyclase and the formation of cyclic AMP.

Aldosterone secretion increases when plasma **potassium** concentration increases and decreases when plasma K^+ levels fall. The rate of aldosterone secretion changes significantly in response to quite small (0.5 mEq/L) variations in plasma K^+ concentration. Potassium acts directly on the zona glomerulosa and appears to affect the aldosterone biosynthetic pathway at three steps; 1) the conversion of cholesterol to pregnenolone, 2) the conversion of deoxycorticosterone to corticosterone, and 3) the conversion of corticosterone to aldosterone.

Changes in plasma **sodium** concentration lead to reciprocal changes in the rate of aldosterone secretion. However, large changes in Na^+ concentration (10-20 mEq/L) are required to produce this effect. Plasma Na^+ concentration by itself probably is not an important factor in the control of aldosterone secretion. The order of importance as regulators of aldosterone secretion is renin-angiotensin > plasma potassium > sodium > ACTH.

Aldosterone binds to an intracellular receptor of high affinity and specificity. Aldosterone reduces the level of receptors, whereas adrenalectomy increases it. Its effects on electrolyte flux are via transcription and subsequent translational synthesis of protein(s) essential to the transepithelial movement of Na^+ and K^+. Its target tissues are the distal tubule and collecting duct of the renal nephron, the digestive tract, salivary and sweat glands, the heart, salt "drive" or appetite, and perhaps the CNS.

Atrial natriuretic peptide (ANP) is a 28 amino acid peptide from the atria of the heart. It is a potent diuretic and natriuretic, but provokes mild kaliuresis (Fig. 7-9). It has hypotensive effects, decreasing systolic and diastolic blood pressure, cardiac output, heart rate, and peripheral vascular resistance. It decreases renin and aldosterone secretion. Thus, it antagonizes the renin-angiotensin system at the level of renin and aldosterone production, angiotensin II, vasoconstriction, and aldosterone-induced sodium excretion. Overall, atrial natriuretic peptide improves hemodynamics of the heart in case of overload. Its secretion is induced by blood volume expansion and subsequent increase in atrial pressure.

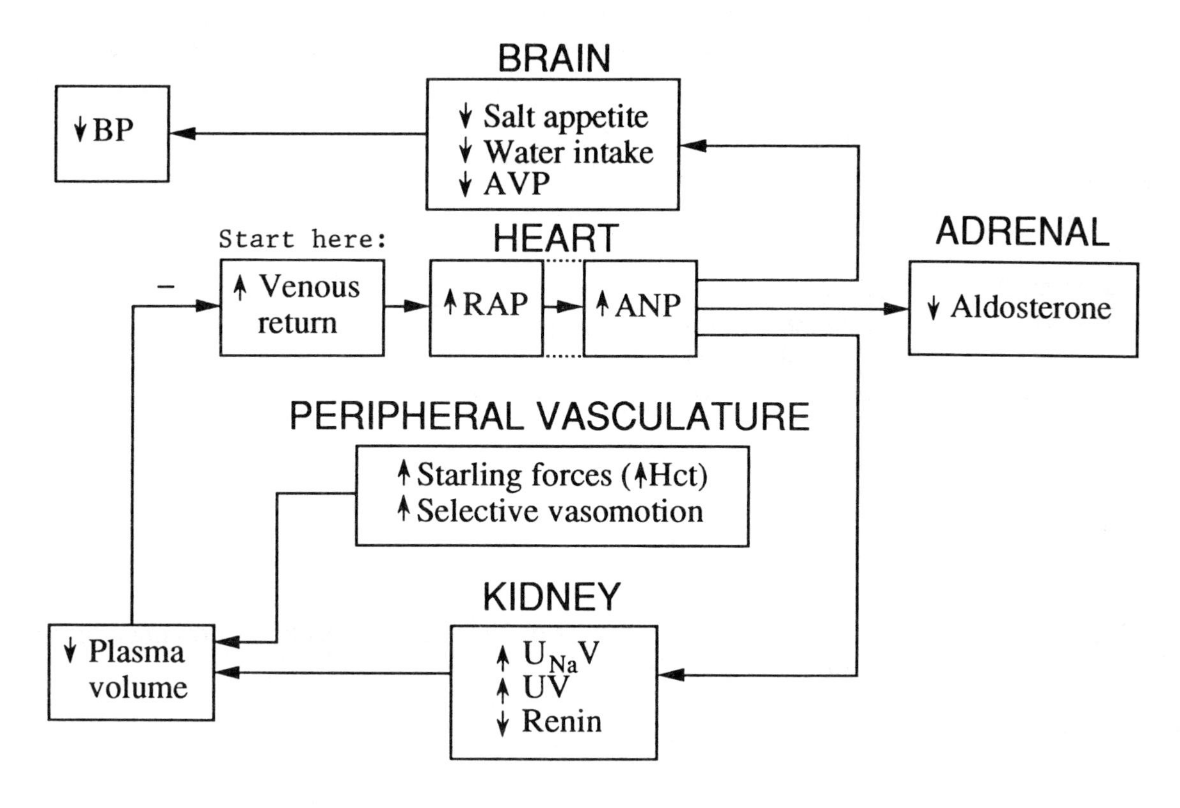

Figure 7-9. The release and actions of ANP. Symbols: AVP, vasopressin; Hct, hematocrit; RAP, right atrial pressure; $U_{Na}V$, urine sodium excretion; UV, urine volume.

Pathophysiology of Mineralocorticoids

Primary hyperaldosteronism is hypersecretion of the hormone due to a tumor or bilateral hyperplasia of the adrenals. It is marked by moderate hypertension associated with potassium depletion and absence of edema. The absence of edema is characterized by the escape phenomenon. Plasma renin activity is low. The elevated aldosterone secretion is not suppressed by blood volume expansion or by the administration of converting enzyme inhibitor, demonstrating the autonomous nature of the tumor. Spironolactone, a competitive inhibitor of the aldosterone receptor, corrects all abnormalities. **Secondary** hyperaldosteronism is due to overproduction of aldosterone due to an excess of stimuli. It may be due to extra-adrenal stimulation resulting from an excess of angiotensin, potassium, or ACTH.

When edema is associated with hepatic cirrhosis, nephrotic syndrome, or heart failure, there is a reduction in the effective circulatory volume. Hyperaldosteronism is secondary to a reduction in effective blood volume, but because there is a low level of sodium ions in the distal tubule (proximal tubular reabsorption of sodium chloride is increased), potassium depletion does not occur. Edematous states are characterized by the absence of escape. Elevated aldosterone in hepatic cirrhosis and heart failure is due to adrenal hypersecretion associated with excessive angiotensin secretion, reduced metabolic clearance of the hormone, and reduced metabolism of aldosterone.

Review Questions

For Questions 50-55. A forty-year-old female presented with the following symptoms: weakness, fatigue, orthostatic hypotension, weight loss, dehydration, and decreased cold tolerance. Her blood chemistry values were as follows:

Serum sodium	120 mEq/L
Serum potassium	6.9 mEq/L
Fasting blood glucose	70 mg/dL
BUN	54 mg/dL
Serum creatinine	1.6 mg/dL
Plasma cortisol	1.5 ug/dL (4pm)
	1.5 ug/dL (8am)
Plasma ACTH	310 pg/ml (normal 30-120 pg/ml)

Hematology tests resulted in the following values:

Hematocrit	50%
Leukocytes	5000/mm^3

She also noticed increased pigmentation of portions of her body and back. Following administration of ACTH plasma cortisol did not rise significantly after sixty and ninety minutes.

50. What endocrine organ is the site of the malfunction? Is this a primary or secondary disturbance?

51. What electrolyte disturbances result from this disorder?

52. Discuss the metabolic disturbances resulting from this disorder.

53. What is the cause of the pigmentation?

54. What type of replacement therapy would you advise for this individual?

55. Diagram the feedback loop for this endocrine disorder.

56. In humans total adrenalectomy is fatal without replacement therapy, whereas hypophysectomy is not. This is explained by

A. the adrenal cortex undergoing compensatory hypertrophy
B. the secretion of aldosterone to normal
C. adrenal catecholamines compensating for the metabolic actions of cortisol
D. Tissue requirements for corticosteroids falling to low levels
E. Plasma concentrations of angiotensin II not increasing to values that are toxic

57. A drug that prevents the binding of dexamethasone to its receptors would cause

A. a negative nitrogen balance
B. decreased concentration of ACTH in the blood
C. decreased concentration of cortisol in the blood
D. increased concentration of insulin in the blood
E. decreased concentration of cortisol in cell nuclei

58. If an inhibitor of converting enzyme in the renin-angiotensin system is given to a salt-depleted person, there will be a decrease in

A. Na^+ clearance
B. K^+ clearance
C. secretion of renin
D. plasma concentration of Angiotensin I
E. thirst drive

59. Treatment of a normal person with large amounts of cortisol causes

A. decreased hepatic glycogen content
B. skeletal muscle anabolism
C. increased adrenal size
D. central deposition of fat
E. resistance to peptic ulcers

60. Cortisol differs from growth hormone, since it

A. increases the transport of amino acids across some cell membranes
B. increases excretion of a water load in adrenocortical insufficiency
C. can cause hyperglycemia and ketonemia
D. is produced in increased amounts during surgical stress
E. is produced during acute hemorrhage (10% of blood volume)

61. All of the following statements about the adrenal cortex are true **EXCEPT** which one?

A. After hypophysectomy few changes occur in the zona glomerulosa and in the secretion of aldosterone.
B. The zona fasciculata cells respond to circulating angiotensin II by producing aldosterone.
C. In both sexes dehydroepiandrosterone is a major product, and small amounts of testosterone and estradiol are also formed.
D. Inadequate conversion of 11-deoxycorticosterone to corticosterone results in hypertension, salt retention and adrenogenital syndrome.
E. The secretion of aldosterone is sensitive to change in $[K^+]$ and in $[Na^+]$ of plasma.

ANSWERS TO ENDOCRINOLOGY QUESTIONS

1.
A 1. Acetylcholine
A 2. Glucagon
B,C,D 3. Testosterone
B,C 4. Triiodothyronine

2. Chronically elevated plasma levels of a hormone, like that of neurotransmitters, causes a down-regulation of the hormone's receptor on target cells. Thus, the cells become hyporesponsive to the hormone. An example is type II diabetes mellitus.

3. Corticotropin is controlled by CRH.
Growth hormone is controlled by GHRH and somatostatin.
Thyrotropin is controlled by TRH.
Prolactin is controlled by dopamine (a PIH) and several PRF factors.

4.
A 1. Cortisol
C 2. Estradiol
B 3. IGF-1
D 4. Triiodothyronine

5. D. GH, FSH, prolactin, and ACTH are anterior pituitary hormones controlled by releasing and/or release-inhibiting factors from the hypothalamus. Vasopressin is synthesized in the paraventricular and supraoptic nuclei of the hypothalamus and stored in the posterior pituitary.

6. C. Proopiomelanocortin is a single precursor molecule synthesized in anterior pituitary corticotrophs. It is enzymatically cleaved in the corticotrophs to an amino-terminal peptide, ACTH, and β-LPH. Only ACTH has known biological activity.

7. D. Inadequate sleep due to emotional disturbances associated with abuse is the indicated answer since GH secretory capacity (pituitary reserve) (A) and IGF levels are normal (C,E).

8. B. Hypophysectomy would remove the trophic pituitary hormones TSH, ACTH, GH, and GnRH, leaving the individual with hypogonadism.

9. E. Growth hormone enhances lipolysis.

10. B. Lactotrophs secrete prolactin (A). Gonadotrophs secrete LH and FSH (B). Corticotrophs secrete ACTH (C). Thyrotrophs secrete TSH ((D). Somatotrophs secrete GH (E).

11. D. TSH, LH, and FSH are glycoproteins secreted by gonadotroph and thyrotroph cells of the pituitary and by the placenta which secretes hCG, an LH-like protein (A,B,C).

12. B. The actions of growth hormone which are insulin-like include amino acid transport across membranes and protein synthesis.

13. B. Cortisol secretion is highest in the morning between 6 and 8 am. Vasopressin may be higher in the morning than the previous evening, depending on the state of hydration (E).

14. Somatomedin stimulates **chondrogenesis** in the epiphyseal plate of long bones, secretion is **enhanced** by androgen, and feeds back to negatively inhibit **GHRH** at the hypothalamic level.

15. C. The posterior pituitary hormones are synthesized in the supraoptic and paraventricular nuclei of the **hypothalamus** and are transported down long axons to their endings in the posterior pituitary. Vasopressin has CRF-like actions (E).

16. B. Oxytocin stimulates contraction of an "estrogenized" uterus (B), and the myoepithelial cells of the mammary glands (A). It is a prolactin-releasing factor (D) and participates in, but does not initiate, labor (E).

17. E. ADH insufficiency does not decrease the number of vasopressin receptors.

18. Plasma osmolality; angiotensin II; stress = blood pressure; blood volume.

19. C. Vasopressin secretion is stimulated by hypovolemia, pain, and anxiety (A). Its actions, which include increasing water permeability of the nephron (D) and increasing peripheral resistance of the vasculature (E), are inhibited by alcohol (B).

20. E. Vasopressin is synthesized in the supraoptic nucleus.

21. It appears to be a secondary (tertiary) disorder of the hypothalamus. TRH and/or ability to respond to exogenous TRH should be verified to confirm this.

22. The feedback of T_3 and T_4 is at the level of the pituitary to diminish the sensitivity to TRH and to inhibit the secretion of TSH.

23. An insufficiency of TRH production or a lack of TRH receptor on the thyrotrophs could cause this problem.

24. A goiter may be present because of the elevated TSH.

25. Replacement therapy with thyroid hormone is indicated.

26. D. T_4 is more than 99% bound to transport proteins in plasma (A) and has a half-life of less than one hour (B). The hormone is stored as thyroglobulin extracellularly in the lumen of the follicle (C).

27. B. Thyroid peroxidase catalyzes several key reactions in the synthesis of thyroid hormones, but has no effect on their intracellular binding (A). Thus, a reduction in their synthesis may result in goiter (B).

28. The endocrinopathy is a lack of parathyroid hormone.

29. Vitamin D increase the fraction of oral calcium that is absorbed by the gut.

30. Tetany and muscle spasms were caused by hypocalcemia, which increases Na^+ conductance in skeletal muscle membranes at rest.

31. What is the renal enzyme that activates 25-hydroxy-cholecalciferol to 1,25-dihydroxy-cholecalciferol?

32. A relative lack of PTH receptors.

33. A. Plasma phosphate is important to bone mineralization (B). Low levels activate renal 1-hydroxylase (C), but hyperphosphatemia is not common (D), because both its gut absorption and renal disposition are hormonally regulated (E).

34. A,B,C Bone — Parathyroid hormone has equally important actions on bone and kidney.
C Gut
A,B Kidney

35. Ketoacidosis. If insulin shock were the case, blood and urine glucose would be low, and ketones would not be produced.

36. The serum bicarbonate has been depleted by buffering keto acids. The individual is in an acidotic state.

37. The acidotic state is responsible for the irregularities in breathing, heart rate and blood pressure.

38. Fluid, bicarbonate, and insulin replacement and monitoring the plasma potassium status.

39. C. Plasma glucose levels would decline due to insulin action, then return to normal due to glucagon action.

40. A. Plasma glucose levels would decline due to insulin action, but not return to normal due to the inhibition of glucagon secretion by somatostatin and the blockade of epinephrine action by the beta-blocker.

41. B. C-peptide levels in plasma. This measurement is a reflection of secretion of pancreatic insulin.

42. Cortisol interferes with insulin binding to its receptor. Thus, the beginning plasma glucose level may be somewhat higher, as will the glucose peak attained, and the time to normal will be prolonged.

43. E. Gluconeogenesis increases for the first few days of starvation, then declines to a much lower rate.

44. Plasma insulin levels of type II diabetics are normal to high and the degree of insulin resistance due to diminished insulin receptors is incomplete. Therefore, there is enough insulin to forestall pronounced ketone production.

45. D. Plasma epinephrine is actively taken up into non-neuronal tissues and metabolized by the enzymes COMT and MAO.

46. B. Epinephrine is formed from norepinephrine by the action of the enzyme PNMT which is induced by cortisol.

47. D. Epinephrine, eating, and cold serve to increase metabolic production of heat. The BMR may actually decline by as much as 50% during prolonged starvation (beyond 1 week).

48. A. Stimulus-secretion coupling involves depolarization of the membrane, followed by elevated cytoplasmic calcium ions which initiate the secretory process.

49. B. The binding of epinephrine to the β-adrenergic receptor results in the activation of a Gs protein which binds GTP and subsequently activates adenylate cyclase conversion of ATP to cAMP.

50. This disorder appears to be a primary malfunction of the adrenal cortex. The late onset (in age) suggests adrenal cortical loss.

51. The laboratory values reveal hyponatremia and hyperkalemia, symptomatic of hypoaldosteronism.

52. The metabolic status is due to hypocortisolemia and hypoaldosteronism. Weakness, fatigue, and dehydration are directly traceable to the electrolyte imbalance. Weakness, hypotension, and hypoglycemia are results of low cortisol.

53. High plasma levels of ACTH can result in patchy pigmentation of the skin and mucous membranes.

54. After laboratory confirmation of suspected hypoaldosteronism, both cortisol (glucocorticoid) and aldosterone (mineralocorticoid) replacement should be instituted.

55. The feedback loop involves the inhibition of CRH and ACTH by cortisol.

56. B. Adrenalectomy is most often fatal from electrolyte imbalance brought on by the loss of mineralocorticoid. Additionally, such an individual may succumb to stress because of the loss of glucocorticoid.

57. E. Steroids act upon target cells by binding to a cytoplasmic receptor which is transported into the nucleus where it induces or represses gene transcription. A glucocorticoid receptor blocker could be expected to prevent cortisol from having its normal mechanism of action.

58. C. A converting enzyme inhibitor will result in less aldosterone secretion. The loss of aldosterone will result in Na^+ excretion (increased Na^+ clearance, A) and K^+ retention (decreased K^+ clearance, B).

59. D. Cortisol promotes liver glycogen deposition (a), muscle proteolysis (B), inhibition of ACTH secretion (and thus decreased adrenal size; C), and peptic ulcers (E). Truncal obesity is one hallmark of Cushing's syndrome (D).

60. B. Glucocorticoids acts by an unknown mechanism, probably at the glomerulus-Bowman's capsule, to enhance the excretion of water, which may account for the frequency of urination among Cushing's patients.

61. B. Angiotensin II acts upon cells of the glomerulosa layer to stimulate aldosterone secretion.

INTEGRATIVE TOPICS IN PHYSIOLOGY

REPRODUCTIVE PHYSIOLOGY

Sexual Differentiation

Chromosomal Sex

The presence of the Y chromosome in the male dictates the development of a testis; the hormonal secretions of the testis then impose male development on the phenotypically indifferent fetus. In the absence of a Y chromosome a testis fails to develop and a female phenotype results. Thus, a central concept in sexual differentiation is that the male is the induced phenotype, whereas the female develops passively in the absence of male determinants. The genotype of the female is 46,XX and that of the male is 46,XY. All oocytes have a 23,X complement, and spermatozoa have either 23,X or 23,Y. The presence of a single Y chromosome ensures testicular differentiation and the development of a predominantly male phenotype.

Gonadal Sex

The fetal testis begins to differentiate when müllerian-inhibiting substance, a glycoprotein hormone from the fetal Sertoli cells, is expressed by the Y-specific genes. The first event in male development is regression of the müllerian ducts. Leydig cells differentiate and begin to synthesize testosterone. Estrogen formation by the fetal ovaries occurs about the same time that the fetal testes begin to synthesize testosterone. Cells multiply in the ovary, but follicular development does not occur until the 15th week when primordial follicles begin to form. Thus, endocrine differentiation in terms of steroid hormone formation occurs at the same time in fetal ovaries and testes.

Phenotypic Sex

The transformation of the wolffian ducts begins after the onset of müllerian duct regression. Various efferent ducts develop during this time, of the testes, epididymis, vas deferens, seminal vesicles, urethra, and prostate.

The internal reproductive tract of the female is formed from the müllerian ducts as the wolffian ducts degenerate. The müllerian ducts give rise to the fallopian tubes, uterus, cervix, and vagina.

From the ninth week through the twelfth week development of the external genitalia of the male and female diverges. In the male the genital swellings enlarge to form the scrotum, and the genital folds fuse over the urethral groove to form the penile urethra. Two additional aspects take place during the latter two-thirds of gestation - descent of the testes into the scrotum and the growth of the external genitalia. After 10 weeks the genital tubercle begins its transformation to the clitoris, and the genital swellings enlarge to form the labia.

Development of the female phenotype apparently does not require secretions from the fetal ovaries; that is, female phenotypic development occurs in the absence of gonads. In the male testosterone combines with its intracellular receptor to mediate differentiation of the wolffian duct. In the cells of the external genitalia, testosterone is metabolized to dihydrotestosterone, which subsequently binds to its receptor and mediates differentiation of those structures. Females have the same androgen receptor system and thus can be virilized when the appropriate androgenic signal is present.

Androgens secreted by the testes during the periods of fetal and neonatal development are responsible for the acyclic pattern of gonadotropin secretion characteristic of male rats. Administration of testosterone to neonatal female rats produces acyclic gonadotropin secretion and sterility and masculinization as adults. Conversely, castration of neonatal male rats eliminates androgen production, and adults exhibit cyclic female gonadotropin secretion. The same is thought to be true of human development.

Puberty

The reproductive systems of males and females are not dormant during childhood, but they demonstrate a low level of activity throughout the period between birth and 8 to 9 years of age. Removal of the small amount of steroid hormones by castration produces a rise in plasma gonadotropin levels. Contrariwise, administration of very small amounts of steroid hormones suppresses gonadotropin secretion. Thus, an intact feedback system operates during childhood. LH and FSH are secreted at low levels and fairly evenly over 24 hours up to age 9. From age 9 to about 13 an increase in LH and FSH secretion occurs during sleep. From age 13 to 16 or 18 gonadotropin secretion increases during the awake portion of the day/night cycle, with mean 24 hour levels at two to four times the prepubertal levels. In the adult gonadotropin secretion is random and episodic throughout the day.

Three changes can explain this progression, but their relative contribution is unclear. 1) The neuroendocrine feedback is altered during brain maturation, so that there is less sensitivity to negative feedback by steroids. 2) The responsiveness of the anterior pituitary to GnRH increases. 3) CNS inhibition of the hypothalamus diminishes. The system operates at a higher level, ie, greater circulating gonadal steroids are required to negatively feedback upon the hypothalamic-pituitary axis. The female hypothalamus-pituitary-gonadal system operates cyclically, but the male system does not. What instigates cyclicity of the female is not known.

These changes during the period from childhood to adulthood cause physical and sexual maturation. The onset of puberty occurs at 10-11 yrs in females and 12-13 yrs in males, thus males are generally about 10 cm taller at the beginning of puberty than females. Females attain the peak of their pubertal growth spurt at 12 yrs and males at 14 years of age. In females menarche occurs just after the peak of the growth spurt, meaning that their growth rate reaches a peak before sexual maturity. Since the male growth spurt is delayed, their sexual development is much farther along by the time they reach the peak of their growth spurt. That is, their peaks of growth rate and sexual development coincide more closely.

In males the increase in testosterone secretion results in growth of the testes and penis, development of body, facial and pubic hair, increase in muscle mass, broadening of the shoulders, and thickening of the vocal cords. In females cyclic reproductive function begins with the first menstrual period, menarche, occurring between 12 and 13 years of age. Regular ovulatory cycles commence up to several years later with intervening cycles being anovulatory. The increases in estrogen and progesterone secretion during puberty bring about growth of the pelvis, increase fat formation over the pelvis, buttocks and thigh, breast enlargement and pubic hair formation.

Physiology of the Male

The testes serve both as reproductive and endocrine organs in the male. The male reproductive system is designed to produce, store, and deliver large numbers of gametes (spermatozoan) to the female, in addition to producing several hormones important to reproductive function. The hypothalamus and anterior pituitary gland regulate these processes via the secretion of gonadotropin releasing hormone (GnRH) and gonadotropins (LH and FSH). In turn, the secretion of GnRH, LH, and FSH is modulated by the circulating concentrations of testicular steroid and peptide hormones (negative feedback mechanisms).

The Hypothalamic-Pituitary Axis

GnRH is a decapeptide synthesized by neurons in the medial basal hypothalamus. Like many other hypothalamic peptides, GnRH is secreted in a pulsatile fashion (high amplitude and short duration pulses) and transported by the hypophyseal portal circulation to the anterior pituitary gland. GnRH binds there to its receptor and stimulates the synthesis and episodic secretion of LH and FSH by a calcium-dependent mechanism(s) in the gonadotroph cells. Although the secretion of both gonadotropins is dependent on GnRH stimulation of the gonadotrophs, the pattern of LH secretion (one pulse every 1 or 2 hours) more closely corresponds to GnRH secretory episodes. Prior to puberty GnRH-induced FSH secretion is substantially greater than in adulthood when FSH secretion is only about one fifth of that observed for LH.

Regulation of Testicular Function

LH and **FSH** are glycoproteins composed of two polypeptide subunits, designated alpha and beta. The alpha subunit is identical for LH, FSH, TSH, and hCG, whereas the beta subunits differ and establish the unique biological activity for each hormone.

The processes of **steroidogenesis** and **spermatogenesis** are anatomically compartmentalized within the testis and are differentially regulated by LH and FSH. **Testosterone** production by **Leydig cells** (or interstitial cells) is largely dependent on the amount of **LH** secreted into the circulation and delivered to the testes. Once LH binds to its membrane receptor on the Leydig cell, adenylate cyclase is activated and catalyzes the formation of cyclic AMP. This second messenger activates protein kinase A which promotes the synthesis of steroidogenic enzymes, increases cholesterol availability, and increases testosterone production. The majority of testosterone in the circulation is bound to **sex steroid binding globulin** (30%; **SSBG**), albumin and other proteins (67%). The remaining 3% of testosterone is unbound or free. Testosterone can be metabolized to **dihydrotestosterone (DHT)** by the enzyme **5 alpha-reductase** in skin, prostate gland, seminal vesicles, and epididymides. The androgenic potency of DHT is some 30 to 50-fold greater than testosterone. Testosterone is also available for metabolism to **estrogens** by the **aromatase** enzyme system in testes, adrenal, fat, liver, brain and pituitary gland. Normal **spermatogenesis** is dependent upon 1) adequate LH-mediated testosterone production by Leydig cells and 2) the direct actions of both **testosterone** and **FSH** on the seminiferous epithelium to promote germ cell development and **Sertoli cell** production of **androgen binding protein (ABP)**, **inhibin**, **estradiol**, and other factors. The concentration of testosterone within the male reproductive tract is considerably higher than blood concentrations due to the presence of ABP.

Only the Sertoli cells and spermatogonia contact the basement membrane of the seminiferous tubules. The so-called **Blood-Testes Barrier** is formed by the tight junctions between the basal portions of adjacent Sertoli cells within seminiferous tubules. Sertoli cells envelope and separate the developing germ cells into unique, stage-dependent environments which are completely partitioned from everything outside seminiferous tubules. Production of mature spermatozoa from quiescent stem cells requires about **74 days**. There is a potential for the production of 64 mature spermatozoa from each committed spermatogonia entering spermatogenesis. As the germ cell differentiates and matures, it moves in a basal to luminal direction in the seminiferous tubule.

Endocrine Feedback Mechanisms

Synthesis, storage, and secretion of GnRH by the hypothalamus and gonadotropins by the anterior pituitary are negatively regulated by gonadotropin-induced testicular steroids and peptides. Testosterone and its metabolites, estradiol and DHT, inhibit LH secretion by various mechanisms; 1) reducing hypothalamic **secretion of GnRH** (testosterone, estrogen, DHT); 2) reducing the pituitary **response to GnRH** (testosterone, estrogen); and 3) reducing the **LH pulse frequency** (testosterone) **or amplitude** (estrogen). All of these examples are **long-loop, negative feedback mechanisms. Inhibin** preferentially **inhibits FSH** secretion, but the regulation of FSH is not as well understood. Some steroids may also play a role in the negative feedback inhibition of FSH by 1) **reducing the biological half-life of FSH** (estrogen) and 2) reducing the FSH pulse amplitude (DHT).

Accessory Reproductive Structures

Once spermatozoa are released into the lumen of the seminiferous tubules, they are transported out of the testes via the **efferent ducts** and into the **epididymides** where **spermatozoa maturation**, **concentration**, and **storage** occurs. During the processes of sexual excitation, penile erection, emission, and ejaculation, some 300 x 10^6 spermatozoa are rapidly moved from the epididymides, through the paired **vas deferens** and converge at the **ejaculatory duct** system. The secretions of the **seminal vesicles**, **prostate** and **bulbourethral (Cowper's) glands** are deposited into the ejaculatory ducts in a sequential manner during ejaculation and are expelled through the **urethra** with the sperm cells.

The function of all the accessory sex glands is dependent upon **androgens**, since castration and androgen deprivation results in the involution of these structures.

The constituents of seminal plasma are shown in Table 8-1.

Table 8-1. Composition of human ejaculate (3.0 ml).

	Volume	From	Substances
First fractions	<0.6 ml	Epididymides	Spermatozoa, phosphorylcholine, alpha glycerophosphorylcholine
	0.5 ml	Prostate	Zn^{2+}, acid phosphatase, citric acid, liquefaction factors
Later fractions	1.5-2.0 ml	Seminal vesicles	Fructose, prostaglandins, secretions for coagulation

Erection, Emission, and Ejaculation

Parasympathetic and sympathetic innervation of the reproductive system are intimately involved in normal sexual function in the male. **Erection** of the penis is controlled by **parasympathetic** nerve activation resulting from tactile stimulation of the genitalia (sacral spinal reflex arc). The neural pathways mediating psychogenic erection are not well understood. Penile erection is largely a **vascular event.** Parasympathetic activity produces vasodilation of penile arteries and occlusion of veins, causing the two lateral corpora cavernosa and ventral corpus spongiosum to become engorged with blood.

Emission (deposition of semen into the posterior urethra) is dependent on **sympathetic** activation of smooth musculature of the ductus deferens, ampulla, prostate, and seminal vesicles, and partial closure of the bladder neck. **Ejaculation** occurs following further **sympathetic** input to completely close the bladder neck and **parasympathetic** input to cause contractions of the striated perineal musculature. **Orgasm** is a cortical sensory experience with efferent input from smooth muscle contractions of the accessory sex glands and pelvic striated musculature.

Review Questions

1. Development of male external genitalia requires

   A. a single Y chromosome
   B. androgens
   C. prior regression of the wolffian ducts
   D. that there be no estrogen in the fetal circulation

2. Müllerian-inhibiting hormone is

   A. secreted in response to human chorionic gonadotropin
   B. synthesized and secreted by Sertoli cells
   C. required for differentiation of the wolffian ducts
   D. synthesized and secreted by interstitial cells of Leydig

For Questions 3-5. A 16-year-old male was concerned because he did not exhibit deepening of his voice, had scanty pubic and axillary hair, absence of beard and mustache, small penis, poor muscular development, and psychosocial immaturity.

Clinical evaluation indicated the following:

Serum testosterone	137 ng/dL
Sperm count	13 million/ml semen

The following tests were performed:

Clomiphene (a weak estrogen agonist which stimulates gonadotropin secretion) given at 100 mg/day for seven days: 0% increase in LH.

GnRH (100 ug iv): 0% increase in LH in twenty minutes

HCG (5000 I.U. iv): 50% increase in plasma testosterone one to three days after injection

The boy was subsequently treated with 25-75 of units FSH three times/week and HCG as described above. Sperm count and testosterone levels were both near normal after two months of treatment, and primary and secondary sex characteristics appeared.

3. What is the endocrine disorder in this individual? Is the disorder primary or secondary?

4. Why is HCG used in the treatment?

5. Why would both FSH and HCG be needed in the treatment?

6. Testosterone is

   A. produced primarily in the Sertoli cells of the testes in response to FSH.
   B. converted to DHT by 5α-reductase in the cells of the prostate gland and some other reproductive tissues.
   C. present in the plasma as a free hormone not bound to plasma proteins
   D. synthesized from estradiol-17β in the testes
   E. not required for the initiation and maintenance of spermatogenesis

7. In the male FSH

   A. concentrations in plasma decrease after castration
   B. concentrations in plasma are independent of hypothalamic GnRH secretion
   C. binds to receptors on primary spermatocytes, causing them to enter meiosis
   D. stimulates Sertoli cells to synthesize inhibin and androgen-binding protein

Physiology of the Female

The female reproductive system exhibits regular cyclical changes which optimize the chance of fertilization and pregnancy. Like the male, female reproductive function is dependent upon the hypothalamic-anterior pituitary axis. The gonadotropins, FSH and LH, regulate ovarian cyclicity and steroid production which in turn regulate cyclical changes in the uterus, cervix, vagina, and breasts.

The Menstrual Cycle

As its name implies, the menstrual cycle begins with the first day of menstruation (day 1 of the cycle) and is complete when the following menstrual period occurs. The average menstrual cycle length is 28 days with a normal range of between 23 and 33 days. Events which occur during the menstrual cycle may be subdivided into three interdependent cycles related to hypothalamic/pituitary, ovarian, and uterine functions.

The Ovarian Cycle

The ovarian cycle includes the **follicular phase** (days 1-13), **ovulation** (day 14), and the **luteal phase** (days 15-28). During the follicular phase a group or **cohort of follicles** begin to grow within the ovary largely in response to **FSH stimulation**. Each follicle contains an **oocyte** which also grows and matures in synchrony with the follicle. The **number of oocytes** contained within each ovary is finite and continues to decline over the reproductive life of the female. **Follicle growth** is characterized by accelerated **mitotic activity of granulosa** cells, formation of an **antrum** (follicular fluid accumulation within the follicle), and increased **estradiol production** by the granulosa cells in response to both LH and FSH. Elevated circulating **estradiol** concentrations exert a **negative feedback regulation** on the endocrine secretions of the hypothalamus (GnRH) and anterior pituitary (LH, FSH) throughout much of the follicular phase. The larger antral follicles also produce a peptide hormone, **inhibin**, which specifically **inhibits FSH** secretion. During the latter half of the follicular phase a dominant follicle emerges from the cohort. The subordinate follicles and their oocytes become **atretic** and are resorbed by the ovary. Maximal estrogen production by the **mature graafian follicle** triggers the midcycle **LH** (and FSH) **surge** via a **positive feedback** mechanism on the hypothalamus and anterior pituitary. The LH surge is responsible for initiating **ovulation** of the graafian follicle and subsequent formation of the **corpus luteum (luteinization** of the follicular **granulosa** and **theca** cells). An elevated **progesterone** concentration is the hallmark of the **luteal phase**, although the primate corpus luteum also produces **estrogen** and various peptide hormones (ie, **relaxin, oxytocin**). The functional lifespan of the corpus luteum is approximately 14 days and requires **LH support** throughout. In the absence of oocyte fertilization and pregnancy, the corpus luteum undergoes complete **regression** by day 28 of the cycle. Secretory products of the corpus luteum decline precipitously during regression. However, the mechanisms regulating the regression process are poorly understood.

The Uterine Cycle

The uterine cycle includes the **menstrual phase** (days 1-4), **proliferative phase** (days 5-14), and **secretory phase** (day 15-28). The uterine cycle is directly related to the ovarian cycle and is dependent upon steroid production by the ovary. Although **menstruation** marks the start of a menstrual cycle, it actually represents the **end of the uterine cycle**. By the end of the menstrual period, all of the upper endometrial layers (functionalis I and II) have sloughed. Under the influence of **follicular estrogen,** these endometrial layers are regenerated during the **proliferative phase**. As the endometrium thickens, **elongation of the uterine glands** and **growth of coiled spiral arteries** into the stratum functionalis occurs. At this point the glands are not convoluted and do not exhibit appreciable secretory activity. The proliferative phase is also accompanied by the secretion of **clear, elastic, watery, cervical mucus.** Peak secretory activity by the cervical sebaceous glands occurs during the preovulatory rise of estradiol. The consistency of the cervical mucus during this phase facilitates the **passage of spermatozoa** into the uterine lumen.

Following ovulation the influence of **progesterone** (and **estrogen**) from the **corpus luteum** initiates the **secretory phase** of the newly regenerated endometrium. These steroids are responsible for **preparing the uterus for pregnancy** should fertilization occur. During this phase endometrial glands become branched and highly convoluted, and the spiral arteries project throughout the stratum functionalis. Secretory activity by the glandular epithelium is increased coincident with the rate and extent of corpus luteum development and steroid secretion. The **cervical mucus** becomes **opaque, thickened,** and **dehydrated** under the influence of **progesterone**. This type of mucus acts as a **barrier to spermatozoa**. If pregnancy does not occur during the secretory phase, the endometrium is shed **(menstruation)** following corpus luteum regression and **progesterone** (and estrogen) **withdrawal.**

Fertilization and Implantation

There are several events which must occur within the female reproductive tract to both gametes (spermatozoa and oocyte) prior to successful fertilization and development of the embryo. First, the **timing of insemination is critical** due to the limited functional lifespan (generally 24 to 36 hours) of the ejaculated spermatozoa and ovulated oocyte. Spermatozoa are deposited within the vagina and must penetrate the cervical mucus in order to reach the uterine lumen. The seminal plasma acts to neutralize the acidic environment of the vagina and facilitates motility of sperm, thereby allowing sperm to move as a convoy through multiple mucus tracks. Spermatozoa undergo **capacitation** within the uterine lumen. This process occurs by removal of seminal proteins from sperm cell membranes, promoting spermatozoa to a **fertile** state **(freshly ejaculated spermatozoa are infertile)**. Within minutes thousands of spermatozoa are swept up the reproductive tract toward the oviducts, aided by **uterine contractions** and ciliated uterine epithelium. Only a few hundred of the 300 million spermatozoa ejaculated actually make it to the **ampulla** (upper third of the oviduct) where **fertilization** occurs.

At ovulation the **primary oocyte** expels the **first polar body** as it undergoes the first of two **meiotic** divisions. The resulting **secondary oocyte** remains arrested at the metaphase II stage until fertilization occurs. The finger-like projections of the **fimbria** collect the ovulated secondary oocyte and direct it into the lumen of the oviduct. As the capacitated spermatozoa contact the surrounding **corona radiata** cells of the oocyte, the membranes of the **acrosomal cap** on sperm heads begin to perforate, allowing **proteolytic enzymes** (eg, hyaluronidase) to be released. This **acrosome reaction** allows the sperm cells to digest their way through the corona radiata cells and mucopolysaccharide layer **(zona pellucida)** surrounding the oocyte. **Oocyte activation** occurs in response to the first spermatozoon that makes contact with the oocyte's plasma membrane. This initiates the release of protein-containing, **cortical granules** from the cortex of the oocyte into the **perivitelline space** (space between the zona pellucida and oocyte membrane). This causes biophysical changes in the zona pellucida and oocyte plasma membrane which effectively block fertilization by more than one spermatozoa (**block to polyspermy**). During the process of **fertilization** the sperm and oocyte membranes fuse, and the spermatozoa is engulfed into the oocyte. The oocyte completes its **second meiotic division** and extrudes the **second polar body** into the perivitelline space. The sperm head, containing the male complement of DNA, decondenses to form a **male pronucleus** and a nuclear membrane is synthesized around the female complement of DNA to form a **female pronucleus**. The **pronuclei then fuse** to complete the fertilization process (12 to 24 hours after ovulation).

Cleavage or **mitoses** of the one-cell **zygote** into a two-cell, four-cell, eight-cell, and early **morula** (16-cells) **pre-embryo** occurs within the **oviduct**. At the late morula stage (32-cells) the pre-embryo reaches the **uterine** lumen where **blastocyst** development occurs (approximately 128 to 256 cells). The anatomy of the blastocyst includes an outer layer of trophectoderm (**trophoblast**) which will become the **fetal placenta**, an inner cell mass (**embryoblast**) which will become the **fetus**, and a fluid filled cavity or **blastocoele**. The blastocyst must "**hatch**" out of the zona pellucida before **implantation** into the endometrium can occur (about 144 to 192 hours after ovulation). The hatched blastocyst first attaches to the surface epithelium. Trophoblast cells in the **attachment zone** then differentiate into **cytotrophoblast cells**. Prior to penetration cytotrophoblast cells **fuse** together and form a syncytial (large multinucleated cells) layer called the **syncytiotrophoblast**. This alters the function of cytotrophoblast cells so that **penetration into the endometrium** can occur. The syncytiotrophoblast layer is also the source of **human chorionic gonadotropin** (**hCG**) which maintains the function of the **corpus luteum** (**progesterone secretion**) throughout the first trimester of pregnancy. It should be noted that as a species, human fecundity is very poor. This is due largely to an estimated 60% to 70% embryonic death prior to and during implantation. Most of these early pregnancy failures occur prior to any extension of the menstrual cycle and are therefore not detected.

Pregnancy

Gestation lasts 37 to 41 weeks in humans and is subdivided into **trimesters**. **Implantation** is completed by the end of the second week of pregnancy, marking the end of the **pre-embryonic stage**. With formation of the **primary germ layers** (ectoderm, mesoderm, endoderm) and **extraembryonic membranes** (amnion, yolk sac, allantois, chorion), the **embryonic stage** begins (weeks 3 through 8). During the **first trimester** the **placenta** becomes firmly established, and **embryonic/fetal organogenesis** occurs. The critical nature of **conceptus** (embryo/fetus plus extraembryonic membranes) development during the first trimester is shown by the high rate (approximately 25%) of spontaneous pregnancy losses during this period. This contrasts with 2% to 3% fetal death during the second trimester and less than 1% in the last trimester.

Human placentation is **hemochorial**; the chorionic villi are in direct contact with maternal blood. The **chorionic villi** are the **sites of nutrient**, **waste** and **gas exchange** between fetal and maternal systems. The placenta is also a **transient endocrine organ** producing several steroids, prostaglandins, and peptide hormones. The corpus luteum is the major source of progesterone secretion during the first 6-8 weeks of gestation. However, the developing trophoblast takes over by about 8 weeks, making the placenta the main source of progesterone during pregnancy. The placenta has very limited capacity to synthesize cholesterol from acetate. The principal precursor for biosynthesis of progesterone is low-density lipoprotein (LDL) cholesterol. Progesterone is essential for the maintenance of pregnancy, including the ability to inhibit T lymphocyte cell-mediated responses which allow the body to tolerate the "foreign antigen" status of the fetus.

The rate of estrogen production during pregnancy increases markedly; urinary estriol can increase up to 1000-fold. The corpus luteum is the major source of estrogen during the first few weeks, but afterward nearly all of the estrogen formed is from the placental trophoblast (Fig. 8-1). However, the placenta is unable to convert progesterone to estrogens because of a deficiency of 17α-hydroxylase. The placenta has to rely on androgens produced in the maternal and fetal adrenal glands. Estradiol-17β and estrone are synthesized by the placenta via conversion of dehydroepiandrosterone sulfate from both the maternal and fetal blood.

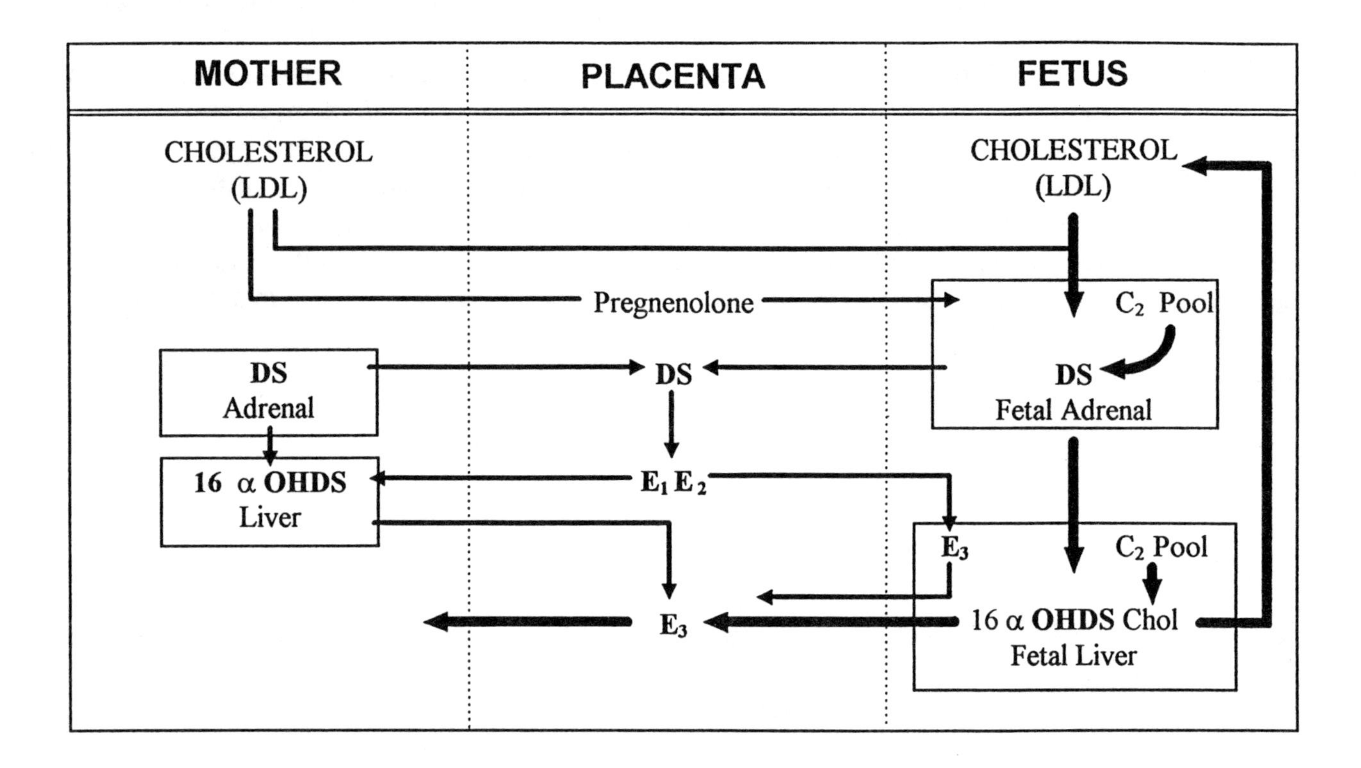

Figure 8-1. Sources of estrogen biosynthesis in the maternal-fetal-placental unit. Symbols: C_2 pool, carbon-carbon unit; DS, dihydroepiandrosterone; E_1, estrone; E_2, estradiol-17β; E_3, estriol; LDL, low-density lipoprotein. (Modified with permission from Carr BR and Grant, NF, *Clin Perinatol* **10**:737, 1983).

The maternal compartment is the ultimate destination of estrogen and progesterone secreted by the placenta.

Human chorionic gonadotropin (hCG) secretion by the placenta is measurable in maternal blood by about two weeks, reaches peak concentrations by ten weeks of pregnancy, and declines thereafter. Early production of hCG is essential for maintenance of progesterone production by the corpus luteum. After the fifth week of pregnancy production of progesterone by the corpus luteum gradually declines as placental steroidogenesis increases, described as the **luteoplacental shift**. Placental steroidogenesis (**progesterone, estrogens**) continues to increase until term. During the period when hCG production is declining, **human placental lactogen (hPL)** production increases, reaching peak blood concentrations during the third trimester. The structure of hPL is very similar to both growth hormone and prolactin, and its functions reflect these similarities. **Maternal intermediary metabolism** is modified by hPL, especially during the last trimester, to increase lipid mobilization, enhance carbohydrate-stimulated insulin secretion, and increase insulin resistance in some maternal tissues. Collectively, these maternal alterations in metabolism are thought to **enhance maternal free fatty acid utilization** while **sparing glucose for use by the growing fetus**. Human placental lactogen may also play an important role in **mammary gland development**. Several other maternal systems which are affected by the altered endocrine environment of pregnancy are listed in Table 8-2 on the next page.

Fetal growth rate increases most dramatically during the third trimester and then levels off towards term. Maturation of the **fetal hypothalamic-pituitary-adrenal axis** and increased production of **fetal adrenal steroids** (cortisol, dehydroepiandrosterone) during the second half of pregnancy alter **placental steroidogenesis**, especially during the last month of gestation.

Parturition

Prior to parturition placental estrogen production is increased relative to progesterone, thereby **increasing** the **estrogen:progesterone ratio**, attenuating the so-called "**progesterone block**" on the uterus. This allows estrogen to exert effects on the myometrium and cervix which facilitate parturition (birth of the baby and placenta). **Gap junctions** become more abundant and facilitate **coordinated contractions** during labor and fetal/placental expulsion. **Estrogen** induces myometrial cells to **synthesize receptors** for **estrogen, oxytocin** and **prostaglandins**. In response to myometrial stretching and pressure exerted on the cervix by the fetus, oxytocin is reflexly released from the posterior pituitary and binds to its receptors on the myometrium. **Oxytocin stimulates** the production of uterine and placental **prostaglandins** which increase **intracellular calcium** and promotes **myometrial contractility**. Estrogen also affects the **cervix** by increasing its responsiveness to **relaxin** (corpus luteum, placenta) and prostaglandins (uterus, placenta). In response to these three hormones, the cervix undergoes biochemical changes which result in **structural remodelling** characterized by increased water content, vascularization, and mass, and decreased organization and mass of collagen and dermatan sulfate. These changes begin well before term and are responsible for **cervical dilation** and **effacement** during labor.

Menopause

Fertility in women declines rapidly after the age of 35 years until **menopause** (reproductive senescence), which is characterized by the **cessation of menstrual cycles**. The vast majority of women experience menopause between 45 and 55 years of age (mean of 51 years). The onset of reproductive senescence is gradual and results from a **reduction** in the ability of the **ovaries to respond to gonadotropins**. This lack of ovarian responsivity to gonadotropins is thought to accelerate as the

TABLE 8-2. Maternal changes associated with pregnancy.

GENERAL
- Increased body mass (25-30 lbs)
- Changed posture (lordosis and kyphosis)
- Fatigue and nausea

HEMATOLOGICAL
- Increased blood volume (about 50%), RBC mass (20-30%) and WBC's (neutrophils)
- Decreased hemoglobin (15%) (physiological anemia of pregnancy)

CARDIOVASCULAR
- Heart moves upward and forward from growth of the uterus.
- Increased HR (10-15%), SV (30%), CO (up to 60%) and cardiac volume (10%)
- Unchanged blood pressures
- Increased venous pressures in legs, edema

RESPIRATORY
- Increased O_2 consumption (30-60 ml/min)
- Increased tidal volume (200 ml) and alveolar ventilation
- Unchanged respiratory rate
- Decreased residual volume (20%) and expiratory reserve volume
- Dyspnea is common.

RENAL
- Increased GFR (50%) and renal plasma flow (50-70%)
- Glucosuria (reabsorption is exceeded)
- Cumulative retention of 2 oz of NaCl

URINARY TRACT
- Increased bladder capacity, residual volume and frequency of micturition
- Decreased bladder and urethral tone
- Increased risk of urinary tract infection because of urinary stasis

GASTROINTESTINAL, due to increased progesterone
- Decreased esophageal tone, gastric acid, and colon motility
- Increased mucus production
- Rectal stasis and hiatal hernia
- Hemorrhoids are common.

MUSCULOSKELETAL, due to relaxin and progesterone
- Softening of ligaments (pubic and sacroiliac joints)
- No loss of bone density

SKIN, due to estrogen, progesterone and MSH
- Midline hyperpigmentation and dry skin
- Chloasma (mask of pregnancy)
- Striae gravidarum (mechanical stretch)
- Spider angiomata and palmar erythema

REPRODUCTIVE SYSTEM
- Vulvar swelling and varicosities
- Thick vaginal secretions
- Eversion of cervix and mucus plug
- Increased uterine mass (16-fold) and uterine blood flow (5 to 10-fold)

METABOLIC
- Increased ingestion of 300 kcal/day (for fetal and placental growth)
- Sex steroids increase tissue glycogen storage and increase hepatic glucose production.
- Corticoids reduce liver glycogen storage and increase liver glucose production.
- Increased insulin secretion, glucose utilization and pancreatic B-cell hyperplasia
- Increased production of ketones and FFA
- Reduced glucose tolerance from increased production of human placental lactogen
- Insulin resistance from increased prolactin

ENDOCRINE
- Progesterone produced in placenta instead of corpus luteum at 6-10 weeks.
- hCG production peaks at 10-12 weeks and then declines.
- hPL production increases from 8 weeks until term.
- Estrogens and corticosteroids increase to term
- Prolactin comes from decidua, fetus and mother.
- Labor is induced by maternal oxytocin acting on myometrial receptors and uterine production of prostaglandins.

numbers of ovarian follicles and oocytes decline during the aging process. Several signs associated with the **premenopausal period** include **irregular menstrual cycle lengths, lower circulating ovarian steroid concentrations**, and **elevated gonadotropin concentrations**. As ovarian production of estrogens and inhibin dwindle, so to does their negative feedback on the hypothalamus and anterior pituitary, thus GnRH and gonadotropin (especially FSH) concentrations begin to rise. At **menopause** the **ovaries** are **functionally inactive**, resulting in **estrogen deprivation** and symptoms associated with this condition. **Atrophy of the vagina** occurs and may be associated with **atrophic vaginitis, dyspareunia** (discomfort during intercourse), and **vaginal bleeding** as the vaginal epithelium thins. **Calcium absorption** by the gut is **reduced** in the absence of estrogen which facilitates **accelerated bone loss** (up to 1.5% bone mass lost each year after menopause) and predisposes postmenopausal women to **osteoporosis**. **Hot flushes**, and associated symptoms such as **anxiety, depression, irritability**, and **fatigue** are reported in approximately 75% of menopausal women. **Hot flushes** are thought to be the result of vasoactive and CNS effects exerted by **elevated GnRH, LH** and **catecholamines**. All of these menopausal symptoms may be controlled, reduced, or eliminated by **estrogen replacement** therapy.

Contraception

There are several methods of contraception, including the rhythm method, withdrawal, barriers such as condoms, foams and diaphragms, intrauterine devices or IUDs, sterilization (gonadectomy, tubal ligation, or vasectomy), abortion, and oral steroid contraceptives.

The least effective methods are the rhythm and withdrawal methods, followed by the barrier methods, and IUDs, with oral contraceptives being the most successful (greater than 99%).

Oral contraceptives prevent pregnancy by inhibiting ovulation through suppression of gonadotropin secretion. There are three types of oral contraceptive pills: a combination of estrogen-progestin, phasic estrogen-progestin, and progestin-only pills. The ideal pill is one that contains the least amount of steroid to prevent pregnancy with the least amount of adverse effects.

Other than the condom, the only proven effective means of fertility control in men is surgical interruption of the vas deferens (vasectomy). There is no readily reversible and effective pharmacological contraceptive for men. Research to date has included inhibition of hypothalamic-pituitary function, of spermatogenesis, and of epididymal function. A significant toxicity of all compounds studied has preclude their clinical trials in men on a large-scale basis.

Review Questions

8. Progesterone

   A. binds to plasma membrane-bound receptors that the transport the hormone to the nucleus of the cell
   B. decreases the minimal stimulus necessary for contraction of myometrial smooth muscle.
   C. reduces myometrial contractility
   D. prepares endometrial cells to respond to estrogen by increasing estrogen receptor

For Questions 9-11. A 26-year-old female complained of severe, dull pain and cramping in the lower abdomen. There were no other physical findings. A laparoscopy revealed the presence of ectopic endometrial tissue on the uterine wall and ovaries. Danazol (a synthetic androgen and inhibitor of gonadotropins) 600 mg/day, was prescribed for up to nine months to inhibit ovulation, suppress the growth of the abnormal endometrial tissue, and achieve appreciable symptomatic relief, with a 30% possibility of conception after withdrawal of the therapy.

9. What is this condition called? What causes it?

10. What is the rationale for using danazol?

11. Could oral contraceptives be used as effectively? Why do you think so?

12. Which of the following is not a precursor in the synthesis of 17β-estradiol in the ovary?

    A. Testosterone
    B. Cholesterol
    C. Pregnenolone
    D. Estriol
    E. Androstenedione

13. FSH and LH

    A. are glycosylated proteins that bind to membrane-bound receptors
    B. concentrations in the plasma peak during the early follicular phase of the menstrual cycle
    C. are both required for the development of ovarian follicles
    D. secretion by the anterior pituitary gland is modulated by inhibin

14. Menopause occurs because

    A. the pool of ovarian follicles is depleted
    B. pituitary gonadotropes synthesize excessive quantities of FSH and LH
    C. excessive plasma inhibin levels suppress FSH production
    D. the uterus is no longer capable of sustaining a fetus

15. Propulsion of spermatozoa through the female reproductive tract to the fallopian tubes

    A. is due to rhythmic motion of the flagellum of the spermatozoon
    B. is retarded by contractions of the vagina and uterus
    C. is retarded by the cervical mucus
    D. is retarded by the negative vaginal pressure generated during orgasm

For Questions 16-20. A twenty-something woman stated that it had been six weeks since her last menses. Her pregnancy test result was positive. By the sixth month of pregnancy, she felt irregular contractions of the uterus but no complications were present. After nine months, a healthy 7 lb, 3 oz girl was delivered with no complications. Breast feeding was planned.

16. What hormone is the basis of pregnancy tests?

17. What hormonal mechanism prevented spontaneous abortion of the implanted embryo?

18. What prevented the uterus from initiating labor before the designated delivery time?

19. What maintains milk production after birth?

20. Is it possible to get pregnant while breast feeding? Explain.

21. Implantation of the conceptus

    A. occurs most often in the ampulla of the fallopian tube
    B. occurs at the blastocyst stage when it consists of approximately 200 cells
    C. is preceded by decidualization of the stromal cells of the endometrial epithelium
    D. is inhibited by progesterone

22. Which of the following statements is correct?

    A. The maternal-side and fetal-side trophoblast cell membranes form the principal elements in the transplacental exchange of nutrients.
    B. Free fatty acids must be esterified to triglycerides prior to being transferred across the placenta to the fetus.
    C. Transfer of D-glucose from mother to fetus occurs by active transport across the placenta
    D. Prolactin is synthesized by amniotic and chorionic cells and secreted into the amniotic fluid
    E. During pregnancy, maternal oxygen-carrying capacity increases so that, at a given P_{O_2}, maternal hemoglobin can carry 20 to 30 percent more O_2 than fetal hemoglobin, thereby facilitating transfer of oxygen from the mother to the fetus.

23. Suckling

    A. inhibits oxytocin secretion
    B. increases estradiol secretion
    C. develops the alveoli and lactiferous ducts of the mammary gland
    D. increases prolactin secretion

24. Biosynthesis of steroid hormones during pregnancy

A. is regulated by hCS
B. is inhibited by DHEA and DS of maternal and fetal origin
C. involves conversion of progestins to androgens by the placenta
D. involves rapid 16-hydroxylation of estradiol to estriol in syncytiotrophoblast cells of the placenta
E. involves a shift of progesterone production from the corpus luteum of the ovary to the chorion during the eighth to ninth weeks of gestation

PHYSIOLOGICAL RESPONSES TO EXERCISE

This section will review the cardiopulmonary adjustments associated with maintained light to moderate physical activity. Every day examples are jogging, bicycle riding and swimming. Exercise may take other forms which will not be considered here, such as weight lifting with very brief efforts of robust activity.

The change from the resting state to exercise requires substantial and coordinated adjustments to several physiological systems. The respiratory and cardiovascular systems are central to this response, but changes of neural, metabolic and endocrine systems are important roles themselves and also alter the respiratory and cardiovascular systems. This section focuses on how these two dominant systems respond to aerobic exercise. In general, the physiological factors discussed represent values for an unconditioned young adult male. However, the specific response to exercise depends on many factors, such as the length and intensity of exercise, the conditioning or previous exposure to exercise, and sex, age and other factors. We will describe responses to sub-maximal intensity exercise of a moderate duration at sea level; for example, jogging for 2 miles at a moderate speed (8 minutes/mile) or similar activity. Where appropriate, data will be given for maximal exercise. This level of exercise is typically defined by the maximal level of oxygen consumption (VO_2 max) achieved during exercise testing (12-16 times from basal levels).

The Respiratory Response to Exercise

A resting person requires a VO_2 of 250 ml/min for basal metabolic needs. This requires a tidal volume of 0.5 L of air inspired at a respiratory rate of about 10/minute for a minute ventilation of approximately 5 L. Mixed venous blood returning to the lungs has a relatively high P_{CO_2} (46 mm Hg) and a low P_{O_2} (40 mm Hg). Exchange across the lungs produces an arterial blood lower in CO_2 (40 mm Hg) and higher in O_2 (90 mm Hg). At the onset of exercise there is a rapid increase in the depth and frequency of breathing followed by a slower secondary rise ventilation. As exercise is maintained, a ventilatory plateau is reached representing a steady state. During moderate exercise ventilation is in the range of 25-50 L/min. Respiratory rate may increase 2-3 times and tidal volume 4-5 times, thereby increasing minute ventilation (tidal vol. x resp. rate = minute ventilation) 10-fold, but ventilation can increase to over 150 L/min during maximal exercise. As O_2 consumption increases progressively from light to moderate exercise, arterial P_{O_2}, P_{CO_2} and pH are remarkably well maintained because of increased ventilation. This means that increased in ventilation is well matched to increased metabolic demands of exercise. This virtual constancy of arterial blood gas values argues that very efficient control systems maintain these values within narrow ranges, although how control mechanisms interact to produce this effect is uncertain.

If the exercise level approaches VO_2 max, the anaerobic threshold is reached. This occurs when significant amounts of metabolic acid begin to accumulate in the blood. Measurement of the rate of CO_2 elimination and blood lactic acid indicates when the anaerobic threshold occurs. Work performed beyond the anaerobic threshold results in a disproportionate amount of CO_2 elimination and blood lactic acid accumulation.

At rest and even during heavy exercise the work and O_2 consumption for the mechanisms of breathing is minimal. At rest inspiration is the active component, primarily of the abdominal diaphragm, while expiration is completely passive due to elastic properties of the respiratory muscles. Increased ventilatory requirements require recruitment of accessory respiratory muscles particularly during expiration.

During exercise, the ventilation/perfusion (V_A/Q_C) progressively increases. As pulmonary ventilation increases, cardiac output does not increase as much, and V_A/Q_C may reach 3-4:1 from the original proportion of approximately 1:1. During maximal exercise ventilation may increase to over 150 L/min, yet cardiac output is maximal at 40 L/min (both in highly trained individuals). Therefore, cardiac output and tissue perfusion are the limiting factors during exercise rather than ventilation. During exercise both pulmonary capillary distension and recruitment occur as pulmonary blood flow increases, maintaining a sufficient time for adequate red blood cell oxygenation.

Ventilation does not immediately decrease to resting values upon the termination of exercise. Both frequency and depth of ventilation decrease some at the end of exercise and then slowly return to normal. VO_2 follows the same pattern. This decrease in ventilation and VO_2 over many minutes, is largely explained by the "oxygen debt" incurred by exercise. Oxygen debt is the amount of O_2 consumed above basal levels after exercise is finished. It is due to replenishment of the body's high energy phosphate and O_2 stores, removal of exercise-induced anaerobic waste products, and the O_2 needed to supply the elevated ventilation and cardiac output still present during the recovery period. The coronary effect is seen at the beginning of exercise, when the respiratory and cardiovascular systems do not fully supply the O_2 demands of muscle, since the systems have not increased their respective transport capacities. This leads to an O_2 deficit, where high energy phosphate, O_2 stores, and anaerobic metabolites are utilized.

Cardiovascular Response to Exercise

The cardiovascular response to exercise is an immediate increase primarily in heart rate with a lesser increase in stroke volume which together increase cardiac output. The change in cardiac output is similar to the respiratory response; a rapid increase within seconds and a slower increase throughout exercise. This cardiac response is a neurally-mediated increase in the ratio of cardiac sympathetic/parasympathetic activity coordinating SA nodal firing and AV junctional conduction. This increased ratio favors adrenergic-mediated increases in cardiac contractility and heart rate. Central motor control, baroreceptor input and ascending afferent neural activity from joint and muscle spindles all have projections to the medullary autonomic control areas. Additionally, the sympathetically-innervated adrenal glands release catecholamines which circulate in the blood and augment this cardiac sympatho-excitation.

Exercise causes a decrease in total peripheral resistance, with the largest decreases occurring in the most metabolically active muscles. Increases in cardiac output provide exercising tissues with additional blood flow to meet oxidative mitochondrial requirements. Increased tissue blood flow, hyperemia, results through arteriolar and pre-capillary sphincter dilation via release of tissue autocoids (adenosine, K^+, prostaglandins) in the exercising tissues. This increased blood flow serves two purposes; one is to increase O_2 delivery, and another is to facilitate removal of metabolic waste in those exercising tissues. Recruitment of previously closed capillaries occurs to further minimize both the diffusion distance from the supply capillary to the energetic mitochondria and the O_2 delivery time. The specific exercising muscle blood flow may increase many-fold as capillary dilation and recruitment of these vessels occur throughout the muscles, thereby decreasing vascular resistance. This hyperemia is met by a similar increase in cardiac output thereby maintaining a normal arterial pressure and gradient for tissue blood flow. In well-trained athletes cardiac output can increase by 35 L/min above basal output (5 L/min). Almost the entire blood flow increase is consumed by exercising muscles, since mean arterial pressure remains in the normal range. If there were vasoconstriction in non-metabolically active tissues and no compensatory increase in cardiac output, then arterial pressure would immediately fall, thereby decreasing tissue blood flow and quickly stopping the exercise process.

In the untrained person maximal exercise increases cardiac output by 20 L/min, principally by a 3-fold increase in heart rate (to about 200 beats/min) with a much smaller increase (less than 50%) in calculated stroke volume (to 100 ml/beat). Mean arterial pressure is maintained near 90 mm Hg due to simultaneous increases in systolic pressure (contractility related) and decreased diastolic pressure (peripheral resistance related). Pulse pressure increases, activating the arterial (high pressure) baroreflex and contributing to arterial pressure stability during exercise. At the start of exercise substantial fluctuations in arterial pressure are sometimes observed and may be minimized by the arterial baroreflex. However, peripheral control of blood flow in exercising tissues is dominated by local factors mentioned above and is not influenced by peripheral baroreflex-activated sympathoconstriction. Additionally, arterial pressure stability results from exercise limitations. As near-maximal exertion is performed, even more capillaries dilate (decreasing peripheral resistance), and O_2 extraction from the blood into tissues is maximized. However, cardiac output has a maximal upper limit. As vascular resistance decreases, arterial blood pressure may decrease if cardiac output cannot maintain tissue perfusion. Maintenance of excessive exercise further decreases blood pressure, leading to decreased tissue flow and decreased muscular performance. Continuation of exercise would then reduce cerebral perfusion, resulting in dizziness. The normal response is to stop exercising before fainting, a clear physiological adaptation that ultimately limits exercise.

During exercise hyperemia does not occur in all organs, but it is partitioned by O_2 demand to oxidative exercising tissues (Fig. 8-2). Cerebral flow is maintained over a wide exercise range, but both splanchnic and renal flow decline. Coronary blood flow increases directly in proportion to increased cardiac metabolism. The largest increase (amount and %) of cardiac output is to exercising skeletal muscle as a function of exercise level. Skin blood flow is variable and will change according to the body thermoregulatory needs, but it is shown in the Fig. 8-2 to be relatively constant. Skin blood flow may increase tremendously to cool an elevated core body temperature during exercise. This increase in skin blood flow may obligate up to 5 L/min of cardiac output, thereby decreasing blood flow available to active muscles and limiting performance.

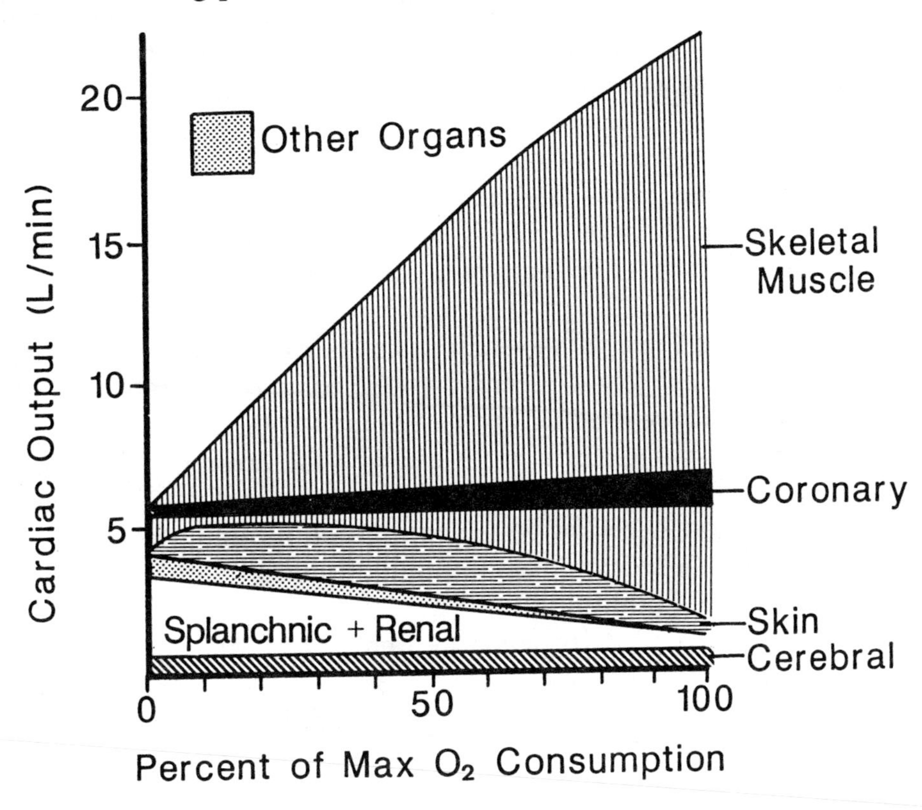

Fig. 8-2. Redistribution of cardiac output during exercise. (Modified with permission from Clausen, JP, *Scand. J. Clin. & Lab. Invest.* 24:306, 1969, Blackwell Scientific Publs., Oxford, and from Ross, J., Best and Taylor's *Physiological Basis of Medical Practice*, 11th ed, West, JB [ed]. Copyright 1985 by Williams & Wilkins, Baltimore).

Exercise results in large increases in hormone concentrations, altering both cardiac dynamics and intermediary cellular metabolism. Release of adrenal catecholamines stimulates cardiac sympathetic beta receptors. Subsequent adenylate cyclase activation increases heart rate, stroke volume and contractility, together augmenting cardiac output. This catecholamine response is sufficiently effective to increase cardiac output even in individuals with heart transplants, allowing them to do marathon runs despite having no direct sympathetic innervation to the heart. Additionally, sympathetic activation shifts the heart pressure-volume Starling curve upward and to the left. Peripheral tissue adenylate cyclase is also activated, increasing intracellular free calcium and stimulating glycogenolysis in both the liver and muscle, thereby increasing available glucose for oxidative metabolism. Catecholamine-induced lipolysis increases free fatty acids - an essential fuel source for cardiac metabolism. Cortisol and thyroid hormone also increase. However, much lower levels of insulin are required for glucose transport into cells during exercise. During exercise the untrained person primarily relies on stored tissue glycogen. The physiological control mechanisms liberating fatty acids and glucose are not well regulated in the untrained individual and exhaustion occurs before adequate amounts of these compounds are available.

Responses of Trained Persons

The response to exercise in trained athletes reflects primarily changes in the cardiovascular system, since the respiratory system is not limiting. The major differences are that cardiac output is now much more dependent on stroke volume than on heart rate and that the flow distribution to the muscles is more adapted to minimize diffusion distances, principally via increased capillary density.

Conditioned athletes are characterized by lower heart rates (physiological bradycardia). Some well-trained athletes showing heart rates below 50 beat/min, requiring a 50% greater stroke volume than sedentary persons. The ejection of large stroke volumes with low heart rates requires much less myocardial O_2 consumption than smaller volumes at higher heart rates.

The immediate increase in heart rate with exercise is not as great as for sedentary persons. During systole the heart empties more completely; both end-diastolic volume and pressure are decreased when compared to sedentary persons. There is a Starling shift to the left on the pressure-volume curve (Fig. 3-5), producing a greater cardiac output for a given end-diastolic pressure. Maximal cardiac output during exercise in world class athletes is approximately double that of sedentary persons, about 40 L/min despite their lower maximum heart rate (150 beats/min). Stoke volume can be 265 ml/beat in these athletes compared to 100 ml/beat in sedentary exercising individuals.

Trained persons have an altered neural response to exercise and "anticipate" its onset. Heart rate, contractility and blood flow all increase in response to the expectation of immediate athletic performance. This neural response, called central command, is not understood, but it allows a more rapid transition from rest to sub-maximal and maximal exercise.

A greater amount of aerobic work can be performed for a longer period of time in trained individuals before exhaustion occurs. This is a function of both the increased cardiac output and microcirculatory changes in exercising tissues. The effect of increased collateral circulation is particularly effective in the heart, since potential blockage or spasm of one artery does not limit blood flow and cause myocardial infarction. Tissue hypoxia is minimized, because blood flow has several routes to the affected tissue, and there is increased capillary density.

The endocrine response to exercise are also altered in well-trained athletes. Both plasma epinephrine and norepinephrine still increase in response to exercise, but the magnitude of the response is highly damped. Glycogen mobilization and lipolysis remain working or may even be more effective, suggesting increased catecholamine receptors and/or efficacy. The effect of exercise on insulin is even stronger after training. Diabetics who exercise regularly can often reduce or completely eliminate their exogenous insulin needs and achieve much better control of blood glucose.

Any increase in daily exercise is associated with health benefits. Exercise training produces decreased blood cholesterol, a positive shift in blood lipids and a more efficient heart, all contributing to a lower cardiovascular risk. In many cases in hypertensive individuals exercise returns arterial pressure to normal levels without additional medication. Most people who exercise on a regular basis report the need for fewer hours of sleep, more daily energy and feeling better.

Review Questions

25. The maximal level of exercise performance is increased after several weeks of exercise training. When compared to the untrained state, this is due to

    A. increased cardiac output, increased arterial pressure and increased muscle capillary density
    B. increased cardiac output, decreased arterial pressure and increased muscle capillary density
    C. increased cardiac output, the same arterial pressure and increased muscle capillary density
    D. the same cardiac output, the same arterial pressure and increased muscle capillary density
    E. increased cardiac output, the same arterial pressure and decreased muscle capillary density

26. A patient with asthma (chronic obstructive lung disease) can perform anaerobic exercise (weight lifting) quite well, but at a much lower level of aerobic exercise (jogging) has severely restricted athletic performance. Why is this true?

    A. Asthma decreases pulmonary blood flow, thereby limiting cardiac output.
    B. Anaerobic exercise does not increase $\dot{V}_{O_2}$, therefore the lungs do not participate in the anaerobic exercise response.
    C. Aerobic exercise is associated with increased pulmonary ventilation to meet a continued need for increased oxygen in the tissues.
    D. Aerobic exercise is associated with increased cardiac output, which is decreased by asthma increasing total peripheral resistance.
    E. The patient is unable to adequately increase pulmonary ventilation.

27. On hot and humid days a normal person's exercise response is limited by which of the following?

    A. Sweating is ineffective as a cooling mechanism, since skin evaporation is decreased, and a large amount of cardiac output is going to the skin.
    B. High humidity decreases the P_{O_2} in alveoli, thereby decreasing hemoglobin saturation.
    C. Skin blood flow decreases to minimize skin exposure to high environmental temperatures.
    D. Body core temperature increases due to decreasing cardiac output.
    E. Increased temperature in the alveoli shifts the oxyhemoglobin saturation curve to the right, thereby decreasing blood P_{O_2}.

ANSWERS TO INTEGRATIVE TOPICS QUESTIONS

1. B. Exposure of the undifferentiated urogenital tubercle to androgen causes it to develop into scrotum and penis. Testosterone is thought to be metabolized to dihydrotestosterone, which is the active agent in these tissues.

2. B. Müllerian-inhibiting hormone causes regression of the müllerian ducts, an active process in male development.

3. The diagnosis is hypoandrogenism secondary to hyposecretion of pituitary gonadotropins.

4. HCG is an LH-like hormone and thus stimulates steroidogenesis.

5. FSH promotes sperm production, while HCG stimulates testosterone secretion.

6. B. Testosterone promotes the differentiation of the Wolffian ducts and is metabolized to DHT which causes differentiation of the external genitalia.

7. D. FSH binds to its receptor on Sertoli cells to initiate the series of reactions leading to secretion of inhibin (which feeds back to inhibit FSH secretion) and the intracellular production of ABP.

8. C. Progesterone increases the minimal stimulus necessary for myometrial contraction, reducing myometrial contractility.

9. A. Endometriosis is due to the appearance of endometrial tissue in the abdomen.

10. B. Danazol is used to suppress stimulation of the misplaced endometrial tissue by gonadotropins.

11. C. No. Estrogen and progestin stimulate endometrial tissue.

12. D. Estriol is not a precursor for 17β-estradiol synthesis.

13. A. FSH and LH belong to the family of pituitary hormones that includes TSH (A). They act by binding to membrane-bound receptors on target cells and increase dramatically at mid-cycle (B). FSH alone can cause development of follicles (C), and it is inhibited by inhibin from the ovary (D).

14. A. Nearly complete loss of oocytes leads to cessation of estrogen and progestin production.

15. A. Propulsion is aided by contractions (B) and negative vaginal pressure (D). Sperm pass more easily through watery mucus (C).

16. A. Human chorionic gonadotropin.

17. B. High levels of progesterone maintain pregnancy and prevent myometrial contraction.

18. C. The relative E/P ratio is less in the quiet myometrium than in the contracting myometrium. That is, maintained progesterone levels help prevent myometrial contractions.

19. D. Prolactin from the anterior pituitary.

20. E. It is not likely. The secretion of prolactin inhibits ovarian function either directly or by inhibiting gonadotropin secretion.

21. B. Fertilization is in the fallopian tube (A).

22. A. Free fatty acids and glucose are transferred by simple diffusion (B,C)

23. D. Suckling has the effect of increasing both oxytocin and prolactin secretion.

24. E. This changing focus of progesterone production is referred to as the "luteal-placental shift".

25. C. Cardiac output is increased due to strengthening of the cardiac muscle (hypertrophy and increased capillary density) and more capillaries in the peripheral muscles. This couples the increase in cardiac output to peripheral tissue perfusion. Such matching results in the same arterial pressure.

26. E. Anaerobic exercise can be performed while holding your breath, but some period of time between exercise allows the oxygen debt to be repaid. By contrast, aerobic exercise is maintained, and asthma decreases alveolar ventilation, directly decreasing the intensity of exercise.

27. A. The large amount of sweat on the skin indicates that evaporative cooling is not working as efficiently as when it is drier. The body slightly heats up and a larger amount of cardiac output is directed at the skin in an attempt to decrease core body temperature. This skin blood flow therefore cannot go to the muscles for exercise.

CLINICAL LABORATORY VALUES

Test		Value
ACTH		10-50 pg/ml
Aldosterone (Serum)	Recumbent Upright	3-9 ng/dL 4-30 ng/dL
Alkaline Phosphatase (Adult)		20-70 U/L
Antidiuretic Hormone (Serum Osmolality >290)		2-12 pg/ml
Bicarbonate (Total CO_2)		23-29 mmol/L
Blood Pressure		120/80 mm Hg
Blood Urea Nitrogen (BUN)		7-18 mg/dL
Calcium		4.1-5.3 mEq/L or 8.5-10.5 mg/dL
Catecholamines (plasma)		
Norepinephrine		110-410 pg/ml
Epinephrine		<50 pg/ml
Chloride		98-106 mEq/L
Cholesterol		140-250 mg/dL
Cortisol	(8 am) (4 pm)	5.0-23.0 mg/dL 3.0-15.0 mg/dL
Estradiol (Adult)	Male Female	20-50 pg/ml 20-350 pg/ml
Glucagon		50-200 pg/ml
Glucose	Fasting 2 hours postprandial	65-95 mg/dL <120 mg/dL
Growth Hormone (Adult)		<5 ng/ml
Insulin (Fasting Adult)		5-25 μU/ml or 0.2-1 ng/ml
Iron	Male Female	65-175 μg/dL 50-170 μg/dL
Lactic Acid (Lactate)		0.5-2.2 mmol/L
Osmolality (Serum)		278-298 mOsm/L
Oxygen (Arterial)		75-100 mm Hg
Parathyroid Hormone		11-54 pg/ml
Phosphorus (Adult)		2.7-4.5 mg/dL
Potassium		3.6-5.1 mEq/L
Progesterone	Male Female	0.12-0.3 ng/ml 0.3-20 ng/ml
Protein		6.0-7.8 g/dL
Sodium		136-145 mEq/L
Sperm Count		40-250 million/ml of semen
T_3 (Triiododthyronine) Free		230-660 pg/dL
T_3 RIA (Triiodothyronine)		120-200 ng/dL
T_3 RU (Resin Uptake)		26-35%
T_3 Total		80-220 ng/dL
T_4 Free		0.8-2.4 ng/dL
T_4 RIA (Adult Thyroxine)		5-12 μg/dL
>60 years	Male Female	5-10 μg/dL 5.5-10.5 μg/dL
T_4 Total		5-12 μg/dL
Testosterone (Adult, total)	Male Female	437-707 ng/dL 25-45 ng/dL
Thyroid-binding Globulin (TBG)		15-34 μg/ml
Thyroid-stimulating Hormone (TSH)		<10 μU/ml
>60 years	Male Female	2-7.3 μU/ml 2-16.8 μU/ml
Triglycerides	Male Female	40-170 mg/dL 35-135 mg/dL
Vitamin D		20-76 pg/ml